The Last Drop

ALSO BY TIM SMEDLEY

Clearing the Air

The Last Drop

Solving the World's Water Crisis

TIM SMEDLEY

PICADOR

First published 2023 by Picador
an imprint of Pan Macmillan
The Smithson, 6 Briset Street, London EC1M 5NR
EU representative: Macmillan Publishers Ireland Ltd, 1st Floor,
The Liffey Trust Centre, 117–126 Sheriff Street Upper,
Dublin 1, D01 YC43
Associated companies throughout the world
www.panmacmillan.com

ISBN 978-1-5290-5814-7 HB
ISBN 978-1-5290-5815-4 TPB

1 3 5 7 9 8 6 4 2

A CIP catalogue record for this book is available from the British Library.

Diagrams by ML Design Ltd

Typeset in Warnock by Jouve (UK), Milton Keynes
Printed and bound by CPI Group (UK) Ltd, Croydon, CR0 4YY

Visit **www.picador.com** to read more about all our books
and to buy them. You will also find features, author interviews and
news of any author events, and you can sign up for e-newsletters
so that you're always first to hear about our new releases.

'Walls of stone will stand upstream to the west
To hold back Wushan's clouds and rain
Till a smooth lake rises in the narrow gorges.
The mountain goddess if she is still there
Will marvel at a world so changed.'

'Swimming', Mao Tse-tung

'Water's precious. Sometimes it can be more
precious than gold.'

The Treasure of the Sierra Madre (1948)

Contents

Troubled Waters

Karameh Dam, Jordan, August 2021

I'm standing on the shores of a large lake in one of the world's driest regions. The lake is calm and peaceful, offering some respite, to the eyes at least, from the burgeoning heat of an August morning. We're 400 m below sea level in the Jordan Valley, the lowest land point on Earth, where the sun is punishing from the moment it rises and rainfall doesn't come for months on end. On the far shore, bleached-white banks blur into sky, with the hills of the West Bank ghostly apparitions through the haze. A lone, thin white waterbird briefly makes an appearance, barely able to break the silence or its own lethargy, before giving up and landing again. Near me on the foreshore, a handful of stunted brush trees stand unhappily in the water, a telltale white crust on their lower branches. For there is a fatal problem with this lake. The Karameh Dam, the second largest in the Kingdom of Jordan on its construction in 1995,[1] was built on salty ground. Its water, desperately needed for drinking and irrigation, is entirely unusable.

A car appears, kicking up white dust, and out gets Eshak Al-Guza'a, project manager for the NGO EcoPeace Middle East. He's an engaging thirty-something with a ready grin and readier grasp of figures. 'This is one of our biggest failures,' he laments, as we stare out at the disconcertingly beautiful yet dead lake. 'Almost 52 million cubic metres [52M m³] of unusable water. In the 1990s we had all the expertise, all the technology to build a good-quality dam, but they did not do enough research about the saline springs and rocks – so we confined this fresh water, made it saline, and now there's no technology to fix it.' Adding insult to injury, a desalination plant was built by its shores in 2010 but closed just months later, unable to cope with the now extraordinarily high salt content – saltier even than the Dead Sea, just kilometres away. Meanwhile, Nasser – our driver for the day, and EcoPeace's private-investment manager – sizes up the shoreline for a potential tourist park with eco-lodges and water sports. But, in truth, the lake is receding by around a metre a year through evaporation, as climate change sees the heat intensify and rains lessen. The dam's main sluice gate now stands many metres above the waterline; opening it would only unleash a slow tumble of bone-dry pebbles.

The Karameh Dam was built, my hosts tell me, to fulfil a dream. A minister's young son allegedly had a dream one night that a dam should be built on this site.[2] So it was – for between £43 million and £75 million – in one of the driest, saltiest valleys on Earth. It appeased his son, a few influential farmers and doubtless the UK-based engineering firm that won the tender. A heady mix of hubris and greed persuaded the government to spend millions destroying what precious little surface water existed in a valley carved by the once-mighty River Jordan (today just a mere trickle of its former self, due to over-abstraction and over-allocation). 'During the investigation phase of the dam site some scientists and specialists advised the government not to construct the dam,' writes Elias Mechael Salameh, professor of hydrogeology and hydrochemistry at the University of Jordan. 'But the Jordan Valley Authority of the Ministry of Water and Irrigation (JVA MoWI) ignored all criticism and warnings.' In an interview with me, the JVA MoWI's former secretary general, His Excellency Saad Saleh Abu Hammour,[3] confirms they had such a report, and lays the blame solely with the then ministers in charge: 'This is corruption,' he says, adding that one minister 'owned land over there.'

Abu Hammour now describes Jordan's water situation as its 'very worst' historically: 'We are looking for every single drop of water everywhere.' Meanwhile, the Karameh Dam lies silent and useless, full of trillions of valuable, wasted drops. It serves as one of the modern world's great emblems of water mismanagement – but, as we will see, it's merely one of a long and sorry list. As Salameh wrote in 2004, the Karameh Dam 'proved to be a hard lesson, not only learnt by Jordan, but also by other areas of the world, where decent scientific work defending the common good is ignored for the greed of a few beneficiaries.'[4]

The water capacity of the Karameh Dam, around 50M m³, is significant in more ways than one. The summer I visited, in 2021, Jordan was forced to purchase additional water from neighbouring Israel to stave off its latest water-shortage crisis. The amount bought – which came with an admission of national water insecurity and a side serving of regional humiliation – was the very same, 50M m³.[5]

Lake Mead, USA, December 2021

Minutes after leaving the National Parks Service (NPS) HQ in Boulder City, Nevada, we crest the hill, and I gasp – the 'bathtub ring' that encircles Lake Mead is revealed in its bleached-white ignominy, a forlorn witness to water lost. I'd read that the lake had dropped by 43 m in twenty years of drought – but seeing it in person brings home the scale. The Statue of Liberty is only slightly taller, at 46 m. If Lady Liberty's feet touched the current water level, only her torch would rise above the original line showing when the lake was full. The whiteness of the 'bathtub ring' is caused by calcium carbonate in the water attaching to the red sandstone: as the lake level descends, tidemarks form on the rock like suds on an emptying tub. Even from a distance, each one is as visible as tree rings, distinguishing the relatively good years from the relatively bad. Looking down from a viewing platform near the Hoover Dam, my host, Justin Pattison, acting deputy superintendent of the NPS, points out the ring showing his past year of service, closest to the water level. It's a wide one, marking a drop of perhaps 3.7 m. The four intake or 'penstock' towers for the hydroelectricity rise out of the water like skyscrapers, at least 61 m tall. They look skinny and exposed, showing their frilly underwear – the metal grills that would prevent underwater debris being sucked into the turbines, were they actually under water. Now they stand embarrassed before a photo-taking public. The intake towers, perhaps even more than the bathtub ring, make clear just how much closer the reservoir is to dead pool – when the water level becomes too low to run through the dam – than it is to ever being full again. I take some bemused selfies, feeling like a disaster tourist. On the other side of the dam, a little launch jetty for observatory boats hangs uselessly from the rock, high above water it would have floated on perhaps just a year before.

As we drive around the lake, the scale of America's largest reservoir dawns on me, too – even this one small corner is vast, reaching to the Red Mountains on the horizon. How much vaster it should be, however, is made clear by new islands of white rock that litter the lake itself, previously little known to anyone but pleasure boaters' depth monitors. Hemenway Harbor, where a hundred or so pleasure boats are moored and the tourist steamship the *Desert Princess* sets off – or used to – edges ever further towards a new archipelago of islands that breached only recently. Just ten years ago, this would all have appeared an invincible expanse of water.

Bankside at Hemenway Harbor, the launch-site road – as wide as a UK motorway – comes to an abrupt end some 50 m before the water. Each 30 cm drop in water level – which is occurring on average every month – equates to several metres of bank incline. Boaters need a concrete launch ramp to avoid sinking into the mud, but the NPS can't extend the road fast enough. 'The problem we're dealing with,' says Pattison, 'is that by the time we get the engineering done, do the bid reviews and actually get something constructed, it's useful for a couple of months before it needs extending again. We literally can't keep up with the rate at which the water's declining right now.' To our left is a brand-new, $3-million section of road, built this year to reach the water. But it already doesn't reach the water and stands idle. A $3-million road to nowhere. 'It becomes a definition of insanity after a while, right?' admits Pattison. 'To continue to dump millions of taxpayers' money to extend these launch ramps with the same end result of being unusable.' Keeping these boat sites open would need investment 'to the order of $20 million in the next two years. And you know what that gets us at the end of those two years? What we're looking at right here.' If (when) the lake reaches a 305 m elevation – which could happen as soon as 2024 – Hemenway Harbor will have to be abandoned entirely. Dead pool is just 273 m.

As Pattison frames it, the amount of water coming into Lake Mead won't change much within the next year. Whatever the size of drop, he says, drop it will: 'the water is only going down'. It's dropped 43 m in twenty years, so the trend only points in one direction. 'We often hear, "Well, we just need that miracle May," or, "Just one good winter and this will all turn around . . ." What we try to impress upon people is that this is a twenty-plus-year downward trend; regardless of what happens in the next two, three, four, five years, it will continue to go down. It's just how quick that happens.' If the 4.6 m year-on-year decline continues, as is currently projected, the lake will reach dead-pool elevation sometime around 2029, after which the hydroelectricity turbines will no longer turn, and the Colorado River will no longer flow beyond the Hoover Dam into Arizona, let alone reach Mexico (which it barely does now) or the sea (which it hasn't in years). If and when that happens, 40 million people – plus another million or two by then, given population growth and the never-ending south-west housebuilding boom – lose their ninety-year source of water supply.[6]

Akosombo Dam, Ghana, October 2021

I visited a third great reservoir in 2021. The greatest, in fact: Lake Volta, Ghana, is the world's biggest man-made reservoir by surface area. It takes a whole-day trip from the capital, Accra, so we set out

early. Emmanuel, our driver, puts on a CD of early 2000s R&B –
Craig David, Usher, and the jarringly expletive-filled (in a devoutly
Christian country) 'F**k It (I Don't Want You Back)' by Eamon.
After its fourth rotation, and still struggling through the Greater
Accra traffic, I politely request a change. When we're finally released
from the grinding cogs of the city, we reach the Ghanaian countryside: flat green vistas pockmarked by steep, tree-covered hills that
rise jaggedly from the ground and just as sharply disappear. Near
the Shia Hills, baboons sit on the roadside, accepting snacks thrown
from passing vehicles.

At midday, we reach the town of Akosombo, home to the start of
the lake and the source of the country's (and much of its neighbours') hydroelectricity: the Akosombo Dam. Our request for an
official site tour with the Volta River Authority (VRA) remains
stuck in bureaucratic purgatory – a paper in-tray on a desk in the
VRA's Accra headquarters. The security checkpoint doesn't let us
through, citing 'national security' and heightened tensions with
Togo over a disputed electricity-supply deal. Defeated, we go to the
nearby Volta Hotel Akosombo, where Emmanuel knows of a glass-
panelled restaurant with an excellent view over the dam. Inside, we
take some unsatisfactory photos. It was a long drive just for a selfie
behind glass. On my phone map, I spot a public road that passes
beyond the dam and touches the shore of the mighty Lake Volta
itself, at a recreation centre marked 'Dodi World'. I persuade a sceptical Emmanuel to drive there instead. Here, we find an empty,
off-season tourist ferry, and a view of the lake that unfolds all
around us, filling me with happiness that verges on the emotional,
on this long day in a long week. Steps lead down to the water, where
a lone fisherman is launching a wooden dugout canoe. He waves to
say it's OK for us to come down, so we do. Spontaneously, I reach
down and touch the water, then wash my face and neck in it. It is
good, clean water. I needed to see and feel it. And then I needed to
understand how this vast volume of fresh water, covering 3.6 per
cent of Ghana, isn't enough to stop the country from struggling
with a water crisis of its own.

Answers come not at Lake Volta, but at the nearby holding reservoir for water supply, Lake Kpong. 'Water shortage isn't a problem for us at all, not at all,' argues Yaw Adjei, manager of Kpong Water Treatment Plant (KWTP), one of five such plants built between 1954 and 2014 in the Volta region. Adjei oversees them all. At KWTP alone, he tells me, with the printed figures to hand, 'monthly, we abstract 4,482,433 m^3 of water for treatment. That's for a month. And we supply 3,630,275 m^3 [of that] to Accra.' According to Adjei, Lake Volta's water quality is so good that they only need to use two chemicals to treat it: lime, to adjust the pH, and chlorine. It then travels via 1.07-m-diameter pipe to the next pumping station, 48 km away, before going another 30 km or so to the capital for consumption. So, water quantity isn't the problem. There are no 'bathtub rings' here, the mighty lake looks full to the brim. Adjei's biggest headache is power. Pumping water in high-pressure pipes over 80 km requires a lot of power, and Adjei has no control over that. Despite being so close to the 1-GWh hydropower turbines of the Akosombo Dam, sometimes the power goes off, and so do the pumps. The largest reservoir within Accra itself, Weija, on the other hand, has the opposite problem: it is close enough to the people to distribute cheaply, but it's heavily polluted, from both industry and illegal gold mining (known as *galamsey*), making it very expensive to treat to drinking-water standard. 'Weija is something like stagnant water,' complains Adjei, 'and anything that is dumped into it remains.' The solution in 2014 was to build a new treatment and pumping plant here at Kpong. Built by the Chinese and consequently known locally as 'the China plant', it can produce 149,000 m^3 per day. But, says Adjei, Accra is growing so fast that demand still outstrips supply. He thinks they should build a second plant here at Kpong, 'Even a third. Because this river, this lake, will never run dry. Never. We can put as many plants as we can build here.' But this comes down to money and 'the powers that be'. According to the figures of state utility Ghana Water Company (GWC), the country's water demand is 1,131,820 m^3 per day, whereas they can only supply 871,500 m^3 per day – a deficit of 23 per cent.[7]

Unofficially, others give a much higher deficit figure. For many people in this, one of Africa's wettest countries, that means no potable water at all.

As we drive away from Kpong back to Accra, the air cools and black clouds roil in the middle distance. The heavens duly open. Roadside drains explode into brown fountains and the roads turn into muddy rivers. I fear for our return as potholes become ponds, but Emmanuel drives calmly on, unfazed. Closer to Accra, the clouds part, and we cross a small bridge where I see a group of water trucks lined up. These are the private tucks that supply water to fill the rooftop 'polytanks' found on most Accra properties, with capacities ranging from 1,000 to 10,000 litres (L). On Yaw Adjei's desk, a company calendar proudly declared that GWC's water 'is five times less costly' than water from such trucks. I crane my neck as we pass to see what the trucks are doing. Below the bridge runs a small river, flowing fast from the storm shower, its water a deep red-brown with mud and detritus from the side streets and storm drains. The trucks are filling up their tanks with it, to be sold to fill polytanks. The further we drive from Lake Volta's bounty, it seems, the less accessible it becomes.

Oxfordshire, southern England, midsummer 2022

During the summer months in the Oxfordshire town where I live, I go swimming in the local 50 m lido. With my inelegantly slow breaststroke, from time to time I accidentally gulp some of the pool's opulent, chlorine-clean 5.9M L (5,900 m^3)* of water. Sometimes I swim while it's raining, when fewer people brave it, alone in my lane with the strangely comforting feeling of having water both above and below me. I stand a bottle of water at the end of the lane, to drink from halfway through my swim. I normally have a shower afterwards, even if I've showered that morning. I live a wet, drenched, quenched existence. But, as I discovered when

* One cubic metre (m^3) = one thousand litres.

researching this book, this won't last. I am living on borrowed time and borrowed water. Water stolen from nature, drained from rivers and lakes and returned polluted, allows me to live this way. It will have to stop – not through some altruistic hand-wringing desire to do better, but because even in England this amount of water will soon be unavailable. Like many parts of the world, we are now using more water than we can sustainably supply. As surface-water and groundwater levels dwindle year by year, a crisis awaits. It's simple maths. Here, too, demand is outstripping supply.

The UK's average annual rainfall is around 1,200 mm, compared to less than 400mm in Afghanistan or double figures in Egypt.[8] However, despite our winter storms, significant parts of the UK are staring down the barrel of empty water butts. Much of that four-figure rainfall average is propped up by the rainy highlands of Scotland, Wales and northern England. In south-east England, where I live, the average annual rainfall lingers around 600 mm – comparable to Lebanon or Kenya, and far drier than Sydney, Australia.[9] This also happens to be the UK's most populated area, with some 18 million inhabitants packed into just 19,000 km², including London. And it's drying up, fast. The government's Water Abstraction Plan shows that, in England, 28 per cent of ground-water aquifers and up to 18 per cent of rivers and reservoirs are unsustainably abstracted, with more taken out than is put back in.* Not a single one of England's rivers are classified as being in good ecological health – this includes chalk streams, a delicate habitat almost entirely unique to England. However, much of the public remains oblivious to a problem that we are all, at least in part, responsible for causing. Over half of the fresh water abstracted in the UK is for household use. The average Brit happily uses 153 L of water per day, through showers, toilets, dishwashers, washing machines and garden hoses. Yet climate-change projections show that dry summers in England are set to increase by up to 50 per

* 'Aquifers' being the layers of porous sand and rock that hold water underground, while 'abstracted' simply means taking water out.

cent, with the amount of water available reduced by at least 10–15 per cent by 2050.

Freshwater scarcity, once considered a local issue, is increasingly a global risk. In all ten annual risk reports since 2012, the World Economic Forum (WEF) has included water crisis as one of the top-five risks to the global economy. Half of the global population – almost 4 billion people – live in areas with severe water scarcity for at least one month of the year, while half a billion people face severe water scarcity all year round. Global water demand has increased by 600 per cent over the past one hundred years, while available fresh water has fallen by 22 per cent over the last twenty.[10] Something has clearly got to give.

Given that 70 per cent of our blue planet's surface is water, such a problem seems impossible. However, 97.5 per cent of that is sea or salt water, unfit for human consumption. The remaining 2.5 per cent is fresh water, but almost two thirds of it is trapped in the ice caps and permafrost. One lake alone – Baikal, in Russia – holds 20 per cent of the world's drinking water (and even that is being polluted by tourism). As the charity WaterAid puts it, 'If a bucket contained all the world's water, one teacup of that would be fresh water, and just one teaspoon of that would be available for us to use.' Rising demand and climate change are putting that teaspoonful of fresh water into shaky hands.

There's only ever the same, finite amount of water churning around in our water cycle. Every drop of water on Earth has been here since the beginning of time, constantly recycled again and again. Up to 60 per cent of the adult human body is water (even bone is a surprisingly splashy 31 per cent). When you die and are cremated or buried, that water will be released again, to the atmosphere or the earth, and those water molecules will eventually be reunited with old friends back in the clouds. We are as intimately connected to the water cycle as rivers and lakes are. Yet, from the Yellow River in China to the Colorado River in the United States, many rivers no longer reach the sea. Often artificially straightened and dammed, water is sucked out and channelled off to supply farms, industries and households. Great lakes, from the Aral Sea in

Central Asia to Lake Urmia in Iran, have nearly disappeared. Groundwater aquifers, from the Ogallala and Central Valley in the US to India and Pakistan's Upper Ganges and Lower Indus, are being depleted faster than they can refill. The remaining fresh water is increasingly polluted with fertilizers and pesticides causing algal blooms that smother and choke ecosystems. Streams in Pakistan and Bangladesh run a rainbow of colours, dyed by the apparel industry that caters to Western fast-fashion trends. In Europe, three quarters of Spain is at risk of desertification, as are parts of Portugal, southern Italy and Cyprus, while centuries-old olive groves wither away.[11]

'Climate change *is* water change; the climate crisis *is* water crisis,' stresses Torgny Holmgren, executive director at the Stockholm International Water Institute (SIWI). 'The bottom line is, there is a huge increase in the demand for water. We have a growing population, growing economies, more urban populations. The OECD* predict that, if these trends continue, we will need 50 per cent more water in 2050 compared to twenty years ago. And of course, that is impossible, because water is a finite resource . . . This will impact all of us.'

The COVID-19 pandemic, which unfolded in parallel to my book research, brought such water issues into sharper focus. 'The first thing everyone was told to do was stay home and wash your hands vigorously,' recounts Gary White, CEO of Water.org, who typically flies 300,000-plus kilometres a year to visit water projects, but in 2020 was grounded like the rest of us. But for the 2.3 billion people in the world without safe drinking water and sanitation at home,[12] 'if your daily routine is walking to collect water, you don't have enough water to increase your handwashing by five times without making more trips out or paying more money to water vendors . . . It's not like COVID woke us up to the need for water for hygiene; we already knew that. But I certainly think we hadn't seen the lack of access to water and sanitation as a *global* crisis before. When

* Organisation for Economic Co-operation and Development.

somebody [being unable to] wash their hands in one country becomes the critical link to the spread of disease, then suddenly water and hygiene becomes a global risk.' The pandemic widened the cracks already riven throughout the global water system. In June 2021, Mami Mizutori, the UN secretary-general's special representative, added: 'Drought is on the verge of becoming the next pandemic and there is no vaccine to cure it.'[13]

Much of this book was written – and the research trips took place – across 2021 and early 2022. This preceded major climate events, from the catastrophic flooding in Eastern Australia and Pakistan to the European drought that saw the Rhine no longer able to carry coal barges to power stations, and the Yangtze in China all but drying up.[14] But my research foretold their grim inevitability. As sure as June follows May, so every year will now see new climate disasters that outstrip those that came before. The purpose of this book isn't to catalogue such climate shocks or stand as a snapshot in time. Rather it's to tell the stories with universal relevance, that can equally influence individuals and decision makers now and in decades to come. It is also to offer hope. The good *and* bad news is that, as I saw on the banks of the Karameh Dam, water crises are usually caused by all-too-human mismanagement, not climate. But, as climate change bites, precipitation patterns change and climate refugees are forced to move, the timeframe to get our act together is becoming ever shorter. We are currently using up the water sources on which our very existence relies. We can continue doing that, until the very last drop. Or we can decide to change our approach before it's too late. The world isn't running out of water – people are. This book is the story of how this crisis happened and how we can still avert disaster.

PART 1:
RUNNING OUT

CHAPTER 1

Approaching Day Zero

Driving from Amman to the Jordan Valley, we descend 1,400 m in a matter of minutes. To my surprise, maybe because I only associate it with aeroplanes, my ears pop. Once down at the valley floor, we pass farmland with row after row of bare metal hoops – the ribs for plastic polytunnels. The soil is full of scraps of black plastic, remnants of sheets laid to retain moisture during growing season and simply ploughed back in afterwards. Nasser slows down to point out his favourite date farm. I wonder whether I've ever met anyone before who *has* a favourite date farm. We spend a lot of the time lost. Nowhere in Jordan has a precise address, and there's no formal postal service. The valley is Jordan's Wild West; rural communities cluster around a single main road that cuts north to south, eking a living out of agriculture, shops and services. Streets bustle with more energy than seems right in the heat. The drive is punctuated by long expanses of irrigated agriculture, street sellers, fig plantations, the occasional shepherd corralling sheep, boys sharing donkey rides, parched desert and, every so often, the King Abdullah Canal.

The King Abdullah Canal is Jordan's equivalent to Israel's National Water Carrier; both built in the early 1960s to divert the water of the Yarmouk River and Jordan River to the respective country's major cities. The 1994 Israel–Jordan peace treaty gives Jordan an annual supply of 50M m^3 of water, including 75 per cent of the water from the Yarmouk River, a transboundary river it shares with Syria, plus 20M m^3 transferred by pipe from Lake Tiberias (Sea of

Galilee), all of which goes into the King Abdullah Canal. As the country's main water source, from the roadside it looks decidedly modest – as if the entire North of England relied on the Leeds–Liverpool Canal for all its water supply.

Yana Abu Taleb, Jordanian director of EcoPeace Middle East, wants to show me the canal's source – the diversion weir, from where each country takes its share. Being so close to the Syrian and Israeli borders, security is tight; we are scrutinized by armed guards, then waved into no man's land, where we stand looking down at the Yarmouk Valley. On the horizon sits the Sea of Galilee; below, the sun-bleached hills and canyon give way to a thin strip of lush green that follows the river. A local later tells me that the Yarmouk is only a quarter of what it once was, ravaged by the abstractions upstream (and the 1987 Jordan–Israel water-allocation agreement), the Syrian civil war and climate change. Further up the valley, a broken, rusting Ottoman-era bridge* that once served the Hejaz railway stands high above the river – far higher than would be necessary now. Seeing the three together, the narrow canal and two stream-like transboundary rivers, makes it hit home just how precarious the region's water resources really are.

Jordan's water deficit is such that it needs to import considerably more to ensure a continuous supply, but it's set to pay an increasingly high price. The water it gets annually via the 1994 agreement is sold by Israel at a 'cost price' of 4 cents (US) per cubic metre. In 2021, Jordan requested a further 50M m^3 for the same price. Israel said that price was too low, Yana tells me, but she thinks they'll come round on the understanding that 'our national border is a national security issue . . . it's in their own interests'. Much of EcoPeace's work promotes these mutual interests. Its very existence, with an equal balance of Palestinian, Israeli and Jordanian staff and

---

* In 1913, you could have had coffee in Damascus (Syria) in the morning, before taking the Ottoman Hejaz railway to Amman (now Jordan), and continuing west to Haifa (now Israel) on the Mediterranean coast for lunchtime, or on to Medina (now Saudi Arabia).

a director representing each country, is emblematic – showing what can be done through collaboration.

By November 2021, King Abdullah himself* personally instructed his government to devise a better long-term water strategy. According to the Water Ministry, demand in 2022 was expected to reach around 555M m³, with only around 510M m³ available. Dr Raya Al-Masri, a Jordanian academic at Surrey University's Centre for Environment and Sustainability, told me that Jordan's dams are at 40–60 per cent capacity most years, 'which is around 340M m³ [in total]. But this year [2021], it's less than 30 per cent capacity. That is a huge deficit.' The plan devised by the king's water minister to fill the deficit is to increase pumping from the Disi fossil aquifer shared with Saudi Arabia, from 12M m³ up to 14M m³. However, researchers like Al-Masri are not confident that this will work in the short-term, let alone the long-term. She tells me that the improvement to Amman's water supply after the Disi water began pumping in 2010 lasted only until 2013. Initially, it was claimed that the Disi aquifer held enough water to last for fifty years.[1] But now it seems the quantities are not sustainable, and other sources will be needed before 2025. That means it's only lasted for fifteen years. The difference, Al-Masri says, is the demand calculations were based on a much smaller population, and didn't fully take into account the consequences of climate change. Jordan's latest census counted 1.3 million Syrians among an overall population of 10.2 million – an increase of almost 15 per cent. 'With the refugees, the population of the north increased, so you need more water,' says Al-Masri. 'The water supplies in that area were already stressed, and the major groundwater aquifers already depleted.'

The costs of pumping even faster and deeper from the Disi aquifer, as the water minister wants to, also become prohibitive because, as the aquifer is depleted, the water quality worsens, becoming highly mineralized, potentially even radioactive, and difficult to treat. Some of the heralded 'fifty-year water supply' was always

---

* The canal was named after his grandfather, not him.

going to have to stay in the ground, Syrian war or not. So, with no other water source, the desert's fossil aquifer is further relied upon. This brings down the timeframe of its usefulness – and brings forward the potential date of Jordan's 'Day Zero', when the water runs out and significant rationing begins. According to World Resources Institute (WRI), the Middle East and North Africa have twelve of the top seventeen most water-stressed countries, with Jordan in the top five. Amman used to receive water twice weekly from the network, Yana tells me. 'Now, we receive it once a week, and sometimes for just eight hours. So, imagine the struggle?' Omar Shoshan, chairman of the Jordan Environmental Union, told the *Jordan Times*, 'We are very close to Day Zero.'[2]

Cape Town's Day Zero

It was the city authorities of Cape Town, South Africa, in late 2017, who first coined the term 'Day Zero': the day the water supply for a city of over 4 million people would be turned off and residents would queue at standpipes for their daily water ration. In Cape Town, this was calculated to happen if the water capacity of the city's major dams fell below 13.5 per cent. With no rain imminent, Cape Town was on a trajectory to reach Day Zero on 12 April 2018. The phrase became a clarion call that both horrified and mobilized all Cape Town residents. Placards in toilet stalls read *AVOID DAY ZERO. Only Flush After 4 No. 1s* or *If it's yellow, let it mellow. If it's brown, flush it down.* Hotels pleaded with guests to take *90-second showers only!* providing them with trays to stand in to collect the waste water for reuse in toilet cisterns. Swimming pools dried up and gardens turned yellow. A city once chosen by settlers for its abundant water resources had suffered two consecutive years of 'once in a millennia' drought. 'If it happens, Cape Town would become the first major city in the world to shut down entirely the supply of running water in all of its homes,' said the *Globe and Mail*. Cape Town's then mayor, Patricia de Lille, told a press briefing, 'We have reached a point of no return.'

On 1 January 2018, the city declared 'Level 6' water rationing: 87 L per person, per day (the WHO recommends a minimum of 100 L for sufficient health and sanitation). By February, that was heightened to 'Level 6B': just 50 L per person, per day. Then the final Day Zero plan was announced for the impending April date: all piped water would be turned off and replaced with 200 water stand-pipes dotted around the city. These standpipes, known locally as 'pods', would give a daily per-person allowance of barely 25 L (for comparison, the average American uses around 380 L*). The pods would have toilet facilities on site, with queues of up to 5,000 people expected.[3] As 12 April approached, Louise Stafford, director of South Africa Water Funds for the Nature Conservancy (TNC), living on the outskirts of Cape Town, remembers the logistics. Plans were made at community level, she says: people found out where local springs were located and showered in buckets. 'It was preparing for the worst . . . How would the people in informal settlements get water? It had a huge ripple effect. Some people really panicked.' At the time, she blogged that 'more than 30,000 people in the agriculture sector have lost their jobs because there's no water to irrigate the crops. Some farmers are cutting the buds off orchard trees because if they fruit and there's no water, it could damage them irreparably. This will have a knock-on effect on the next harvest. People from outside Cape Town are donating food for livestock because there's a shortage of fodder.'[4]

But, in March, the rains finally came, and the reservoirs slowly began to recover. In May 2018, the Theewaterskloof Dam, by far the biggest in the system, did drop below the Day Zero threshold, to 12 per cent. But the five smaller dams fared better, keeping total storage to 21 per cent. The Day Zero date was pushed back to June, and

* Incredibly, despite being one of the most water-scarce countries in the world, Qatar has one of the highest domestic water-consumption rates, around 1,200 L per person, per day, thanks to expensively government-subsidized desalination and free water for nationals.

then, thanks to more winter rain, was delayed indefinitely. The standpipes were never installed. At least, not yet.

Because of the high population density of cities and increasing urbanization, urban water supply is particularly vulnerable. It is estimated that, by 2050, 685 million people living in over 570 cities will face a decline in freshwater availability of at least 10 per cent, due to climate change. According to a UN report, 'Some cities, such as Amman, Cape Town and Melbourne, can experience declines in freshwater availability by between 30 to 49 per cent, while Santiago may see a decline that exceeds 50 per cent.' It adds, drily, 'The societal impact and consequences are likely to be severe.'[5]

The question is not *if* but *where* Day Zero will strike next. There are many potential candidates. This chapter – indeed, this book – doesn't set out to exhaustively catalogue them all. Instead, I offer a collection of stories with universal relevance; regions hanging from the precipice of water scarcity, that together form a global picture.

Cape Town's Day Zero had, in fact, been predicted almost a decade earlier. In 2006, at the request of South Africa's water minister, academics Asit K. Biswas and Cecilia Tortajada – then at the Institute of Water Policy, Lee Kuan Yew School of Public Policy, Singapore – gave a keynote lecture in which they predicted that 'at least one major city would face a very serious water crisis within the next two decades. It would be a crisis of such magnitude that no earlier generation had ever faced.' They co-authored a follow-up piece in 2018 saying that Cape Town's Day Zero 'confirmed our forecast' and added a new prediction: 'within the next 10 years at least 10 important Indian cities will face even worse water crises than Cape Town.'[6] Just one year later, Chennai, India's sixth-largest city, with some 11 million people, ran out of water.

Chennai, India

In July 2019, all four major reservoirs serving Chennai, the sixth most populous city in India, dried up. They didn't just come close to some 12 per cent capacity figure; they became sun-cracked mud. At the start of the month, the total reservoir storage capacity was recorded at just 0.1 per cent.[7] There was no municipal water supply. Day Zero had rudely arrived at the state capital of Tamil Nadu. The streets became clogged with private water tankers and desperate queues. Milk tankers were repurposed and relined to carry water from Jolarpet, some 220 km away. Soon, so were oil and kerosene trucks. Water was being taken from abandoned quarries, treated with chlorine and sent to desperate residents.[8] Alpana Jain, who regularly works in Chennai, tells me, 'People were only getting water once in fifteen days through a tanker; they would line up and fill up several jars. They were skipping baths, dry cleaning themselves, using only what was necessary. It was a challenge.' In wealthier neighbourhoods, where homeowners had private – often illegal – boreholes down to groundwater, 'people started digging deeper. Even 120 m wells had gone dry. So, they were digging deeper, but some just hit rock.'

Alpana Jain is now manager for cities at the Nature Conservancy India, with a focus on water. 'Chennai, in the last ten years, has been going through alternating years of floods and droughts,' she tells me. The city once had around 1,000 lakes, only 200 of which remain – many of them becoming 'like dustbins for solid waste and sewage'. When Chennai lost its lakes and wetlands, it lost its water resilience. 'Wetlands help in absorbing flood waters and then saving them for the drought or the dry season,' explains Jain. With the wetlands destroyed, degraded, or simply built upon, that was no longer the case.

I ask how the Chennai crisis affected India as a whole. Did it shock people that one of its biggest cities could run out of water? 'Absolutely,' says Jain. 'Every time there's a drought or flood, people totally get it and start saving every drop of water. But they soon forget. This time around, it was different. People became aware. Simultaneously, the Central India Planning Commission released a report indicating that, by 2030, almost twenty big cities in India could run out of water. So, that also became a very big thing.' Chennai did have building-code requirements for rainwater-harvesting

tanks to be installed in all new buildings, which could have helped
to avert the 2019 crisis. 'India is very good at making the best policies,' says Jain. 'Enforcement is the biggest challenge. And we are
not so good with that . . . If people's [tanks] got clogged and so on,
no one bothered to fix it. And suddenly, there was a very high
increase of urbanization, people moving from rural areas to urban
areas, and that's when things went awry.'

Vaibhav Chaturvedi, fellow at the Council on Energy, Environment and Water (CEEW), Delhi, argues that Chennai's Day Zero
'forced people to wake up' to the reality of water scarcity. 'It stunned
those living in Chennai; this could never have been imagined. This is
a high-income city, one of the metropolises of India.' But, he argues,
'it's not that Chennai is the only city which could face such severe
water stress . . . Pick a state and there will be a super-dry region.' A
24/7 water supply is almost unheard of anywhere in India, says Vaibhav. The government report Jain cited put it starkly: 'By 2030, the
country's water demand is projected to be twice the available supply,
implying severe water scarcity for hundreds of millions of people . . .
54 per cent of India's groundwater wells are declining, and 21 major
cities are expected to run out of groundwater.'[9]

Jackson, Mississippi

On 4 March 2021, Jazmine Walker took to Instagram in desper-
ation. During an impassioned fourteen-minute post to camera, she
said, 'I am now over the two-week mark of being without clean
water.'[10] Tap water 'has been on and off – this morning it is off, and
when it's on it's not fit to wash with, it's not fit to cook with, it's not
fit to drink.' An historic winter storm had caught the Southern
States by surprise, freezing and bursting many water pipes across
Mississippi's largest city. At least eighty water-main breaks and
leaks had been reported. Storms such as this are freakish historically, but may become the new winter norm – a similar deep freeze
had hit the US in 2020 – linked to a slipping of the gulf stream as
the Arctic warms, pushing previously 'Arctic' weather conditions

further south. 'It's important,' implored Walker, 'for us to not nor-malize the very inhumane thing it is to experience not having clean water, especially in, quote unquote, the richest nation in the world.' She was now reliant on bottled water for drinking and brushing her teeth. Many people were forced to collect and melt down snow and ice for washing. A local newspaper announced that 'Non-potable (flushing) water will be available on Monday from 9 a.m. to 6 p.m. at the following locations', offering a list of school halls and commu-nity centres.[11]

Walker, a Black resident in an 80 per cent Black city, said it was 'no coincidence' that this was happening to one of the country's majority-Black cities. Drinking-water systems that violate federal safety standards are 40 per cent more likely to occur in places with higher percentages of residents of colour, according to the Natural Resources Defense Council (NRDC), due to historic under-resourcing of municipal water infrastructure. Over a quarter of Jackson residents live below the federal poverty line, with obvious parallels to the lead-contamination water crisis in majority-Black Flint, Michigan, nearly four years earlier, which also came about due to chronic underinvestment. When that happens, basic services fail. Community leader Ronnie Crudup told NBC News, 'Infra-structure has been a historic problem, and for years each administration kept kicking that can down the road. This is a long-time issue, but now we're paying a severe price for that neglect.'[12] Jackson mayor Chokwe Antar Lumumba added that the city's wastewater and drinking-water systems require at least $2 billion for repairs. 'It isn't a matter of *if* these systems will fail, it's a matter of *when* these systems will fail,' Lumumba told the *Mother Jones* podcast. 'Pipes were bursting throughout the system because they are over a hundred years old. They're like peanut brittle.'[13]

The requirement for Jackson's residents to boil tap water before using it was lifted on 17 March, having reached almost a month without potable water supply. The *New York Times*, meanwhile, reports an increase in drought and water shortage across the United States: 'By 2040, according to federal government projections, extreme water shortages will be nearly ubiquitous west of Missouri. The Memphis Sands aquifer, a crucial water supply for Mississippi, Tennessee, Arkansas and Louisiana, is already overdrawn by hundreds of millions of gallons a day.' On top of that, 'At least 28 million Americans are likely to face megafires like the ones we are now seeing in California, in places like Texas and Florida and Georgia.'[14] Wildfires are both a symptom and a cause of water scarcity. Many regions rely on trees and forests as nature's water filter, feeding clean water into river basins; following forest fires, that 'natural service' has gone.

Mexico City

The mayor of Mexico City, Claudia Sheinbaum, described the 2021 drought as the city's worst for thirty years. Countrywide, 80 per cent of Mexico suffered an historic lack of rainfall. But Mexico in general, and Mexico City in particular, has long flirted with Day Zero. Ironically, when I talk with Anaid Velasco, an environmental lawyer in Mexico City for the Mexican Centre for Environmental Law (CEMDA), it is raining heavily in North America's largest city. 'But rain in Mexico City doesn't solve the problem of access to water,' she explains. Poor infrastructure means rain 'goes straight to the sewage; it's basically wasted. Seasonal rain doesn't solve the problem of scarcity . . . We go from extreme wet to extreme dry.'

Like many old cities, it is saddled with a debt from its colonial past: 'the Spanish drained the lakes and put rivers into pipes – a crazy way to start a city,' says Velasco. Thanks to over-extraction, Mexico City's ground level is sinking by 50 cm per year on average, as hundreds of boreholes drain the groundwater below like milkshake through a straw.[15] The city was built on the bed of Lake Texcoco, home of the legendary 'floating' Aztec city Tenochtitlán, and becomes more vulnerable to sinking each time the system digs deeper to access those waters. 'We aren't respecting the water cycle; we're unable to inject enough water back underground,' she says.

'So, we're having to take it from other water systems – in a way, stealing it from other communities.'

In Mexico City, says Velasco, around 40 per cent of the water is lost to leakages in the pipe network. This is well above the global average,* but it's an invisible problem that attracts little political attention and public investment, despite water scarcity being a known issue. Velasco, like most Mexico City residents, doesn't drink the tap water. She uses a water filter, but many people don't even trust those, with an estimated 73 per cent relying on bottled drinking water, a rate which is believed to be the highest in the world.[16] Many houses don't have access to even poor-quality water: 'They do not receive *any*,' stresses Velasco. These neighbourhoods, including densely populated areas such as Iztapalapa, rely on water trucks, paying 'at least ten times more' than those connected to the public water supply. The poorer still, on the steep slopes of Acalpixca, rely on water delivered by donkey, each animal loaded with four 20-L jerrycans.[17] When politicians want to win votes in such areas, they promise water access, Velasco says, but when elected they don't actually deliver.

Consistently the least-served districts are those populated by indigenous communities, she says. She has represented the Yaqui community in the Sonora Desert, northern Mexico, whose water source was being drained by a nearby city authority for municipal use. The city's excuse, says Velasco, was that the 'water was needed to fulfil the human rights of the people in the city.' It was a cynical but clever stance that turned many in the city against the Yaqui's struggle. While the court was deciding, an aqueduct to the city was built anyway. Several years on, Velasco is still negotiating a percentage of who gets what out of that aqueduct. She alleges that one of the Yaqui leaders was imprisoned on a fake charge. 'We are the

* Which is hard to put an exact figure on, but the World Bank have it at around 35 per cent, with Denmark among the lowest (7.8 per cent) and Ireland (49 per cent) on the higher side.

sixth-most dangerous country in the world to be an environmental defender,' she says, matter-of-factly.

São Paulo, Brazil

The closest a 20-million-plus megacity came to Day Zero was São Paulo, Brazil, in September 2014. Protesters gathered outside the headquarters of the Basic Sanitation Company of the State of São Paulo (SABESP) over water shortages imposed across South America's – and the southern hemisphere's – largest city, complaining of 'days without water'.[18] SABESP officially admitted that the reservoir levels of the Cantareira system, its largest set of reservoirs, serving nearly 9 million people, were down to just 7.4 per cent capacity on 25 September – perilously close to a Day Zero, and indeed well below Cape Town's threshold. Water was now being pumped from the 'dead storage' – essentially, from the muddy sediment.

Lise Alves, an experienced freelance reporter in São Paulo, tells me that, during the 2014 crisis, residents were told to reduce their water usage by 20 per cent or be fined. She recalls stories of people frantically filling buckets, bathtubs and sinks as soon as the water supply came on. We speak on 7 September 2021, at the start of the dry season, and she says they're running out of water again. 'Brazil is run a lot on hydroelectric plants. The President [Bolsonaro] has already warned there'll be an energy crisis bigger than we had twenty years ago; and water rationing in São Paulo is already expected, because there's not enough for the hydroelectric plants. We're paying three times the price of energy right now.' According to *La Prensa Latina*, the Brazilian government's new 'water scarcity' rate (amounting to a 14.2-reais ($2.60) surcharge on every 100 kilowatt-hours consumed) is designed to encourage households to reduce their electricity consumption.[19]

São Paulo plays the same dangerous game every year. Alves describes it as 'a recurring nightmare'. In September 2016, the reservoir levels again plummeted to 17 per cent capacity.[20] A federal-government water-emergency alert issued in May 2021 stated that Brazil was experiencing the worst drought in ninety-one years, with winter rains in São Paulo 56 per cent lower than average. One SABESP technician projected that the Cantareira system would again reach below 20 per cent capacity the following summer. Alves says she's seen reports suggesting that it could soon be worse than 2014. She adds that the water they do have is unclean: there's a sewage problem thanks to people dumping stuff in rivers and tributaries. Flooding can cause big problems too. 'January and February down here, it takes maybe half an hour for the streets to flood because there's also a drainage problem – the sewers and everything get clogged up.' This year it rained hard, but 'in the wrong places', she laughs. The reservoirs 'just didn't fill up'.

In fact, the rains arrived in October, a month after we talk, just in time to avert disaster, and the Cantareira's capacity reached 29.9 per cent – above the official rationing threshold of 20 per cent, but well below the 60 per cent normal for that time of year.[21] By January 2022, the cycle began all over again, with the *Brazilian Report* stating: 'six out of the 10 most populous states in the country began 2022 with water reservoirs at levels below those of last year.' Whether São Paulo experiences Day Zero remains a yearly gamble.

Santiago, Chile

'There was some rain this year, and people were pretty excited,' Andrea Becerra tells me, raising my hopes. 'But it's still below historic averages. In fact, it's still the second-worst year in the last twenty. So, quite terrible.' My hopes are now dashed. Becerra, former Latin America Project lead for the Natural Resources Defense Council, says that the Santiago Metropolitan Region – one of Chile's sixteen administrative districts – has suffered a decade-long drought, with no more than 300 mm of annual rainfall. Prior to 2010, rainfall was closer to 700 mm a year. Before the crisis reached Santiago city, the impact was mostly felt 'in the rural communities . . . aquifers that dried out, farmers having to dig deeper and deeper wells. Last year, over 35,000 animals died from lack of water in the Metropolitan Region.' But now the drought is biting in the urban metropolis, too.

Santiago's surface water comes from the rain- and glacier-fed El Yeso Dam. NASA satellite images show that this mighty reservoir rapidly halved in size between 2016 and 2020, due to lack of rain and receding glaciers.[22] The glaciers in the Andes are melting at an average of a metre in thickness a year, the fastest loss for a mountain region, relative to its size, in the world. Bolivia's Tuni glacier reduced to just 1 km² in 2021, endangering the water supply to the capital La Paz.[23] In other areas, however, rapid glacial melt has given a false sense of security – when glaciers are melting so fast, they can swell streams and rivers, making up for the lack of rain. But this is fossil water – once gone, it doesn't come back; and once you pass peak meltwater, it is a one-way ticket down. These frozen reservoirs, relied upon for so long, will simply be gone; without rainwater and groundwater, there are no other options. The NRDC report states, bleakly: 'Precipitation [including snowfall] in the Metropolitan Region's Andes mountains has fallen 3 cm every 10 years . . . This has contributed to 8.54 to 15.14 gigatons of glacial retreat – this would have been enough to supply all of Chile's water needs for the next 14 years.'

In 2018, the Aculeo Lake south-west of Santiago, a tourist hot-spot and one of the largest natural bodies of water in central Chile, with a surface area of 11.7 km², totally dried out. It hasn't returned. In just five years, reports Lorena Guzmán in *Diálogo Chino*, Chile's drought has cut water availability by 37 per cent.[24] 'Santiago has a Day Zero coming, there's no doubt about that,' says Becerra, who wrote the NRDC report on the region.[25] Two thirds of its water supply comes from glacier melt, Becerra says; that's the only thing that has spared Santiago from Day Zero over the last decade. So, what happens when the glaciers are mostly melted? 'That's the big question.'

In April 2022, we got the answer. Water rationing was announced for the first time in Santiago city. A green, yellow and red alert system was introduced, based on water levels in the Mapocho and

Maipo rivers. Yellow (a water-pressure drop for all) was implemented immediately. Red alerts – taps being turned off every four,
six, or twelve days – were expected as soon as the summer.[26]

Belgium

'We scientists have been warning for ten years that drought and
water scarcity would become a problem in Belgium,' Marijke Huysmans, professor of groundwater hydrology (Free University of
Brussels) and hydrogeology (KU Leuven, Belgium), tells me. 'And
we were laughed away, literally: "In Belgium it rains all the time!"'
But, during the drought of 2020, thousands of homes suddenly had
no tap water. 'We were on television and in newspapers every day.
It was like a trigger in the public perception that this was something
serious.' Belgium – famous for rain, peeing fountains and muddy
cycling – was recently ranked twenty-second on a list of the most
water-stressed countries, above Mexico and Syria.[27]

Belgium is a case study in how modern urban development affects
groundwater. It's an immaculately hydrologically drained country. If
all roads lead to Rome, then in Belgium all roads, dykes and drains
lead to the sea. But the longer dry periods of climate change mean it
must now hold some water back, or the taps will run dry. In 2020,
Huysmans wrote in her groundwater blog, 'Why does Flanders suffer
from water shortages? The warm and dry summers of the past few
years have made it all too apparent . . . Flanders has water reserves of
1,500 m³ per person per annum. This is "very little" compared to
international norms.' Generally, less than 1,000 m³ per person, per
annum is considered 'a severe shortage.' Huysmans continues, 'Furthermore, Flanders is a densely populated region . . . chaotic town
planning has meant that large areas are built on or tarmacked or
concreted over. 14.5 per cent of the surface area of Flanders is
covered in this way.' Combined, these three factors, plus climate
change, 'create a recipe for water supply issues.'[28]

Huysmans says the situation is 'getting worse . . . our groundwater is not doing as well as it should be this time of year . . . We

complain about rain a lot, as I'm sure you do in the UK,* but object-
ively it doesn't actually rain so much.' Drought and flood are two
sides of the same coin. Dry or tarmacked surfaces also lead to
higher flood risk when heavy rains do come – as they did so
catastrophically in July 2021, when forty-two people died across
Belgium. Climate change is wreaking havoc with the regional rainfall, says Huysmans, and the duration, 'intensity and the frequency
of the droughts we are seeing now, it's even worse than the climate
models predicted . . . there is another drought and another drought,
and groundwater levels are just going down.'

There's no clear distinction between groundwater and surface
water: rivers and streams are fed by groundwater and, to an extent,
vice versa. Huysmans explains: 'Many of our small rivers are mainly
fed by groundwater: 60–80 per cent of the river water in the small
rivers used by farmers, for example, actually comes from groundwater.' Even when we speak, in December, farmers are restricted
from pumping water because the levels are still so low. This has
happened four years in a row, she says, and has been a disaster. A
comparison of shallow groundwater levels between 2000 and 2020
finds decreased groundwater in 70 per cent of measuring stations
across Belgium.[29]

Water demand is also an issue. Non-essential use of domestic
water is typically banned every summer, but Huysmans says certain
regions have come very close to having no tap water at all. During
very dry and hot periods, the demand for water increases by 50 per
cent. 'Obviously, that becomes very, very difficult for the drinking-
water companies.' According to Huysmans, Belgium should be
'prepared for both more long periods of drought and more intense
rains, because this is exactly what we expect.'

---

*   I can confirm.

Pakistan

Pakistan is running out of drinking water. Per-person surface-water availability in the country went down from 5,260 m^3 per year in 1951 to around 1,000 m^3 in 2016 – below the crucial 'severely water-scarce' threshold – and is likely to drop to about 860 m^3 by as soon as 2025.[30] 'The availability of clean drinking water is a huge problem,' explains Maheen Malik, country coordinator at the Alliance for Water Stewardship (AWS), speaking to me from Lahore, Pakistan's second-largest city, in late 2020. Pakistan has a growing population and limited resources. 'Because we share water with other countries, the geopolitics must be taken into account,' says Malik. 'One of the reasons for the major discontent between Pakistan and India over Kashmir is the water that's available [there]'. And then there's agriculture, which 'makes up more than 90 per cent of water used. We are very much an agriculture-based economy.' The Pakistani authorities anticipate a likely deficit of 38 km^3 of water by 2025* (128 km^3 is used annually for agricultural irrigation), with total groundwater reserves of only 30 km^3.[31]

Pakistan's four provinces each have their own distinct but equally acute water problems. In the north, the main issue is access, says Malik: 'Usually women or kids go and fetch it from the nearby river. Whereas in Punjab, we have access, but there's a lot of pollution. In Sindh, the groundwater has a very high TDS [Total Dissolved Solids – a generic water-quality reading] because it's very salty – there's a lot of salt-water intrusion, so you can't utilize the groundwater, and there's limited surface water. Then, in Balochistan, water is as scarce as it gets.'

Pakistan ranks fourth globally in terms of annual groundwater abstraction – an unwanted title it vies for alongside the much larger

---

* The measurement of 1 km^3 of water is exactly as it sounds: picture a 1,000m-sided cube of water.

India, China and the United States. The Indus Basin aquifer, from which Pakistan extracts most of its groundwater reserves, has been ranked the second-most overstressed in the world.[32] With informal, often unregulated water pumping, this is causing a literal drain on Pakistan's water supply. There's no licensing system, Malik says, 'so many people extract groundwater via unlicensed boreholes.' Pakistan is also the world's largest exporter of groundwater, in terms of the water used to produce water-intensive export products such as leather, textiles and rice.[33] The little water Pakistan does have is being packaged up and sent elsewhere. The likely climate consequence of extreme drought is extreme flooding. This happened to Pakistan with devastating consequences in September 2022, when the most widespread flooding in its seventy-three-year history left one third of the country under water. Balochistan, Pakistan's poorest and typically driest province, saw flooding destroy 60 per cent of the houses.[34] As the floodwaters receded, Unicef warned that 'the sheer scale of damage is being revealed. Hundreds of thousands of homes have been damaged or destroyed, while many public health facilities, water systems and schools have been destroyed or damaged.'[35] Al Jazeera reported 1,500 dead, commenting that 'this tragedy has highlighted the desperate urgency of reinforcing climate defences, in Pakistan and elsewhere.'[36]

The Mekong Delta

The Mekong River is the longest in South-East Asia and the tenth-largest in the world, flowing some 4,350 km through China, Myanmar, Laos, Thailand, Cambodia and Vietnam, covering more than 800,000 km^2, its waterside home to around 100 million people. Two capital cities – Vientiane (Laos) and Phnom Penh (Cambodia) – stand on its banks. A failure of the Mekong would cause a domino effect of Day Zeroes for countless cities. The Vietnamese section alone is inhabited by more than 21 million people.

Chi Nguyen was born and raised in the Mekong Delta region of Vietnam and lectures in environmental communication. When we speak in May 2021, she is in her hometown, Can Tho. As the major food-producing region of Vietnam, the Mekong Delta is acutely vulnerable to climate change and transboundary water disputes. Despite being the 'rice bowl of Vietnam', Nguyen tells me, it is 'the poorest region and least-developed in Vietnam'. In Brian Eyler's book *Last Days of the Mighty Mekong*, he situates Can Tho, 90 km upstream from where the Mekong meets the East Sea, within 'one of the most engineered landscapes on the planet, a flat and fertile agricultural plain crisscrossed with a network of 30,000 km of rivers and man-made canals.'[37] China's upstream dams may be located 2,500 km away, but the massive amounts of fresh water they hold back affect the flow, sediment and salinity levels for Vietnam's Mekong Delta. Such transboundary issues are compounded by the suspicion that 'Vietnam doesn't have a lot of voice with the Mekong River Commission', says Nguyen, of the inter-governmental organization that works to 'jointly' manage shared resources.

She says the many dams built upstream in China affect the river's flow downstream. Most people rely on the river as their main income, but fish stocks are depleted; the decreased quality of the soil, sediment and the water itself affect the rice and shrimp industries. Farmers who could previously grow three rice crops per year, now grow only two. Coconut crops have reduced by half. New crop types are being trialled to work with the increased salinity. But the main solution favoured by farmers is abandoning the river entirely and pumping groundwater instead. The rate of groundwater extraction is so high that it has already exceeded the aquifer's ability to naturally recharge, despite the annual monsoons. In 2019, rainfall reached only 60–70 per cent of the annual average, while the dry season of 2020 broke all records.[38] Every day, 2.5M m³ of water are withdrawn, causing the delta to sink by centimetres each year – a dangerous predicament, given that the sea levels are also rising due to climate change. Salt-water intrusion therefore creeps ever further inland, like a cancer. In 2019, salt water reached as far inland as 130 km.[39] Projections show high tides subsuming much of Vietnam – and most of the Mekong Delta – by 2050.

Ironically, Nguyen has helped documentary-makers from Hong Kong to record how bad the situation has become, while mainstream Vietnamese media are 'not allowed' to mention it. Most local journalists know all too well that they can't criticize China. Having already built eleven dams across the upper Mekong, China has helped downstream countries such as Laos and Cambodia to build more. Researchers found that 'China's dams restricted nearly all upper Mekong wet-season flow', leading to the drought in the lower Mekong region in 2019.[40] This makes Beijing an easy scapegoat. But Beijing is itself one of the most water-scarce capitals in the world, with water availability of only 193 m³ per capita.[41] A 2020 study found that, of all the world's urban centres, China's Pearl River Delta, the country's manufacturing hub and home to tens of millions, is the most at risk from rising sea levels – more so, even, than the Mekong.[42]

Sub-Saharan Africa

Cape Town may not be the most likely site for Southern Africa's next Day Zero. Themba Gumbo, a water engineer from Zimbabwe, based in Pretoria, says that the urban region most at risk is Johannesburg–Pretoria. 'I don't think we are much prepared for that kind of scenario . . . There'll be more chaos here', he warns. This urban conglomeration, home to some 15 million, 'is the economic hub, with all the major industries, but that means it also generates a lot of waste.' Waste flows are poorly managed, and end up in the Vaal River system, the original water source. The severe pollution problem is exacerbated 'not only by pollution from human waste, but also from the mines, what we call acid-mine drainage – the city of Johannesburg is full of holes underneath from mining.'

In Harare, Zimbabwe, too, he says, 'probably 80 per cent of the population is off-grid, with boreholes and tanks; they move water around because the centralized system has failed.' Arguably Harare is already living with Day Zero, forced to survive on private, unconstrained groundwater pumping. The Sahel, meanwhile, is among

the most water-stressed regions on Earth. Record-low rainfalls –
comparable to a major dry period in 2011 that led to thousands of
deaths – have caused a sharp drop in crop production. Patrick
Youssef, Africa director for the ICRC (International Committee of
the Red Cross), told ReliefWeb in May 2022, 'In some places in Bur-
kina Faso people wait in line for 72 hours to access boreholes. Their
lives are completely revolving around reaching water. Should the
situation deteriorate further, we are facing the real possibility of
people . . . dying of thirst.'[43] Further south, in Maputo, the capital of
Mozambique, a persistent drought has been ongoing since 2015,
with water rationing introduced in 2018.[44] Meanwhile, Day Zero in
Accra (a city of some 5 million people), Ghana, 'happened as recently
as April' 2021, Accra resident Richard Matey tells me in June that
year. 'The Ghana Water Company closed the taps for a week due to
some maintenance issues,' leaving about half the population with no
access to water.

'In this part of the world, we say, "If you jump into a river, then
dust yourself down,"' jokes Gumbo. 'Because, most of the time, it's
just a dry channel.' The ongoing challenge for water authorities in
Southern Africa – in terms of both demand and supply – is 'the rate
of urbanization'. 'It's a huge problem throughout South Africa,
maybe throughout the whole region; informal settlements are
sprouting everywhere.' In South Africa specifically, the Water Act
provides basic water rights; each household is entitled to 6 m^3
(6,000 L) free per month. But revenue isn't going to the municipality to manage the services. Not everyone that is required to pay
does so – and the subsidized limit is rarely checked or enforced.
Gumbo tells me of his friend in a Soweto township who was unsure
of where his water meter was: in five years, no one had ever looked
at it. 'That is the state of things.'

*

Back in the drenched lowlands of the Netherlands, Rick Hogeboom,
head of the international Water Footprint Network, has a different
take on the Day Zero concept. He visited Cape Town during the

peak of Day Zero, when rationing was down to just 50 L per day. 'It was no fun. I couldn't even take a shower. And in the slums, of course, the situation was way worse.' But, he says, 'it's a management problem. The infrastructure was just lacking. There's enough water in the region to support the population in Cape Town, period.' He cites wineries and wine production as the mainstay of the area's economy. 'There's a lot of wine farms producing grapes in the outskirts of Cape Town. From a water-systems perspective, that's still the same water. Did the wineries suffer from [the same] water shortages during Day Zero? They did not. It's a $4-billion industry.' Policymakers make water-allocation choices and those choices impact society and inequality, he argues: for any city, water supply is a 'preventable and manageable' problem. The problem, he says, is more typically a lack of storage – both in surface reservoirs and underground – and ageing, leaking pipes. There's long been complacency in many cities and countries, who rely on creaking infrastructure and seasonal rains arriving like clockwork to save the day. This system may have worked, just about, in the twentieth century. But now the rains – and the water cycle – are no longer dependable.

CHAPTER 2

The Water Cycle Is Changing

The Prampram-Tekponya community used to live in a remote coastal village in southern Ghana, surrounded by fertile farmland and abundant freshwater streams. Life was good. Until the developers came. First, they moved the villagers closer to the beach; the seaside held no allure for urban types. Still, there was a large stream and a lake, which the community dammed at the coastal end by planting trees, so it wasn't a bad relocation. Most villagers live in makeshift homes of timber and corrugated iron, or with more traditional mud walls and palm-leaf roofs. Only in the past five years did the coast became fashionable, and droves of developers began building up to and in front of the village. Concrete and breeze blocks steadily choked off the streams feeding their dam. The villagers and the lake may have survived even this, thanks to Ghana's two monsoons a year. But the rainy seasons are no longer what they were. The developers built the coffin, but climate change has driven the nails in.

I visit in what should be the end of the wet season, September, but very little rain has fallen – the village dam looks as if it's the middle of the dry season, barely a metre of muddy water clogged with weeds. This, the villagers tell me, means 'it will certainly fully dry up this time. Then, what will we do?'

Before travelling to Prampram-Tekponya, I speak with Richard Matey of the small local NGO Alliance for Empowering Rural Communities (AERC), who helped set up my meeting with the villagers.

'It is a huge challenge,' he tells me. 'We should have been experiencing heavy rainfall from April to July, but there hasn't been rain. When we hit April, May, June, it gets colder, you normally put on your pullovers . . . in 2021, it's very, very warm; 2020 was recorded as our hottest year and it's likely that 2021 will be even hotter.' Climate change is causing wetter regions to get wetter and drier regions to get drier. The question for countries like Ghana, in the tropics of Africa, is which way they'll go. Matey predicts, 'It is going to be drier – even the tropical areas are beginning to experience dryness.'

To visit the Prampram-Tekponya community, I'm told by my host for the day, Felix Best Agorvor, Richard's colleague at AERC, we must first seek permission from the chief. We pick Felix up in the car on the road out of Greater Accra. While the roads get less and less passable, we never really leave a built-up area. As we pass building sites and industrial yards, Felix – who is only in his twenties – keeps exclaiming, 'That never used to be there!' When we reach Prampram-Tekponya, it is a rural oasis within an urban sprawl – the last of its kind, here. The chief's 'palace' is the only concrete-walled building in the village. We find that the chief isn't at home, but his wife, Naadu Mateye, grants us an audience. Plastic chairs are placed under the shade of a barwood tree in the yard. Naadu Mateye is stern-faced, not keen on long answers, while Felix interprets. The village received a connection to the municipal water supply ten years ago, Mateye tells us, but there is only one tap, connected to a water meter. The village monthly water bill ranges from GH₵1,000 to GH₵2,000 (Ghanaian cedi, or £120 to £240, $142 to $285),* while the villagers pay the chief for what they use, at 60Gp (Ghanaian pesewas, or 7p) per gallon. Yellow plastic one-gallon jerrycans can be seen in every yard we pass. I ask what people do when they can't afford to pay. Naadu Mateye points to the stream, the *yoduwekope*,†

---

* Felix, who is local to the area but not to the village, tells me his own household water bill is typically GH₵300–GH₵400 (£36–£48/$43–$57).
† I am told that *yoduwekope* literally means 'a woman must bathe' – a private space for women to bathe in, away from men.

behind the houses – barely more than a muddy ditch. An older lady I meet later remembers it as a clear, flowing river. It used to be a farming village, with abundant land to farm and reliable rain; now, developers have taken most of the land, and the rains don't come. Everyone I speak to tells me the same thing: the rainy season used to begin in May, but last year it simply didn't arrive. This year was the same again. Now, occasional heavy rain might fall in August or September, hitting long-parched ground and immediately flooding, before draining away.

When I ask Naadu Mateye what solution would help her here, she says, 'A dam,' meaning an additional local pond for the women to fetch water from. Then she receives a call on her mobile – it's her husband, Chief Nene Kwaku Abladu. He is granting us an audience after all, but only over the phone. The chief proposes a different solution: currently he has just one polytank that fills with municipal tap water. So, three more would do the trick. Naadu Mateye, then, wants more storage of free river- and rainwater, while the chief wants more storage of the treated water that he sells – an expansion of his business, in effect, and one he has a monopoly over. Nevertheless, the only tap can go for two weeks without water supply, so one polytank really isn't sufficient to store enough for the community's uses.

When the chief kindly gives us permission to walk around the village and talk to some of the villagers, however, we learn just how costly this water is for people living hand-to-mouth. I talk to one fishing family. The wife, let's call her Afi,* is at home with her mother-in-law, Joy. She confirms that buying water from the chief's polytank costs them 60Gp per gallon and, when they don't have money, they take water from the stream. She tells a child to fetch me a plastic glass of stream water from a nearby jerrycan (indicating that they regularly can't afford the treated tap water) – it's too dirty to see even a couple of centimetres down, let alone to the bottom of the small cup. I ask if it makes them sick. Joy nods and makes a retching

---

* I was asked not to use their real names, only the name of the chief.

motion with her hands. Was the stream cleaner when Joy was young? She smiles and widens her hands, signalling abundance.

Such dirty water is 'treated' in various ways. The traditional way, Felix explains, is to burn dried pumpkin or coconut husks and use the smoke to 'purify' the water; this is a taste that many Ghanaians would recognize and like, he says. Another way is to add 'alum', explains Felix, a chemical water treatment readily available in local markets. When added to water, it's an effective coagulant, sticking to impurities and making them drop to the bottom of the container, clarifying the water ready for drinking. It's also used by locals as an underarm deodorant.*

The chief's charge of 60Gp per gallon of tap water is too expensive, says Afi. She and her family buy that water only for drinking, and for washing clothes for special occasions (including, ironically, meetings with the chief). But they can't afford it all the time. Her husband the fisherman, Ekow, arrives and eyes me nervously before sitting down. He says sometimes he returns from fishing with nothing, and therefore earns no money; on very good days, he might earn GH₵200 (£19.50/$23), but bumper catches are getting rarer. I ask how much of their household expenditure goes on water. Afi and Ekow confer and then respond that their spending is typically around GH₵10 a day, and of that around GH₵5 is for water. So, half of all their already very low household income is spent on water. Some of that goes as profit to the chief, yes, but the majority simply goes to Ghana Water Company (GWC), who charge far more for water than many Western utility companies do, to customers earning on average far less. Even Felix tells me he is shocked by this.

When I ask Afi and Ekow what solution would help them,

---

* When I look it up later, I discover that 'alum' is aluminium sulphate – long used for water treatment, but in the wrong doses can form sulfuric acid. Overuse has also been linked with Alzheimer's disease, according to one recent paper, 'both for workers occupationally exposed . . . and for people who drink the water'. Inexpertly scattering it into collected cans of river water is, therefore, risky.

unsurprisingly their answer isn't the same as the chief's. Ekow would ideally like a piped water connection of their own, to cut out the middleman (the chief, although Ekow is far too deferential to name him) and just pay directly for what they use. GWC has quoted him GH₵2,500 (£305/$415) for such a connection, which is a significant part of his annual income and simply unaffordable. Felix tells me that installing his own connection in town cost closer to GH₵1,000 (£120/$165), due to his proximity to the mains pipe. In town, you are always closer to the municipal supply than you are in a rural village, and GWC rates seemingly have no concession for this. 'Water is expensive,' says Felix, in blunt summary. It makes me think, guiltily, of just how cheap it is in the UK.

The village is cut in half by the *yoduwekope* stream, but at some point a road was added and developers moved in, separating the two sides of Prampram-Tekponya by perhaps as much as a couple of kilometres. The other side feels quite different. They share the same chief, but there seems to be more autonomy here – the population is bigger, there's a school, the houses are built more permanently. I sit under the shade of a large mango tree in the neat, spacious courtyard of an elder. His son, we'll call him Kosi, sits with me to answer my questions in English. This side of the village has a separate water connection and polytank, and the elder whose courtyard I'm in appears to have the monopoly. Here, the rate is 80Gp to GH₵1 per gallon. Perhaps they don't know that this is a 33–66 per cent markup compared to the other side of the tracks. Perhaps the chief is making concessions for the poorest families, after all.

For free water, the one dammed pond is on this side of the village too. But this year, 'The rain didn't come,' says Kosi. 'Barely two, three, four months after the rainy season, the dam dried up. So, we really depend on Ghana Water and have only one tap. Fortunately, our MP donated a polytank, so, when the water is flowing, we fill the tank and then we sell it . . . but, because Ghana Water is not reliable, it doesn't flow frequently . . . It can be only two or three days a month, or they can decide to close it for one month, two months.' When the polytank is empty, they are dependent on the dam or must buy water in the town, for GH¢1.5 a gallon (a 150 per cent markup, compared to the chief's price), plus between GH¢10 and GH¢50 for car hire. 'You have no choice because you don't have water.'

Kosi and his older brother Akwasi suggest we go to see the dam itself, a short walk away. The beautifully mud-rendered dwellings give way to half-finished breeze-block buildings. The dam is a large pond, roughly 200 m by 400 m, surrounded by trees and reeds, with a thick cover of water lilies on the surface. A few well-trodden metres down the bank is the small muddy shoreline where the villagers retrieve their water – a large plastic bottle lies abandoned, half-full of muddy water. The water at this end is only 30 cm or so deep; at this time of year, at the end of the rainy season, the dam should be full, even at risk of overflowing the bank that currently stands high and dry, metres above. Less than a kilometre away, a red-roofed house is built almost into the dam itself, and the roofs of several other developments are visible not much further away. The dam, it seems, is simply an inconvenience for the developers. Wealthy housebuilders don't care about free water from dirty ponds; they have their water delivered via truck or GWC pipe. For developers, the sooner the dam dries up and is built upon, the better.

'This is the only reliable source of water we have,' says Akwasi. 'But people are not reasonable . . . they are not thinking of the future of the dam, they are only thinking to build their houses.' Akwasi believes the dam will inevitably dry up this winter. As we drive away, we pass several more houses being constructed. In the shade of a tree, four young men stare at us as we drive past. 'They are *langas*,' warns Felix. 'If you mess with the builders, they will beat you. They are paid to agitate. That's the fear in the community.' Development brings jobs, voters and taxes. But it does not bring water. And it does not bring the rain.

*

Jay Famiglietti is to water academia what Tiger Woods is to golf: you don't write a book about water without spending some time at his court. His back catalogue includes professor of Earth-system science at the University of California, Irvine; senior water scientist at NASA; and water adviser to California governor Jerry Brown and the US Congress.[1] When we talk, he's director of the Global Institute for Water Security (GIWS) at the University of Saskatchewan, Canada. And he's a podcast host: *Let's Talk About Water*, seasons one and two, and *What About Water? with Jay Famiglietti*, both of which open with a pleasingly banging techno intro to pump you up to match Jay's enthusiasm levels.

Jay's biggest hit to date was his part in NASA's Gravity Recovery and Climate Experiment (GRACE), which identified groundwater depletion in the world's major aquifers. The theoretical work behind it was done as far back as the 1920s, when geophysicists worked out that vast accumulations of underground water could affect the gravitational pull in those parts of the Earth. 'They knew that the movement of water underground impacted the gravity field, but back then they had no way of measuring it,' says Jay, from his makeshift home office – they have recently moved, and behind him is a whiteboard with mathematical equations (his wife's) and a guitar (his) propped up against the wall. The GRACE satellite, launched in March 2002, finally provided the measurement. This wasn't a purely environmental mission. If you're going to run a space programme, says Jay, 'You really need to understand the gravity field to get precision orbits and know where a rocket's going to re-enter the atmosphere.'

GRACE works, Jay simplifies, like a 'scale in the sky'. As the satellite passes the Earth's land surface, its equipment is pulled in ever so slightly by the stronger gravitational pull caused by underground water. The data produced has helped to build a highly accurate global map of groundwater and – over time, as GRACE fed back data from 2002 to 2017 – of groundwater depletion. The 2015 paper, 'Quantifying renewable groundwater stress with GRACE', on which Jay is the corresponding author, was groundbreaking. It found that, of the world's thirty-seven largest aquifers, twenty-one were receding faster than they were being naturally replenished. There were two smoking guns. One: human overuse. The aquifers in the worst health were those with the highest levels of irrigation demand and population density. The most overstressed aquifer systems of all, with the highest depletion rates and little to no available recharge, were the Arabian, the Murzuk-Djado Basin (North Africa), the Indus (India and Pakistan) and California's Central Valley (see Graph 1). The second smoking gun, a changing water cycle, caused by climate change. As water disappears from one region, it must reappear elsewhere – and indeed there are some major aquifers with a positive bank balance, including the Northern Great Plains (US and Canada), the Amazon Basin and the Karoo Basin (South Africa).

But having more water doesn't necessarily make these places the 'winners' in a changing water cycle. Wetter can mean increased flooding. Saskatchewan in Canada, where Jay used to live up until recently, is one of the places getting wetter. 'Some of that is good; Canada grows a lot of food and wants to grow more food. So, it's an abundant resource. The bad part is the way [precipitation] is being delivered is more episodic. The variability is changing.' There's more heavy rain, interspersed with longer periods of drought; the oscillations from wet to dry are getting stronger. This means more floods *and* more drought. Jay describes it as an 'intensification of the water cycle'. A warmer world sees more water held in the atmospheric stage of the water cycle,* meaning that the 'swings of the water cycle are getting bigger.' When I speak to Jay, I note that the UK has just had its wettest-ever February followed by its driest-ever May. 'That's it,' he says. 'Like a see-saw from wet to dry – the whole world is seeing this . . . and that's super difficult for water management.'

In his 2016 RBC Distinguished Lecture at the Water Institute, Jay elaborated on his 2015 paper.[2] The most depleted aquifers were 'past sustainability tipping points,' he said. 'Some of them at super-rapid

---

* This doesn't mean there is more water overall, just that more water is being held in the air as water vapour at any given point – for reasons we'll get to.

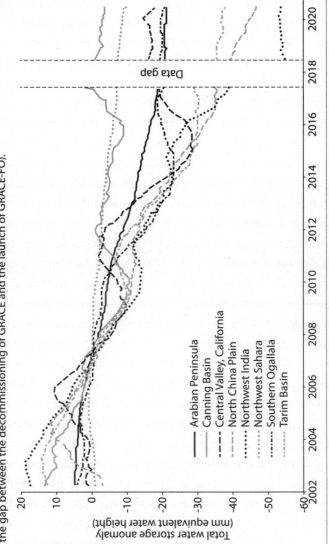

Graph 1 Watching groundwater depletion from the sky

Changes in total water storage across major aquifer systems globally, as shown by the NASA Gravity Recovery and Climate Experiment (GRACE) and follow-on (GRACE-FO) satellite missions (the vertical dashed lines indicate the gap between the decommissioning of GRACE and the launch of GRACE-FO).

Source: Famiglietti, J. S. and G. Ferguson, 2021. The hidden crisis beneath our feet, Volume: 372, Issue: 6540, Pages: 344-345, DOI: (10.1126/science.abh2867)

rates, like India, Pakistan, the Middle East, Central Valley, the Guarani aquifer. These are all our big food-producing regions . . . I don't think we're really ready . . . The world is not ready . . . We will have the haves and the have nots of water.' When we speak in 2020, Jay stresses that his 2015 paper wasn't simply a snapshot in time: 'The trend is linear,' he says, and this is broadly climate-driven. 'The high latitudes are getting wetter, the arid regions drier, the ice sheets and glaciers continue to melt . . . If you were to remake that map every five years, you'd see the same direction of travel.'

According to a 2017 World Bank report, 'Climate change will worsen the situation by altering hydrological cycles, making water more unpredictable and increasing the frequency and intensity of floods and droughts.'[3] Sandra Postel, director of the Global Water Policy Project, appeared on Jay's podcast in late 2021 and proclaimed that 'the water cycle is broken'. Our part in that can be found in how we've managed water and land. Globally, 60,000 large dams are blocking rivers from flowing their natural course, said Postel: 'We have huge levees to guard against flooding, but rivers are meant to flood. They're no longer connecting to their floodplain – that's a break in the water cycle. Groundwater is the base flow for rivers.' The double whammy of man-made climate change and man-made engineering means that 'the heating of the atmosphere is fundamentally changing the water cycle, the warmer atmosphere is holding more moisture, which means that droughts and floods are intensifying.' Postel says this is fundamentally changing water cycles everywhere – 'locally, globally, regionally' – and will dramatically impact 'food security, water security, the health of people and nature in general.'[4] Our 'command and control' approach to water is, she says, a 'Faustian bargain that is breaking the water cycle.'

In 2021, it seemed, the Devil wanted his payback. In July, heavy rain hit western Germany. Cologne recorded 154 mm of rainfall in only twenty-four hours, nearly double its monthly average for the whole of July (87 mm). In Reifferscheid, an incredible 207 mm of rain fell in only nine hours. The world watched as houses and cars flowed down picture-postcard black-timbered streets. Buildings collapsed like matchstick models, and whole fields disappeared into sinkholes. Within a week, Germany recorded forty-nine deaths; there were six more as the storm passed through the Netherlands, Belgium and Luxembourg. At the height of the chaos, in Germany's worst-hit Rhineland-Palatinate state, 1,300 people were assumed missing.[5] On 16 July, Fridays for Future activist Luisa Neubauer tweeted: 'Here in Germany, dozens have died in floods, hundreds are missing, thousands have lost their homes. It's devastating. This is the climate crisis unravelling in one of the richest parts of the world – which for a long time thought it would be "safe". No place is "safe" anymore.'[6]

Barely a week later, Zhengzhou, China, said, 'Hold my beer.' The Zhengzhou weather station recorded a staggering 201.9 mm in *one hour*, between 4 p.m. and 5 p.m. local time, on 20 July. According to weather blogger Minghao Zhou, this was a new hourly record for all 2,418 national stations in mainland China and 'possibly the largest downpour in human history for a city with 10+ million population.'[7] By the end of that day, the Zhengzhou station had recorded 612.9 mm in twenty-four hours. To put that into perspective, I've been keeping a rain gauge in my back garden for this book, diligently updating a weekly spreadsheet. Throughout the whole of 2021, in soggy old England, where it rains every month (the longest dry period I've had is three weeks), I recorded 624 mm. Pretty much the same as Zhengzhou received in that one day. No wonder the hydrologist and climatologist Peter Gleick tweeted on 23 July: 'I'm immersed (so to speak) in the history of extreme weather. But this is unfathomable to me.'[8] It was easily the highest daily rainfall recorded in Zhengzhou: its annual average is only 640 mm.[9] Other parts of the province were also badly affected, with dozens of cities flooded. On 2 August, the Henan provincial authorities reported 302 deaths, 292 of which came in Zhengzhou.

According to UN Water, the number and duration of droughts has increased by 29 per cent since 2000, compared to the two previous decades; but flood-related disasters have increased by 134 per cent over the same period. In December 2021, record rain hit Los Angeles too. Parts of Santa Barbara County saw more than 178 mm in one day, while downtown Los Angeles surpassed its previous record, set in 1888. The *Los Angeles Times* described 'the normally constrained L.A. River . . . sucking vehicles down its surging waters.'[10]

Shortly after the extreme events of Germany and Zhengzhou, I contacted Dr Jennifer Francis, senior scientist at the Woodwell Climate Research Center, Massachusetts, to ask her what the hell was going on. She had just written an op-ed for *The Hill* stating, 'To be clear, summer heatwaves and floods have always happened. What's new is that the intensity, persistence, location, frequency and expanse are beyond what we call normal. Long-standing records are being smashed by large margins . . . the pace of worsening weather is faster than forecasted.'[11] Francis worked in the Arctic in the 1980s with the University of Washington's Polar Science Center. At first, the 'Arctic wasn't changing and climate science wasn't a thing yet,' but then, in the 1990s, ice sheets dating back to the previous Ice Age began melting. She's now, primarily, a climate scientist. She'd rather be back studying the unchanging, pristine Arctic that she still remembers, but it is now, by definition, impossible to be an Arctic researcher without also researching a changing climate. The Arctic is the fastest-warming region on Earth.

'Globally, we're seeing longer wet spells, and longer dry spells. And when it does rain, it tends to rain harder,' Francis tells me, returning to my 'What the hell?' question. 'We're definitely seeing an increase in the frequency of heavy-precipitation events . . . it's really quite straightforward and very directly connected to climate change: as we warm the oceans and atmosphere, evaporation is increasing, which puts more water vapour into the atmosphere. Water vapour is a huge climate story that doesn't get enough attention. Water vapour is a greenhouse gas. It's exacerbating global warming in general. And it's the fuel that storms feed off.' Warming oceans add to this, but so does warming land. 'Extra evaporation from warmer land and air means that the soils are drying out faster, especially during spring and summer when the sun is really strong.

Once you dry out those soils, they can heat up much faster . . . that then feeds into intensifying heatwaves, which further intensifies drought. So, it sets up one of these feedback loops. The shift in the water cycle really means that dry areas are all getting drier and hotter, interspersed with really heavy precipitation events.' Climate change is 'intensifying the water cycle', says Francis, echoing Jay. 'By warming the climate, we're speeding up how much water is evaporated from the oceans, it's going into the atmosphere, and it's raining out faster.'

And then some. Floods devastate water infrastructure and water quality just as surely as – and far more rapidly than – no rain at all. For the last decade, Santiago, Chile, has subsisted with an annual average rainfall below 300 mm. But even those years have seen extreme floods. When heavy rain hits hard, sun-baked ground, it spreads out, not down. Despite far wetter average years from 1912 to 1999, the Santiago Metropolitan Region experienced a 22 per cent increase in the number of floods and nearly six times more landslides between 2000 and 2017; in 2017, in one of its driest years on record, a flood left more than 4 million Chileans without water.[12]

In terms of a changing water cycle, few prospects are scarier than the Amazon. Its rainforest is often referred to as the 'lungs of the world', but just as important is its creation of clouds. A 2019 NASA study* showed that, for the past twenty years, the atmosphere above the Amazon rainforest had been getting drier. Armineh Barkhordarian, the study's lead author, described 'a significant increase in dryness in the atmosphere as well as in the atmospheric demand for water above the rainforest.'[13] Elevated greenhouse-gas levels were found to be responsible for approximately half of this increased aridity, with 'human activity' (namely burning and clearing forests for agriculture) accounting for the rest. Trees and plants need water for photosynthesis and to cool themselves down, pulling water from the soil through their roots and releasing water vapour through

---

* Not GRACE this time, but AIRS – the Atmospheric Infrared Sounder instrument, aboard the Aqua satellite, which launched in 2002.

pores on their leaves into the atmosphere, where it cools the air to form clouds. Rainforests are the perfect circular system. However, as the air gets drier, this system could collapse. 'It's a matter of supply and demand,' said Sassan Saatchi, the study's co-author. 'The trees need to transpire . . . But the soil doesn't have extra water for the trees to pull in.' Much as with the humans, the forest ecosystem is seeing that 'the demand is increasing, the supply is decreasing'.[14] As we saw in Chapter 1, São Paulo almost ran out of water in 2014 and 2021 due to drought and deforestation on the fringes of the rainforest. Were the *entire* rainforest system to collapse, the consequences for the 22 million people of São Paulo, the 200 million people of Brazil, and the world generally, would be catastrophic.

In addition, the jet stream – the core of strong winds, around 8–11 km above the Earth, blowing from west to east – is not as predictable as it used to be, either.* The German and Belgian flooding, the Pacific Northwest heatwave of 2021 (which caused the Canadian town of Lytton to burn to the ground during temperatures of 49.6°C – five degrees higher than anything previously seen anywhere in Canada) and the Mississippi freezing that same winter (causing Jackson's water crisis) were all caused by the jet stream splitting off into 'eddies', much like the currents of a river. 'What happened with the Germany flood was, one of these cut-off eddies in the jet stream, south of the main flow, created this upper-level "low", explains Francis. 'You could think of it like a swirl in a river – in this case, spinning in a counter-clockwise direction. When you get one of these cut-off features that aren't part of the main jet stream any more, they tend to just kind of sit there, sucking

---

* Not to be confused with the Gulf Stream, which is the ocean current of seawater taking warm water from the Gulf of Mexico to north-west Europe. The Gulf Stream is no less relevant, though. Viewed as an underwater energy conveyor belt, it too is being slowed by the melting ice sheets, which could also cause extreme weather, higher tides and stronger hurricanes. (*See* 'The rivers in the ocean mustn't stop', We Are Water Foundation (website), 8 April 2021, https://www.wearewater.org/en/the-rivers-in-the-ocean-mustn-t-stop_339321)

moisture out of the ocean, and basically just transporting and dumping it.' This is the 'dynamics' of the weather event, says Francis; extra water vapour adds the 'thermodynamics'. 'The Pacific Northwest [event] was just the opposite situation, where the jet stream took a big swing northward, and created an eddy known as a "blocking high", which is a very persistent heat creator . . . It was historic, the worst heatwave ever, relative to what the temperatures are supposed to be.' An increase in such extreme weather events is now expected, says Francis. The latest Intergovernmental Panel on Climate Change (IPCC) report (2022) summarizes: 'Climate change affects the occurrence of and exposure to hydrological extremes (high confidence) . . . increases in precipitation intensity (high confidence), local flooding (medium confidence), and drought risk (very high confidence).'[15] But while 'we've known it for a long time,' says Francis, 'the pace of increase in really bizarre extreme events is exceeding our expectations.'

As an Arctic researcher, Francis knows better than most the contribution a warming Arctic makes to such weather events, too: 'The Arctic is warming about three times faster than the globe as a whole,' she says. 'And that means that the difference in temperature between the Arctic and areas south of there is getting smaller . . . that north–south temperature difference is the main fuel that drives the jet stream. Winds are driven by temperature differences. So, as we make that north–south temperature difference smaller, we're weakening the west–east winds of the jet stream. And we know that, when the jet stream gets weaker, it tends to form these eddies that caused the floods in Germany and the heatwave in the Pacific Northwest.' A weakening jet stream with slower winds also causes 'sluggish' hurricanes, that move more slowly, with more devastating intensity. 'When they do strike land, they last longer,' says Francis. Take Hurricane Harvey, which hit Texas and Louisiana in August 2017, and sat in Houston for three days, dumping almost 1.5 m of rain. 'Storms like Harvey, that are moving really slowly and doing much more damage because of it, are more likely to happen.' I ask if the thawing of the permafrost and melting of the ice sheets add further fuel too. 'Oh, absolutely. The number-one contribution to sea-level rise right now is the melting of Greenland. And that's all recent.'

The heaviest rainfall events, known as 'atmospheric rivers' or rivers in the sky, are therefore on the rise too. Atmospheric rivers are narrow columns of water vapour flowing through the sky that pick up moisture from warm areas and deposit it onto colder regions, leading to extreme rain or snowfall. Scientists at the University of Tsukuba, Japan, ran climate simulations in 2021, using nearly sixty years of weather data, and found that atmospheric rivers 'become more frequent in a warmer climate', a finding that consistently appeared across all the models they ran. While focusing on Japan, Taiwan, north-eastern China and the Korean Peninsula, the authors also note that 'the findings of this study should also apply to other mid-latitude regions', including western North America and Europe.[16] In February 2022, an atmospheric river hit New South Wales, Australia. In the three days to 28 February, Greater Brisbane received 676.8 mm of rainfall, the largest three-day (and largest weekly) total ever recorded there – more rain than my back-garden rain gauge recorded over an English year, and more than Brisbane's annual rainfall in some previous years. The Aussie press termed it a 'rain bomb'. The weather station at Mount Glorious, rural Queensland, recorded 1.5 m of rain that week.[17]

Such erratic and violent changes in the water cycle 'will directly affect the terrestrial water budget', says the UN World Water Development Report on Water and Climate Change (2020). Paradoxically, you can have more rain with less available water. Longer dry seasons and shorter, more intense rainy periods mean there might be more annual rain, but it will hit dry ground and skip the crucial groundwater recharge or wetland filtration stage, jumping straight to flooding (as seen in Pakistan). It's the water cycle, but not as we know it. It's speeded-up, super-fast broadband piped in to ageing, analogue infrastructure. 'Climate-change-induced changes in the cryosphere are also widespread, leading to a global reduction in snow and ice cover', the UN report continues. 'Snow cover, glaciers and permafrost are projected with high confidence to continue declining in almost all regions throughout the twenty-first century . . . [This is] expected to have a negative effect on the water resources of mountain regions and their adjacent lowlands.'[18]

Dr Junyan Liu began her career working on the retreating glaciers of China's Qilian Mountains, first visiting Laohugou Glacier No. 12, the range's largest glacier, on 12 July 2018. Water from the Laohugou and neighbouring glaciers irrigate the Hexi Corridor and were integral to the development of oasis cities along the Northern Silk Road.[19] In 1959, the Laohugou Glacier was 10.1 km long. But, from 2006 onwards, it began retreating at a pace of 13 m a year, losing 500 m by 2018. Junyan says she didn't feel the meaning of that number until she went there. On her first visit, in July, a flash-flood caused by melting glacial water had destroyed the road, so they had to climb the last 3 km to the glacier. On her second visit, in August, an even bigger flood of glacial melt meant that she and scientists from the IPCC had to walk the last 20 km. The run-off from Laohugou No. 12 is now believed to be double what it was sixty years ago. Such rapid glacial melt causes two very distinct problems: short-term flooding, and a long-term, total, devastating loss of water. The Qilian Mountains were previously considered relatively water-scarce, despite the reliable seasonal melt. But now, says Junyan, around 40 per cent of the local river basin is fed by glacial melt. 'Because of the increasing mountain water from the glacier, people started to plant a lot of vegetables; the city is vibrant, and the economy is booming, just because the water is increasing. But that's thanks to the melting glacier . . . this area will [soon] suffer from the water shortage, right after the peak-water tipping point.' She describes it as a 'fake boom'. Retreating glaciers may sound like a hyper-local issue, but in fact, she says, 'the Tibetan Plateau is a very important part of the overall global atmospheric system': changes to the ecosystem temperature here will certainly impact the global climate system.

Junyan now heads up the urban climate-risk programme for Greenpeace International, Beijing. 'Dense urban areas, especially in

China, are more vulnerable compared to many other places,' she tells me. 'When there is a flood in our agricultural system, the soil will absorb the water within one or two days, and the agricultural system will gradually go back to normal.' But cities take longer to recover: 'we saw houses destroyed, road infrastructure torn down.' I talk with Junyan in 2020, after floods had hit southern China. China's meteorology administration had just released a report saying that heavy-precipitation events in China have increased by almost 20 per cent since 1960. 'Previously, we thought that climate change is a very slow process,' says Junyan, sadly. 'That it changes slowly, bit by bit . . . But, actually, it's not; it's a disaster right now.' What neither of us knew then was that the floods of 2021 would be worse still.

The changing water cycle is being felt the world over. In Uganda, says environmental campaigner Nirere Sadrach, talking to me from Kampala, 'these days, we have changes in the rainfall pattern . . . The rain comes sometimes too much, very heavy; it destroys villages and crops.' Heavy rains during September and early October 2021 caused floods, landslides and hailstorms, affecting more than 40,000 people across seventeen Ugandan districts, with almost 4,000 people displaced. It was the worst flooding in the country since 1976.[20] 'At the same time,' continues Nirere, 'when the drought kicks in, it's prolonged. There's an extra two to three months of sunshine.' I tell him I'm surprised he's talking in months, not weeks or days. 'Yes, the dry season used to be two months – this year, it was six . . . The dry spell is too harsh, the heat is too much; it sucks up everything. In these times, access to water is a very big challenge.'

Dr Hans Sanderson at Aarhus University tells me that even Denmark's winters are no longer recognizable: 'We've lost a whole season to climate change.' I find that hard to comprehend. I've visited Denmark in winter and it involves ski jackets and snow boots. 'We only have three seasons now: fall, summer and spring. It's the same in southern Sweden . . . The Swedish met office have officially said that all of southern Sweden do not have winter any longer; it doesn't fulfil the criteria.' We're speaking in March 2022, when it should be nearing the end of the long, hard winter. But, says Sanderson, 'The last three winters there's been zero skiing or skating . . . no snow and no ice. It's the first time I would say with, you know, just nothing.'

Much further south, Gidon Bromberg, co-director of EcoPeace Middle East, tells me that, in Israel, 'because of climate change, the Jordan Valley has seen a dramatic reduction in precipitation . . . In the last twenty years, we've had fifteen years of drought. The upper Jordan was at its lowest output in recorded history . . . efforts to rehabilitate the Jordan are so much more difficult today because of climate change.' As an aside, Bromberg says he is astounded by the level of climate-change denialism in the United States and parts of Europe. 'That is very rare in Israel. And it's not a coincidence. Here, people are feeling climate change on their skin. There's already a dramatic impact . . . increased temperatures, increased days of unbearable heat, and, of course, a decrease in precipitation. No one questions that it's happening here.'

Increasingly, especially in the Middle East, this means having to keep an eye on the 'wet bulb' temperature – the thermometer reading taken with a wet cloth over the bulb, as a proxy for sweat and humidity. A wet-bulb reading above 35°C (95°F) means that the human body can no longer cool itself (the 'normal', dry-bulb temperature being over 50°C, 122°F). Such a temperature can be fatal within hours, even to the fittest of people. Jacobabad, Pakistan, first reached the wet-bulb threshold in July 1987, then again in June 2005, and June 2010 and July 2012. Reaching the threshold there is now an almost annual event, and occurred as early as May 2022.[21] South of the border that year, the megacity of Delhi came mighty close, at 49°C. Wet-bulb warnings may soon become common for hot regions. By mid-century, the news website ProPublica report, the Mississippi River valley will see dozens of wet-bulb days each year.[22]

CHAPTER 3

How the West Was Lost

There's something jarringly beautiful about watching the Nevada sun rise gold against the Trump Tower from my Vegas hotel window. The rust-red Las Vegas hills also reveal themselves with the morning sun, standing testament to the bizarre metropolis that has sprung up beneath them in the blink of a geological eye. I raise my very British morning cup of tea in tribute. To what exactly, I'm not sure, but the scene needed acknowledging somehow.

After my tea, I'm collected to join a 'ride-along' with a Southern Nevada Water Authority (SNWA) water-waste investigator. What looks like a cop car pulls up, blue and white with flashing lights on top; branded 'Water Patrol' and emblazoned with the words 'To Protect and Conserve', it's half NYPD, half *Paw Patrol*. Out steps Officer Cameron Donnarumma, a twenty-something Las Vegas native in a smart blue uniform and badge. He says they patrol the neighbourhoods, letting people know they're there, and admits that the cop-style branding 'definitely does stand out; it's not discreet at all'. Despite the tongue-in-cheek showmanship, he's got a serious job to do. Las Vegas has no water to waste. Residents must abide by strict conservation laws: watering your garden outside your assigned weekday or letting water flow off your property are punishable by Nevada law and offenders will be fined. This is the frontline of Nevada's war on water waste.

Within moments of setting off, in the suburb of Summerlin, north-west Vegas, we spot a violation at the entrance to a residential housing authority (HOA). Grass banks are being overly watered by a sprinkler, and water is running into the road. Donnarumma jumps out and immediately starts recording the crime scene. 'That's pretty common, excessive run-off,' he informs me, shaking his head ruefully. Once water goes into the gutter, or down the street, it evaporates and is lost. 'That's why it's really important to document these types of waterways violations.' A first-time offence receives a warning, but repeat misdemeanours are fined $80, which doubles for each subsequent infringement, up to $1,300. Donnarumma's team ply these streets twenty-four hours a day. 'I started at 3.30 this morning,' he tells me. Annually, some 17,000 investigations take place, around 10 per cent of which lead to fines.

We drive on, slowly, another couple of blocks. Pigeons are happily drinking from a puddle at the end of a street. There shouldn't be any puddles. It hasn't rained for three months. Tracking the trickle to its source, we discover a house where a palm tree is being watered by drip irrigation – but the drip nozzle has popped off, causing more of a mini fountain. Donnarumma leaves a door hanger on the garage which reads, 'Don't waste water. IT'S THE LAW!' (their caps, not mine). He leaves a little yellow flag beside the offending garden sprinkler.

'Most people are pretty cool about it,' says Donnarumma. 'We do give out fines, but education is definitely the number-one goal . . . [Sometimes] we'll get violations and then, within a month or two, all the grass is gone. So, it's a motivator.' To clarify, the grass going is a good thing. Lawn is, astonishingly, the largest single irrigated crop in the US. The water used to irrigate the nation's grass lawns would be enough to supply 2 million US households, says Dr Newsha Ajami, director of urban water policy at Stanford University's Water in the West programme. People are, she says, 'often in denial about outdoor water use and how big it is in the US, and how unessential it is.' The green stuff proves hard to kick, however. SNWA tried various voluntary measures and incentives for lawn removal, with limited success. Since 2007, however, they've taken the mandatory approach: development codes now stipulate no grass in the front yard, and a maximum of 50 per cent coverage of the backyard. There was a carrot, too, with households and businesses offered $3 per square foot (93 cm²) of turf removed. That worked. Even the all-powerful HOAs, many of which mandate well-kempt lawns amongst their residents, had to fall into line. On my ride-along with Donnarumma, I see perhaps only one in ten front yards now with any grass at all (installed before the 2007 code). Most people have moved to 'desert landscaping' – a combination of gravel, rocks, cacti and short trees. Beth Moore, public-information coordinator at SNWA, also on the ride-along, explains that 'the bathtub ring out at Lake Mead is a huge indicator. I think people see that and they realize the water situation is pretty serious here. We live in the desert and need to appreciate and value water a little bit more than other places.'

Not everyone has got the message, though. Ajami tells me later that, if Americans used water as efficiently as Europeans, the savings would be enough to supply 4 million households if you removed all the lawns. Per capita water use in the US is 'way, way out there, compared to all other industrialized countries'. The EU average is 120 L per person, per day; in the US, it's approximately 380 L.[1] 'That's partly because of living in suburbs,' says Ajami, 'with large outdoor spaces that we can no longer afford, when it comes to water.'

*

Spend any time talking to water people in the American West and you'll soon learn who got what when the Hoover Dam was built across the Colorado River, forming Lake Mead – America's largest reservoir – in 1935. From that point on, not all Western states' water stories were born equal. (The allocations are given in acre-feet [af], a standard US measurement for large volumes, and also easy to visualize, being the amount you'd get if you flooded one acre of land with one foot of water*.) Among the seven states that share the

---

* An acre foot is equivalent to approximately 1,233 m³, or 1.2 M L, while a million acre feet is just shy (0.81) of a cubic kilometre.

Colorado River, the Colorado River Compact (1922) 'apportioned' half of it annually to the Upper Basin states (9.25 km³/7.5M af to Wyoming, Colorado, New Mexico and Utah) and half to the Lower Basin states (Arizona, Nevada and California). The total (18.5 km³/15M af) was based on the river's average flows of the previous decade – which were, as it turned out, abnormally high. Centuries of data show an actual average closer to 16 km³ (13M af). The river was therefore over-allocated from the very start.* As for the specifics, the Boulder Canyon Project Act (1928) and the Upper Colorado River Basin Compact (1948) decided that California would get 5.4 km³ (4.4M af) per year, Arizona 3.5 km³ (2.8M af) and little, barely populated Nevada just 370M m³ (300,000 af). Congress didn't dictate in 1948 whether states should use percentages or acre-feet to divvy up their allocation. The Upper Basin states chose percentages: 51.75 per cent for Colorado, 23 per cent for Utah, 14 per cent for Wyoming, and 11.25 per cent for New Mexico. In hindsight, this was much more sensible than the Lower Basin's rigid acre-feet method. As David Owen writes in *Where the Water Goes*, 'allocating water that doesn't exist is much less likely to cause trouble later if you don't specify exact amounts.'[2]

The problem for Nevada – and SNWA – is that this dusty little desert state now boasts two of America's fastest-growing cities: Ferguson and Las Vegas. The Las Vegas Valley today is home to over 2 million people and over 41 million annual (pre-COVID-19) visitors, with its mini re-creations of Paris, New York and Venice. The one thing that Sin City *isn't* wasteful with, then, is water. The 300,000 af allocation doesn't stretch far. Even the famous fountains of the Bellagio Hotel on the Strip aren't supplied by the much-depleted Lake Mead, but rather via a borehole tapped into the brackish,

---

* There is some debate about whether they knew this and chose to ignore it. In 1916, six years before the Colorado River Compact was signed, Eugene Clyde LaRue, a young hydrologist with the US Geological Survey, testified to the US Congress to the effect that the Colorado River was not sufficient to irrigate the basin.

non-potable water beneath the city. From the outset, the driest city in the driest state in America had to learn how to cope with little water.

Las Vegas in Spanish means 'the meadows'. A natural spring made it the obvious site for a railroad stop-off point, for steam trains to take on water. The railroad then spawned a town that nobody could have envisaged, and the groundwater was mercilessly pumped to extinction. But the site of the original spring is still preserved by SNWA. In 2007, it was turned into a tourist attraction of sorts, called Springs Preserve, which also homes the Nevada Museum. The car lot is shaded by solar PV panels, and the site itself – half a square kilometre of restored native habitats and archaeological exhibits – is largely shaded by mesquite trees. Visitors can sign up for 'water smart' gardening classes. Admittedly, it doesn't attract much of a spillover crowd from the casinos and strip joints.

My host at the Springs Preserve is Bruno Bowles, a deceptively youthful (being close to retirement), long-haired park ranger. A short stroll from the visitor centre is a small hill, tastefully replanted with desert-native plants. 'This is one of the coolest features we have here,' beams Bowles. 'It's called a spring mound. This is the birthplace of Las Vegas. Back in the late 1800s, early 1900s, the train ran between Salt Lake City and LA. William Clark and his brother bought a bunch of property here and all the water rights, so they could set up a station. In about the 1920s, Vegas started groundwater pumping . . .' The mound would have been reliably wet on the surface, says Bowles, and formed over hundreds of years as dry desert dirt hit the damp earth and stuck, growing upwards. 'There are several spring mounds of this size throughout the valley. We found stone tools, pottery shards, so we know that the indigenous people lived here as far back as 6,000, possibly 8,000 years ago . . . it's always been important water for people in the valley. It's always been a gathering place.' Ironically, it's now one of the quietest, most serene places left in modern Las Vegas. Yet, without this little mound, Las Vegas wouldn't be here. The original spring dried up in

1962. Pointing to a rusted metal structure in the middle distance, Bowles says, 'You see that well derrick? They pumped so much water there, the land just subsided.' Lake Mead's water saved the city and allowed it to grow.

I head from Springs Preserve to the SNWA offices to meet Colby Pellegrino, deputy general manager of resources for both the SNWA and Las Vegas Valley Water District. The region's megadrought began in 2001 and groundwater is no longer the conversation here. The only game in town is Lake Mead – and that's going too. Pellegrino says the reservoir was full when the drought began. 'Obviously, through those early years, you were seeing breaking records, elevations and storage that we hadn't predicted, sequences of dry years. But you're still not, even four years in, going, "Oh, this is a twenty-year drought."' She says it wasn't until maybe 2015 that they recognized things were fundamentally different; there was a dawning realization that the lake level wasn't going to recover. She likens it to a slow-moving freight train: 'It's going to be very hard to put the brakes on at the last minute.' Pellegrino and SNWA have done more than most to hit those brakes: since 2002, the region's population has grown by 750,000, while using 23 per cent less Colorado River water. 'That's not by mistake,' she underlines, emphasizing the hard work behind the scenes. But it's not going to be enough.

When Pellegrino was born, in 1983, the population of Las Vegas city was 300,000; today, it's 2.3 million. Her birth year was also the last time that Lake Mead was in flood control, when the spillways either side of the Hoover Dam were still meaningful infrastructure, not the quirky tour footnotes they are today. When she was a kid, there was plywood on top of both Glen Canyon Dam and Lake Mead. 'They could not release water fast enough; they were worried about the reservoirs overtopping, so there were plywood slats holding back water.' Does she believe she'll ever see the lake full again? 'I think, if this drought is an indication, twenty years of below-average run-off, with no end in sight – it would take a phenomenal flood, or a decade of really wet years, to get us back. You can't plan for the rosy picture, for that coming back.' Forget those who say that there's always been drought, or that they remember worse ones. They don't. Research shows that 2000 to 2021 was the driest twenty-two-year period in the south-west since at least the year 800.[3]

Local rainfall is only a very small part of the picture. More important is what's happening to the snowpack in the Colorado mountains, the river's source. 'Last year, we had 89 per cent of average precipitation, and our run-off was just 32 per cent,' says Pellegrino. 'That's for a host of reasons, but things that are not going to go away if the climate is warmer. The soil is incredibly dry, so you're starting with a drier sponge than you normally would. Snow is melting sooner: that means that all of your natural system is starting to grow sooner; that means your water users, particularly agriculture, are using more water sooner. But the thing that gets overlooked is the carrying capacity of the atmosphere is greater: the warmer the air is, the more humidity . . . So, you have three strikes against you when it comes to dealing with hotter temperatures.' Another important factor is sublimation, she says, whereby increased solar radiation causes more water to go from a solid state to a gaseous state, without ever entering a liquid state. Basins that rely on snowmelt, as the Colorado River does, are especially vulnerable to sublimation, where the crucial 'cool mountain stream' stage is skipped. According to Steven R. Fassnacht, a professor of snow hydrology at Colorado State University, this means 'more snow can disappear into the air than melts into rivers.'[4]

There is a feedback loop with California's ever-expanding wildfires too. Dr Schwartz, lead scientist at the University of California, Berkeley, Central Sierra Snow Lab, writes, 'Massive wildfires, such as those that we've seen in the Sierra Nevada and Rocky Mountains in recent years, cause distinct changes in the way that snow melts . . . The loss of forest canopy from fires can result in greater wind speeds and temperatures, which increase evaporation and decrease the amount of snow water reaching reservoirs.' There's no way of sugar-coating it, he says: 'We are looking down the barrel of a loaded gun with our water resources in the West.'[5]

Despite the freight train slowly crashing as we speak, with today's water level at 325 m, Pellegrino refuses to contemplate dead pool (273 m, 895 ft) – which would cease water delivery to the Lower Basin states and Mexico, and cut hydroelectricity to Los Angeles. 'I think that, absent the type of collaboration and cooperation we have on the river, dead pool is a possibility,' she says. 'But we have a set of policy that will not let the river go below an elevation of 1,020 ft (311 m). That's what the Drought Contingency Plan was built on, not going below 1,020 . . . And we're starting that work now.'

The Drought Contingency Plan, negotiated by Pellegrino, among others, in 2008, set specific usage-reduction triggers when certain lake levels are reached. In 2021, shortly before we meet, the first level – 328 m (1,075 ft) – was reached, triggering the first federally mandated (non-voluntary) Tier 1 shortage. From January 2022, this reduced Nevada's allocation by 25.9M m³ (21,000 af) (see Table 1). Arizona's supply of Colorado River water via the Central Arizona Project canal was slashed by about 30 per cent, or 631.5M m³ (512,000 af). Somewhat amazingly, California, the biggest water user of all Lower Basin states, takes no cut because of its 'senior water rights' until Tier 3, at a lake level of 312 m (1,025 ft) (legally right, perhaps, but morally and practically wrong).

The Drought Contingency Plan has very real consequences, especially for Arizona farmers. But any negotiation that sees California sit it out until a Third Tier is triggered, is just tapping the breaks. When will they be slammed on? Already water-efficient Southern Nevada sure isn't going to, it can handle its reduction of 25.9M m³ (21,000 af) with ease. Pellegrino confirms that they are already 29.6 M m³ (24,000 af) under budget anyway, thanks to their conservation efforts. According to their Drought Contingency Plan commitments, they barely need to brake at all; cruise-control is fine.* Nevada also holds a trump card – were the lake to reach dead

---

* Pellegrino and her team are far too professional to do so, of course. But the point stands that meeting these spreadsheet commitments ultimately means nothing if the water in the basin continues to disappear.

Table 1 Lower Basin Drought Contingency Plan Contributions – cuts to Lake Mead allocation[6]

	Arizona	Nevada	California	Mexico
Original allowance under Boulder Canyon Project Act 1928 and Mexico Treaty 1944	2.8M af / 3.45 km³	300,000 af / 370M m³	4.4M af / 5.43 km³	1.5M af / 1.85 km³
Tier 0: Lake level less than 332 m (1,090 ft) (enacted Jan 2020)	192,000 af / 237M m³	8,000 af / 9M m³	0 af / 0 m³	41,000 af / 50M m³
Tier 1: Less than 328 m (1,075 ft) (reached Aug 2021)	512,000 af / 631.5M m³	21,000 af / 25.9M m³	0 af / 0 m³	80,000 af / 98M m³
Tier 2: Less than 320 m (1,050 ft) (reached June 2022)	592,000 af / 730M m³	25,000 af / 30M m³	0 af / 0 m³	104,000 af / 128M m³
Tier 2b: Less than 319 m (1,045 ft) (reached summer 2022 but not imposed)	640,000 af / 789M m³	27,000 af / 33M m³	200,000 af / 247M m³	146,000 af / 180M m³
Tier 3: Less than 312 m (1,025 ft) (projected 2023–4*)	720,000 af / 888M m³	30,000 af / 37M m³	350,000 af / 432M m³	275,000 af / 339M m³
Total remaining allowance after Tier 3 cuts	2.08M af / 2.56 km³ (-25 per cent of original)	270,000 af / 0.33 km³ (-10 per cent of original)	4.05M af / 5 km³ (-8 per cent of original)	1.23M af / 1.52 km³ (-18 per cent of original)

- Ten-year average inflow into Lake Powell (2001–9), which is in effect the storage reservoir for Lake Mead = 9 km³ (7.313M af)[7]
- Annual water released from Lake Powell to Lake Mead in 2022 = 9.23 km³ (7.48M af)[8]
- Total take from the Colorado River *after* Tier 3 cuts = 9.4 km³ (7.625M af)

*   When I first wrote this table in 2021, the bureau's projection for Tier 3 was 2025–6. By summer 2022, this had been brought forward by a full two years. 'And yes,' wrote Joanna Allhands in *AZ Central*, 'that's as scary as it sounds.'

pool, and California lose its 4.4M af (5.43 km³) overnight, this would not affect Nevada's modest provision. Prior to 2015, SNWA had two water-intake pipes taking water from the lake to serve Nevada, at 320 m (1,050 ft) and 305 m (1,000 ft). Pellegrino may publicly refuse to contemplate reaching dead pool, but SNWA's actions suggest they are not only contemplating this, but have built the contingency plan. In 2015, a third intake was built – not just below dead-pool height (273 m, 895 ft), but at the very bottom of the lake, and capable of draining right down, like a bath tub, if required, to the very last drop. This was no mere contingency afterthought, but a major work of engineering that cost $1.47 billion for 4.7 km of 6.1-m-diameter tunnel and a third pumping station. Many questioned the return on investment, admits Pellegrino. 'Is it five years? Ten? And I said the return is 90 per cent of your water supply or nothing.' That, in effect, was the ROI calculation. Pellegrino calls it the 'largest climate-adaptation measure the United States has seen.'

A few months after I visit, in April 2022, the lake dropped below the level for the first intake, leaving the steel pipe's huge mouth exposed and gaping uselessly towards the sun. Nevada's third-intake investment looked very shrewd indeed. 'This is a crisis. This is unprecedented,' a panicked spokesperson for the Metropolitan Water District of Southern California told CNN. 'We don't have enough water to meet normal demands for the 6 million people living in the State Water Project-dependent areas.' Meanwhile, a far calmer Bronson Mack of SNWA said there was 'no impact' in Nevada, as their third intake was turned on for the first time to maintain a consistent supply. 'Customers didn't notice a thing.'⁹

Despite the lake's ever-plunging levels, in December Pellegrino still talked a good talk as to why dead pool won't happen: 'to think that the federal government in the States wouldn't step in to do something before Mexico, Arizona and California all get completely cut off . . . I think is unreasonable,' she told me. 'The secretary has a lot of latitude to manage the system, in particular for health and human safety and endangered species. You see that in other river systems throughout the West, where the state water project in

California this year has a 0 per cent allocation. They're allocating no water to contracted water users in some of the Klamath Basin, this year: zero . . . I'm fairly confident that the secretary would never let Lake Mead go to dead pool.' I'm struck by the way Americans refer to their secretaries of state in the way that people from other nations might refer to God. As if the natural world would, if push came to shove, inevitably yield to the secretary and the rule of law. Admittedly, if there's evidence anywhere for that possibility, it's in the Western United States, the world's most hydrologically engineered region, where mighty rivers are tamed behind a multitude of dams, and diversion canals meander through hundreds of kilometres of arid desert. In every previous era of this great frontier, including as recently as Nevada's 2015 third intake pipe, grand hydrological engineering projects have worked miracles, and the mighty Bureau of Reclamation has literally moved mountains. But now the two biggest projects of all, from which all those canals and pipelines flow, Lake Mead and Lake Powell, are at 34 and 29 per cent capacity respectively.* More than 40 million people rely on that water. And the two lakes are dying. I wonder what even the secretary can do in the face of that.

One water insider informed me that the Metropolitan Water District of Southern California – which serves Los Angeles, Orange, San Diego, Riverside, San Bernardino and Ventura counties – are already at the stage of 'figuring out how to allocate water for what they call "health and human safety", for flushing toilets, for making sure basic sanitation exists . . . If you're an agricultural user, you're out of business. If you're a municipal user with another source of supply, you're probably soon not getting any more.' In short, when the lakes dry up, you're left with just the water you can find underground, from desalination and water recycling. This seems an

---

* Those levels were at the time of my visiting, in December 2021. At the time of writing this, in January 2023, they are already down to 28 per cent and 22 per cent, respectively. (For up-to-date levels, visit http://lakemead. water-data.com/)

inevitable future for the West as the megadrought bleeds into a permanently changed climate. The water levels at Lake Powell dropped 30 m in the three years to 2022 alone. They need only drop another 10 m before hydroelectricity production would be severely compromised at Glen Canyon Dam. It's not an *if*, but a *when*. And the *when* may be counted in months, not years.

The Drought Contingency Plan, then, is a sticking plaster where emergency triage is needed. It barely even addresses the original, decades-long over-allocation – that original 9.25 km^3 (7.5M af) each to the Upper and Lower Basins that never existed, not even in non-drought years. In 2021, with lake capacity below 30 per cent, to only begin slowly reducing the allocation to what it should have been in the first place, when optimistically assigned in 1928, is a madness that's hard to explain. They are taking cuts out of nothing, bites out of thin air. Even Pellegrino, who is proud of the hard-fought plan, acknowledges this. The over-allocation was identified in the 1960s, she says. The twenty-year period from 1906, during which the river was allocated, was one of the wettest in recorded history, and among the wettest in 1,500 years. 'Sometimes we get so caught up in the policy that we neglect to understand the simple dynamics, which you just pointed out: if the reservoir is going down, your use is exceeding your inflows . . . You'd be surprised how many people forget that. But, yeah, there is an over-allocation problem. And it's going to be exacerbated by climate change.'

I try again. So, why do those allocations *still* exist today? And why aren't the Lower Basin states cutting by more? 'There's a couple of different reasons . . .' says Pellegrino, exiting sober-science mode and re-entering policy mode. Firstly, she says, even up to the early 2000s it was believed that the federal government would 'augment' the river system, that there would be 'some sort of grand water-supply project, whether an interconnection with another river system or a desalination project or something to bring all of that into balance.' But that never happened. 'Big water projects like that have gone by the wayside for many, many years now.' That is at least logical, combined with the fact that Lake Mead was close to full in

2000. The other reason lacks all logic except belief in the flag and due process: 'Is it worth the pain to go back and get seven states and seven state legislators and seven governors and both houses of Congress and the President to agree that those numbers should be different?' she asks. Before I can jump in with, 'Yes, of course!' she answers her own, evidently rhetorical, question: 'Probably not, as long as we can all agree how to manage the reservoir system.' She says the question of going back to renegotiate comes up frequently, but it's 'fraught with legal challenge'. Meanwhile, the waters of Lake Mead continue to quietly recede, heretically oblivious to the secretary and the 'Law of the River'. When it finally dries to a cracked brown lakebed, someone should erect a plaque: 'Here lay Lake Mead. It was fraught with legal challenge.'

*

I visit Lake Mead the very next day with the National Parks Service (NPS). The lake itself is a National Recreation Area for boaters and hikers. The NPS acting deputy superintendent Justin Pattison kindly shows me around. His gentle but steadfast voice, uniform and sheriff's badge add to the handsomely reassuring vibe. He's the type of guy you'd trust in a crisis. And this *is* a crisis. NPS employees are nomadic by nature; Pattison arrived at Lake Mead only in March 2020. I wonder if he's been here long enough to see any changes, but he quickly puts me right: 'Over the last year alone there's been a drastic change.' According to the charts he pulls up, they started 2021 at lake levels of roughly 364 m and are now at 355 m – a drop of 9 m in a calendar year.* His colleague, Dave Alberg, adds that the average annual decline over the past decade has been 3.6 m, but 'that has been accelerating at a rate I don't think any of us were anticipating.' By early 2020, it had become clear that the decline was irreversible. 'This area typically gets 3.8 to 5.08 cm of

* By June 2022, the lake level was 318 m – a further 37 m drop from when Pattinson was speaking in December 2021, triggering Tier 2 cuts earlier than expected.

rain a year. It's not a lot. But even if it gets soaked all day, it won't make much difference in Lake Mead – where it matters is upstream, in Colorado, Wyoming, Utah and that snowpack.' He repeats the dismal snowpack and run-off figures that Pellegrino told me. 'Even if you've got a decent snow year, the amount of water available to the river is going south very fast,' he says.

It's refreshing to hear federal employees speak so openly, with the Trump don't-mention-climate-change years still shudderingly fresh in the memory. 'This is all primarily driven by climate change, of course,' confirms Pattison, matter-of-factly. Echoing Pellegrino again, he says increased temperatures throughout the entire basin are creating higher evaporation demands – so more moisture is lost to the air, as well as to the extremely dry soil. Combined, these things 'essentially result in a decrease of base flow into the Colorado system'. On the fifteenth of every month, the Bureau of Reclamation publish twenty-four-month reservoir-level projection figures. If a 4.6 m year-on-year decline continues, as currently projected (arguably a conservative estimate, against the backdrop of a recent 9 m decline), the lake would reach dead-pool elevation sometime in 2029, after which the Colorado River no longer flows beyond Nevada. And every month, says Alberg, that 'twenty-four-month out projection drops lower'. The latest projections include the Tier 1, 2 and 3 reductions, but already show that they'll be insufficient to even stabilize the decline, let alone halt or reverse it.

The starkness of this chat in the NPS office, surrounded by pictures of water sports taken in the golden, recent, days of the marine-blue lake, still doesn't prepare me for the sight of the lake itself: the bathtub rings that take my breath away; the sorry, naked intake towers; Hemenway Harbor's $3-million road to nowhere. Next, we head to Boulder Beach, 'heavily utilized by the non-boating public' (which is NPS speak for anyone with a crate of Coors Light and a camping chair). Because everyone drives here, the NPS construct rock 'berms' – knee-high drystone walls – along the waterline to stop cars getting too close to the water and getting stuck in the sediment. But the receding waterline effectively leaves more beach, so Pattison and his team must construct a new berm. Pattison stops the car halfway down the beach and indicates the berm he helped construct when he arrived here, just eighteen months ago. We're still hundreds of metres from the water's edge; two further berms stand between us and the lake – the last one raggedy, more of a gesture. I photograph Pattison standing by his berm, barely believing what I'm seeing. From this spot, the lake is halfway towards the horizon. In two more years, it will *be* the horizon.

We carry on to Boulder Harbor, for which Google Images presents the same dreamy family recreation scenes that adorn the NPS office. Any occasional boater from Las Vegas will talk of it fondly, in the present tense: in its (recent) heyday, you'd get here from Vegas and be launching your boat in under an hour. We approach down a motorway-wide stretch of asphalt – like at Hemenway, except this road does (just) reach the water. But there are no boats, and nobody even trying to launch. Because there's no harbour. Two seemingly innocuous islands that were once the wide, welcoming harbour mouth have steadily joined arms, leaving a marooned, isolated pond. A small *Closed* sign, as if it were needed, rests against a concrete bollard. Floating toilets that once served the boating community now stand dry in the desert. High up on the hillside is a cordoned-off NPS visitor centre, now due for demolition. Ten years ago, that was lakeside too.

On the sombre drive back to Boulder city, we talk about family and future. Pattison's wife is stationed elsewhere with the NPS, but will join him soon. They're looking for a house. 'A swimming pool is on my list of must-haves,' he says. Even if it's filled with Lake Mead water, I ask? He laughs, appreciating the irony, then says thoughtfully, 'Yeah. I guess I'll take my piece of the Colorado too.'

*

The neighbouring state of Arizona is far more wasteful with its water than Nevada. The umbilical cord it built to the Colorado River, the Central Arizona Project (CAP) canal, allowed it to grow up in a watery womb. But it may soon be cut, and the state left to fend for itself in a harsh desert reality.

Going back to the original Boulder Canyon Act, Arizona receives almost ten times as much water as Nevada. The problem was always delivery. After decades of legal wrangling – including conservation-minded President Jimmy Carter's attempt to scrap the project – the federal government built a grand canal, the CAP, through hundreds of kilometres of desert, to bring water to its two major cities, Phoenix and Tucson, arriving at the latter – its endpoint – in 1993. Unlike Las Vegas, Phoenix bathes in an excess that would have made the pharaohs blush. 'Our trajectory has been a little different,' deadpans Marco Velotta at Las Vegas City Hall. 'I find it ironic, because the Hohokam native tribes that lived in the Phoenix area survived for millennia.' But now, to finish his thought, Phoenix may boom and bust in little more than a century. Velotta says there's 'a push to make a quick buck and get out, from the development community and home builders. Not a lot of attention is paid to the impact.' He wonders about Phoenix's hundreds of artificial lakes and fountains. 'I mean, the CAP and the Salt River Project [another canalized river] are engineering marvels . . . [but] there's a dark side to it.'

In Phoenix, I see the 'dark side' first hand. Lynda Person, from the Arizona Sustainable Water Network, and her hydrologist husband John, pick me up from Sky Harbor airport. Lynda wrote an op-ed in the *Arizona Capitol Times* calling out the Arizona legislature for being 'in full denial' of Arizona's water-policy needs: 'Flowing rivers are in decline, water levels are decreasing and, in some places, drying up completely . . . For too long Arizona's flowing rivers have been viewed as a resource to tame, exploit, and fuel growth.' In a follow-up article, she wrote: 'Multiple megadairies and farms have recently located to Arizona where they can grow alfalfa and raise dairy cows without much regulation or thought to their water use [agriculture accounts for 80 per cent of Arizona's water use] resulting in the drying of springs, creeks, and rivers, causing massive loss of riparian habitat all over the state.'[10] To see some of this, we don't need to travel far. The Salt River, or what's left of it, runs alongside the airport. There's no water in it at all, merely a large, dry riverbed. 'This used to be a big river, with beavers and cottonwoods and lush, desert forest habitat,' laments Lynda. Photographs from a hundred years ago show exactly that. There even used to be a ferry crossing. Now, only a huge flood event would see any water run down it. She and John remember the last time being January 1993, almost thirty years ago. As I stand taking pictures, a guy out walking asks what I'm photographing. When I explain, he sounds surprised: 'There used to be water here?'

'This is not considered a flowing river at all any more, after the Granite Reef Dam,' says Lynda. The unofficial total length of rivers lost is over 2,700 km, and what remains is not well protected. Both the San Pedro and Verde Rivers are under stress and threatened. Lynda gestures at the former Salt River, now simply gravel, yellow grass and the occasional shin-high creosote bush. 'We're gonna end up with something like this.' Ironically, the template for how to do things sustainably *was* here all along. The Hohokam peoples formerly lived alongside the Salt River, building canals, irrigating farmland and supplying habitations over a geographical range larger even than today's Phoenix. The Salt River was effectively choked off by white-settler over-abstraction in the early twentieth century, and the building of the Roosevelt Dam (a prototype for the Hoover Dam) upriver in 1911. Following the river's brief, unforgiven attempt at an uprising in 1993, Phoenix chose to install two new dams in the city centre, to cut off its head for good. A bit of scenic water was still desirable for leisure and land-value purposes, however, so a 3-km stretch of riverbed was bookended by dams and filled back in. Not with Salt River water, but with CAP water. Now known as Tempe Town Lake, it's little more than an expensive water feature, filled with precious water from the ailing Colorado River and Lake Mead, hundreds of kilometres away. It took over a month to fill, covering 89 hectares[11].

Arizona's inclusion in the original Colorado River Compact (1922) and Boulder Canyon Project Act (1928) wasn't even a given.

It only became a state – the forty-eighth – in 1912. The 3.45 km³ (2.8M af) it was granted – compared to Nevada's lowly 300,000 acre feet, and Mexico's zero (Mexico's share was only increased to 1.5 million acre feet as an afterthought in 1944, by the U.S.–Mexico Water Treaty*) – was exceedingly generous, a bet on Arizona's potential future prosperity. Arizona's allowance would become a self-fulfilling prophecy: give a desert state water, and you will watch it bloom. But to afford this pampered, pumped water from the CAP canal (which remains the state's single largest energy user), Arizona farmers would need up to a 90 per cent subsidy. Grady Gammage, a former president of the CAP board, is quoted in David Owen's *Where the Water Goes*: 'We ended up making a deal with the farmers . . . that we would sell them Colorado River water at significantly less than the cost of getting it to them – and I mean significantly – and that the cities would cover the gap.'[12]

And so the CAP was built: 541 km long, between 7.3 m and 26 m wide, and 5 m deep, with three tunnels blasted through mountains and fifteen pumping plants to push nearly 500 billion gallons of water uphill.[13] Construction began in 1973. It wasn't completed until 1993, at a cost of more than $4 billion. When the CAP finally arrived in Tucson, it wasn't even wanted. Lynda Person worked on the water survey at the time: 'In 1993, Tucson had just started taking CAP water and they just hooked up the CAP pipe to their

---

* A 1928 Colorado River Commission of Arizona report reads: 'One of the questions which makes a settlement of the Colorado River controversy difficult, is the menace of Mexico. American millionaires own over a million acres [404,686 ha] of land in Mexico which can be irrigated from the Colorado River. If [their] land is irrigated, A MILLION ACRES OF LAND WILL REMAIN DESERT IN ARIZONA.' Their caps, not mine. It goes on to urge Congress to make clear to Mexico that 'neither it nor its citizens' have a legal right to expect 'a continuance' of the Colorado River beyond the US border. ('The Colorado River Question', Colorado River Commission of Arizona, 1928, http://www.riversimulator. org/Resources/Controversy/ColoradoRiverQuestionShallNaturalRe sourcesCreateLocalOrDistantDevelopment.pdf)

water-distribution pipe and let it rip,' she remembers. This caused inevitable problems with the water chemistry: after years of ground-water, the new supply was more acidic, 'peeling things off the pipes'. To the outrage of the Tucson populace, their tap water was coming out brown. The solution was to turn the CAP water underground, recharging the aquifers with it, where it would pre-mix with the Tucson water, and then pump it out of the ground again. (According to David Owen, the fifteen pumping plants consume the equivalent energy of bringing the same volume of water to the boil.[14] So, after that, what's one more pump and polish?)

On a desert road, just outside Phoenix, we drive over the CAP canal, and I stop to take pictures. My mind immediately returns to excitedly photographing the King Abdullah Canal in Jordan – another centralized, concrete canal system supplying a desert region. It strikes me that the CAP looks twice as wide as the King Abdullah. Yet, unlike the King Abdullah, from which every drop is desperately needed, the CAP was built bigger than necessary, used to irrigate export crops and top-up town lakes, for farmers who couldn't afford it and city dwellers who didn't like the taste. As Cody Friesen, a professor at Arizona State University, tells me from his office overlooking the CAP: 'You are stood in a remarkable place with a crap-load of water flowing through it – but by no rights should that be true.' Soon, it may not be. The Drought Contingency Plan will cut the CAP's flow by a quarter, and dead pool would cut it completely.

CAP's communications strategist, DeEtte Person (no relation to Lynda), unusually for a PR person, tells it like it is. The CAP, she says, will 'be getting less Colorado River water in general. It's kind of like a bank account: we're taking a pay cut, and so we won't have as much to put in.' Agricultural users will get hit first, having accepted junior water rights in return for government subsidies to make it affordable. In a surprising historical twist, meanwhile, Native American nations were granted senior water rights. In fact, the federal government allocated 382M m^3 (310,000 af) of CAP water to Indian reservations in Arizona, with priority over all urban and agricultural deliveries – more than the entire state of Nevada.

What will happen to all that agricultural water? Mark Silverstein, formerly of the National Surface Water Survey, writes for *The Nevada Independent* that Arizona's cotton industry alone 'consumes perhaps 200,000 af per year (and likely much more) of Colorado River water just to grow cotton in the desert.'[15] Just one crop drinking up two thirds of the total allocated to the entire state of Nevada. Even more alfalfa – an animal-fodder crop, also largely for export – is gown in Arizona than cotton. I ask Justin Conley at CAP what these farmers will do when they lose the CAP water. 'It's a good question . . . I mean, before the CAP, there was just groundwater pumping – that was one of the big reasons the CAP was approved in the first place. I guess they'll go back to groundwater.' Unsustainable groundwater pumping is, then, Arizona's past and future. People with longer memories remember the effects of unregulated groundwater pumping in the state. It averaged around 5.5–6 km³ (4.5–5M af) per year between 1953 and 1968, while the natural recharge was only about 1.85 km³ (1.5M af) per year. Consequently, groundwater levels declined in some areas by over 38 m and many wells ran dry.[16]

*

My ride-along with Lynda and John is very different to the one I had in Vegas. We're in another middle-class suburb HOA, but here residents share rights to a groundwater well connected via huge valves at the bottom of every garden. Larger yards have two valves. Once a fortnight, the 'water guy' visits and turns the valves on for around an hour, literally flooding the back gardens with untold thousands of litres. In Vegas, a damp driveway will see Officer Donnarumma and his colleagues jump out of their patrol car; in Phoenix, rapids of run-off water flow freely over sidewalks, down roads and into storm drains. Forget sensitive desert landscaping; here, lush, abundant lawns are shaded with citrus trees, eucalyptus and pine. No one knows the volume of water being used and lost, because the wells aren't even metered. 'We realize the contradiction,' one friendly homeowner tells me. 'But it makes for *such a nice garden*.' The water doesn't even have a direct cost. The residents pay a monthly fee of $83 to their housing authority, with garden flood irrigation just one part of the service.

In Arizona, groundwater overuse is as much of a problem as surface water, perhaps more so, considering the CAP-sized sticking plaster they've had to cover up the lost rivers. Lynda notes that, in 1980, Arizona passed regulations to discourage groundwater mining, designating certain areas of the state as active groundwater management areas (AMAs). Unfortunately, though, loopholes and lack of enforcement mean groundwater management hasn't been strong. Residential backyard flooding happening totally legally (the HOA collectively owns the historic water rights) within a Phoenix AMA is just one such loophole. All five AMAs (Prescott, Phoenix, Pinal, Tucson and Santa Cruz) are subject to the 1980 Arizona Groundwater Code, and each, according to the Arizona Department for Water website, 'carries out its programs in a manner consistent with these goals while considering and incorporating the unique character of each AMA and its water users'. Meanwhile, also in 1980, Irrigation Non-Expansion Areas (INAs) were implemented by the Groundwater Management Act: 'which would suggest irrigation non-expansion, right?' laughs Lynda. 'In reality, it's a layer of hoops to jump through that might slow things down a bit.'* Meanwhile, most areas aren't an AMA or INA, 'So, you can pump away.' There are just three INAs, spanning a tiny proportion of the state. Even within INAs, wells with a pump capacity under 132 L per minute are exempt[17] – and farms with non-exempt wells are

---

* The Groundwater Management Act, rather than signifying a sudden Damascene moment of enlightenment, was more the case of jumping before being pushed. In 1979, the Carter administration threatened to pull the plug on the CAP unless Arizona adopted a stringent groundwater code.

required only to report their usage, not to actually reduce usage. Loopholes are the norm, not the exception.

As the water-news website Circle of Blue tells it, 'With surface and groundwater supplies more secure than ever before [because of the CAP and the 1980 Act], Arizona raced to the head of the national pack in economic growth, housing starts, and new jobs. In 2000, Arizona counted 4.8 million residents, 70 percent more than in 1980. The state has added 2.3 million more residents in the two decades since.'[18] But if groundwater levels were plummeting in 1980, only to be propped up by the steroid injection of the CAP, what will happen with a population three times the size and no CAP?

In 2003, the US Geological Survey were already reporting that groundwater pumping in south-central Arizona had resulted in water-level declines in aquifers of 92 to 153 m.[19] They found in 2013 that base-flow levels on the Verde River around Camp Verde, a town 80 km north of Phoenix, had decreased by 12 m^3 (10,000 af) annually between 1910 and 2005.[20] All this before the CAP water started to be cut in 2022, leaving many farmers and house developers with only groundwater to draw from again.

Later, Lynda and John take me to Pinal County to see the effects first hand. We drive over the Verde River, as dead as the mighty Salt River. But it's not rivers we're here to see. Huge cracks have appeared out in the scrub, due to aquifers collapsing beneath the surface. As a trained geologist and hydrologist respectively, Lynda and John have identified some likely sites to seek out. I'm not entirely sure what to expect – maybe some cracks by the side of a road to point at and shake our heads. Round the back of a deserted industrial estate near Apache Junction we find our spot marked out with a yellow 'road closed' barrier and a pile of wooden crates. To the left is a junkyard of 1960s trucks and buses, rusting without expectation of salvation. To the right is desert scrubland. And there, as if pointing to the barrier, is a jagged crack in the earth, like a thunderbolt that fell to earth horizontally. It's huge, perhaps three metres deep in places, though rarely more than a metre wide, and runs for perhaps a full kilometre, with multiple smaller radial fissures running from it like branches. Fans of the films *Tremors* or *Dune* could imagine a giant sandworm about to leap out. We stand, gawp and take photos beside our mini-Grand Canyon that no one knows about. An online groundwater map run by the Kyl Center for Water Policy confirms that the nearest well to this site is indeed in slow and steady decline, with water levels dropping 30.2 m between 1978 and 2020.[21] If water stays out of an aquifer for long enough, the waterless sections will be crushed by the weight of the earth above, like a sponge cake beneath a brick. These huge ground fissures are the direct result.

The next fissure we find, near Queen Creek in the San Tan Valley, is being built upon by property developers. Again, there are long, thin fissures, this time two running parallel, but deeper, perhaps four to five metres in places. Rather than being out the back of a rusting, forgotten yard, these ones are also close to the busy four-lane West Hunt Highway. We follow the fissure to the roadside and find it has been covered with a pile of lose stones. Just metres away is an advertising hoarding for new homes. I take a photo of John standing and pointing at the spot, metres away from rushing traffic. On the other side of the road is a building site for hundreds of new homes. Google Maps satellite images, taken before the building work started, show similar-size fissures on that side too. The rumour is that these have simply been 'bladed over'. JCBs, earth-movers and bulldozers stand expectantly by the existing fissures on our side of the road too. Again, the Kyl Center's map shows the nearest well to the Queen Creek site as a red spot, denoting declining water levels, falling 12.2 m between 2015 and 2020.[22] What this man-made roller-coaster of a groundwater table does to the surface is shown in the fissures. I wonder if any of the homebuyers will be told of this – but I think I know the answer. I later find an article in which a home-owner in nearby Cochise County complains that his well levels are dropping by 1.5 m a year and that land subsidence has split apart two highways and rangeland. 'Earth fissures are showing up all over the county,' he told Circle of Blue. 'You're thinking, "What do we do if one shows up in our driveway?" The thought of moving is always around. But who's going to buy?'[23]

We walk up a mound behind a church to view the site. From our vantage point, we look down at the earth fissures, the endangered road, the house-building sites on perilous ground. Seemingly all of outer Phoenix is a building site for new homes. They are all the same – large-footprint, single-family homes – and will all need hooking up to water and electricity supplies, with no certainty about where those will come from in future nor how expensive they could become. I stand in the desert sun and try to take it all in. Lynda shows me a small flowering cactus. I will soon be flying home for Christmas.

CHAPTER 4

Losing Ground

Iran is an even likelier host than the US for the next Day Zero. Temperatures of up to 50°C (passing that 'wet-bulb' threshold) have already pushed the country into a national water crisis. The Iranian meteorological service reported that October 2020 to June 2021 was its driest period for fifty-three years. From 15 July 2021, people in dozens of towns and cities in Khuzestan Province, with a large ethnic Arab population, took to the streets every evening to protest about their lack of water access. Three protesters were killed.

Iran is a hard place to get to, with or without a pandemic. To tell its story authentically, I recruited an independent Tehran-based researcher, Rastynn Radvar. When we first speak, in August 2021, he tells me, 'In the last ten years of global warming, every year in Iran is warmer than the previous year.' Rastynn is a photojournalist and writer with a track record covering water crises. He's young and immediately engaging, and talks with speed and enthusiasm. It's a story he wants to tell, in part *because* it is suppressed within Iran. Rastynn assures me that it is safe for him to do so. 'This is what I do', he says, with a smile. In the previous month, Human Rights Watch reported more excessive use of force by Iranian authorities against water demonstrators, including the use of live rounds.[1] But he talks me out of doing a simpler story about water issues in Tehran: the city doesn't have many problems, he says, because it's where all the politicians and officials live. If we want something deeper, 'we should go elsewhere than Tehran'. He suggests Yazd Province.

Yazd, I later discover, is at least a six-hour, 600 km drive from Tehran. The next time we speak Rastynn is driving back through Fars Province – even further away – returning from another assignment in the Persian Gulf. He's already sent me his first interview transcripts, and I want to ask him about his trip to Yazd Province. 'Yazd is the centre of tile-making,' he tells me, adding that there are 'lots of factories there – like, not very big factories, but a lot of them.' What really made an impression, though, were the mines: 'for iron, sand, this kind of stuff. And they're mining, like, in the middle of nowhere. It's, like, totally empty desert, but there are lots of machines and mines and . . . *tornadoes* of sand.' The mountain communities used to have enough water, but 'since these mines, even the water in the mountains is drying.' There are water shortages everywhere in Iran now, even in Fars Province, which is greener and close to the sea. 'But, in Yazd, it's totally different. It's, like, disaster. So many empty villages.'

Ardakan was Rastynn's first stop in Yazd Province, chosen over Yazd city as more representative of the problems people face. As he set out to find a local village, he stopped and chatted to an old man walking along the desert road. 'It was crazy he was out there. I could not stand like five minutes outside, it's like the sun being right next to your head!'* The man turned out to be a subsistence farmer called Hossein, who was looking to hitch a lift back to his village. 'I told him what we were doing, we took him to his village, and then he took us to his farm. And we talked.'

Hossein's village of Toot, home to nearly eighty families just fifteen years ago during the last census, is mostly empty now. In a video Rastynn sends me, it looks like a ghost town – not a single person or animal can be seen or heard, just the silent white facades of houses hunkering from the sun, with dusty red mountains like a Martian landscape – nothing obviously growing on or near

---

* I ask Rastynn what the temperature might have been. 'I don't know, forty-something. It's forty-three where I am right now, and it's way cooler than there.' (43°C is 109°F.)

them – looming in the near distance. Toot has electricity but no running water. Every household has a tanker or water pump, buying water from suppliers in Ardakan. Hossein is one of the few remaining farmers, but his farm has dwindled in size to something like an urban allotment. It's hand-to-mouth survival, with barely any surplus to sell at the local market. He speaks to Rastynn in Farsi: 'Each year, the snow and rain is less than the previous years. Our region has been a desert since ancient times, and the lack of rainfall is nothing new for the people of this region. But, compared to the past, you can really see and feel these changes.'

Natural springs are usually the main source of water for agriculture in this area, Hossein says. The village has a working *ab anbar*, a traditional water storage tank, likely hundreds of years old; it is largely underground, with a white conical dome topping it off above ground. Hossein says farmers build ponds to collect spring- or rainwater and 'each village has an underground water reservoir. But, with the day-to-day operation of the mines around this village and city, many of these springs dry up . . . The government is the main owner of these stone, soil and iron mines and, by digging in the mountains and deserts, they cause many problems for farmers and residents of arid areas such as Yazd Province.' Hossein takes Rastynn to see the pond – a square, concrete-lined pool, like a municipal paddling pool, with just a few centimetres of water. A ground-level pipe maintains a trickle, barely more than a drip.

'It was so interesting,' Rastynn enthuses afterwards. 'Saving water is in their culture. They understand very well the value of it, because they don't have it a lot in their lives.' However, it was clear that this way of life is dying out. As the groundwater supply is slowly strangled, even the small vegetable plots will soon go. Hossein says ranching used to be popular here too – there were '500 goats and sheep in the village' twenty years ago, but now 'there is not one four-legged animal in the village'. These days, ranching is not financially viable, and a rancher cannot quench his animals' thirst. 'Agriculture in these areas is left for older people,' says Hossein, who is sixty-four but looks much older. 'No youngster is interested . . . The village is very quiet and lifeless, and this makes me very sad. I am a child of this land and I cannot sell or give up and move to the city.' The rural population here is decreasing every day, he says. The region was once famous for pistachios and pomegranates, with large, green orchards irrigated by channels of groundwater. Now, Hossein can grow only turnip and coriander, sometimes aubergine and tomatoes if he's lucky, and he has only enough water to irrigate twice a week by hand. Interestingly, there is a water market for farmers in Ardakan, where they can sell what pitiful water rights they may still have and move out. The wealthier farmers buy what little water there is and send it to their fields, while the sellers either lose their land or reduce their productivity. At least that water is put to some use, Hossein says: some large pistachio farms are still clinging on by these means. Hossein mostly blames the tile industry and mines, which have profited their owners but seized most of the available water, while creating a lot of pollution, he says sadly. 'The water that flows here from Zayandeh River and Isfahan Province is used only by factories . . . But the government is thinking of more profitable jobs that will make them money . . . and they will completely ignore the rest.'

In the town of Ardakan itself, Rastynn speaks with Kobra, a schoolteacher in her forties who recently moved to the town to build a better future for her daughters. Water access was one of the main factors in her move; much as Hossein said of Toot, many people are leaving her village for urban areas. Even so, she says, in cities such as Ardakan, Meybod, Yazd, Ashkzar and Taft, 'quality water is not found in all neighbourhoods, and many people use tap water with purifiers . . . Water may be cut off two or three times a week. Outages in different areas are usually reported in advance so that people can prepare, but sometimes water is cut off without prior notice.' Rastynn sends me a photo of Kobra's kitchen tap, held together with black duct tape, running at a trickle despite being turned fully on. 'The water [pressure] is so low, I wanted to show you,' he says. 'Can you believe it? But, even with this, she was happy, compared to the village.'

Kobra never thought she'd have to leave her father's village in search of water. Soon, she says, 'it's not unlikely that . . . the residents of the cities of Ardakan and Yazd will have to migrate to more water-rich provinces too.' Again echoing Hossein's words, she says drought has left many farmers and ranchers in the area unemployed and that manufacturing is exacerbating local problems. 'My cousins were forced to sell their land for a small fee to move to the city, and, after losing their agricultural jobs, they now work in a tile factory. These factories also have high water consumption, and, with the intensification of the drought crisis, the closure of factories in this important industry in Yazd may cause many to lose their jobs again.' It's a sad, escalating, inevitable demise that everyone is being enlisted into speeding up, not slowing down.

I ask Rastynn how he felt, as an Iranian, visiting these places. 'I wasn't expecting it,' he admits. Despite having reported on water shortages in many different places, the problems here are severe even for him. 'It was totally different from what I saw in Baluchistan . . . It was so sad to see in the villages, but even in a city like Ardakan . . . I know, in our job, it's not too good to be sensitive, but it really makes you sad, man. I mean, I feel ashamed for the amount of water that I have, and it's not even in my hands.' He tells me people in Tehran are very insulated from the realities in the provinces. There is an Iranian saying, he says, that 'if the dog is not biting them, they don't care about other people's wounds'. I don't know why, but that brings tears to my eyes. And he's right – in our job, it's not too good to be sensitive.

Concreting over the cracks

Over 600 dams have been built in Iran since the Islamic Revolution in 1979, with hydroelectric power now vital to the nation's economy. Yet, as *The Times* reports, 'experts say that reservoirs in such hot and arid areas lose so much water to evaporation – 2B m^3 of water a month in Iran – that they have become part of the problem.'[2] From the early twentieth century onwards, the world's approach to water management has largely relied on one material: concrete. We have concreted over our flood plains, dammed our rivers and canalized our streams. What we gained in engineering efficiency, we lost in long-term sustainability. The landmark World Commission on Dams report (2000), with input from 947 submissions from over eighty countries, found that the world's twentieth-century mega-dams were not built on solid scientific or economic foundations. 'In too many cases an unacceptable and often unnecessary price has been paid', it states, 'especially in social and environmental terms, by people displaced, by communities downstream, by tax-payers and by the natural environment.'[3]

From the 1930s to the 1970s, the construction of large dams became, says the report, synonymous with development and eco-nomic progress. Viewed as symbols of modernization, dam construction accelerated dramatically. By the 1970s, on average two or three large dams were being commissioned every day some-where in the world. While some benefitted from this, most did not: 'Global estimates of the magnitude of impacts include some 40–80 million people displaced by dams while 60% of the world's rivers have been affected by dams and diversions . . . Large dams display a high degree of variability in delivering predicted water and electri-city services – and related social benefits – with a considerable portion falling short', in particular 'the impoverishment and suffer-ing of millions' downstream.[4] At least 20 per cent of the world's freshwater-fish species became extinct thanks to the twentieth-century dams, even though a billion people relied on river fish for protein. But, setting aside the 80 million displaced and one fifth of species lost, did dams work from a purely water-storage perspec-tive? Also no. The authors estimated that between 0.5 and 1 per cent of the total fresh-water storage capacity of existing dams is lost each year to sedimentation, meaning that 25 per cent of the world's exist-ing fresh-water storage behind dams may be lost between 2025 and 2050.[5] Half of active storage had already been lost in 10 per cent of the projects surveyed in 2000. Furthermore, half of all dams were judged as unprofitable in economic terms.

Before her role with Stanford's Water in the West programme, Dr Newsha Ajami trained in civil engineering and began her career working on Iran's big dams. She tells me her interests moved from pure engineering towards understanding policy and social aspects of water. 'The saddest part of this whole story is that the field is still very much dominated by engineers, who have that mindset of, "We need to build the next big thing to solve our problems." That's why we are having this problem – because we don't *have* that next big thing.' And this is from a civil engineer. Who worked on big dams. In Iran. 'In our major metropolitan cities, the majority of people don't even know where the water is coming from,' complains Ajami. Nor do they know where it's going or what happens to it. 'That disconnection is a huge problem, because people don't even know what they're paying for.'

Aquatic-conservation biologist John Waldman has described dams as 'blood clots in our watersheds' due to the 'tremendous harm' they have done. If so, there has never been a greater 'clot' than China's Three Gorges Dam. Completed in 2012, it submerged large areas of the Qutang, Wu, and Xiling gorges, creating an immense deep-water reservoir stretching some 2,250 km, from Shanghai on the East China Sea to the inland city of Chongqing. It generates 22,500 MW (megawatts) of hydroelectricity, making it the biggest power station of any type in the world. A Communist Party official in Beijing memorably told me, for my previous book, not to get 'too carried away by numbers when talking about China.' Everything about China is simply much bigger than everywhere else. But it's hard not to get carried away by this statistic: in the three years from 2011 to 2013, China poured more concrete than the United States did in the entire twentieth century.[6] Much of which flowed into dams.

In the summer of 2020, tens of millions of people across China were affected by torrential rains, floods and landslides. It was the worst flooding to hit China in decades, affecting more than 37 million people and leaving 141 dead or missing. The *South China Morning Post* reported, 'experts have questioned whether massive dams can effectively control flooding downstream . . . [a] geologist from Sichuan said the dam could partially intercept flooding upstream, but it had a limited effect on controlling floodwaters in the middle and lower reaches of the Yangtze.' According to hydro-climatologist Peter Gleick, one of the lessons from the Three Gorges was 'that no dam – no matter how large it was – could prevent the worst floods from occurring.'[7] If the $24-billion Three Gorges Dam, described as 'China's largest single flood-mitigation project', can't mitigate floods, then what are dams good for? Many do nothing. According to the National Inventory of Dams, only 7 per cent of US hydroelectric dams still produce any power. Brian Richter, the river scientist and conservationist, put it neatly on Twitter: 'Building a new dam because you don't have enough water is like opening a new checking account because you don't have enough money.'

A lot of dams 'actually lose water', says Maggie White, senior manager of international policy at Stockholm International Water Institute (SIWI). Much of this is through evaporation, but many dams are very old and leaking too, she says, and need re-solidification and proper maintenance to avoid catastrophe. 'The big hydropower dams have really come to the end of their life cycle.'

In Pakistan, two of the world's largest dams – Tarbela and Mangla – are in need of costly repair to fix sedimentation issues and are running out of capacity. In India, Vaibhav Chaturvedi, at the Council on Energy, Environment and Water (CEEW), Delhi, says, 'There is a realization that the investment in dams has kept on increasing but has not delivered.' Yet still megaprojects continue to excite politicians and investors. In South Asia, the National River Linking Project (NRLP) promises to interlink thirty rivers, approximately 3,000 storages sites and 9,600 km of canals to form a gigantic South Asian water grid. Kangkanika Neog, a water consultant formerly with CEEW, describes it as a poor idea 'hydrologically and ecologically . . . And also economically; it's a very heavy investment.' A pan-India programme transferring water from the potentially water-surplus Himalayan rivers to the water-scarce western basins is an immediately attractive idea, but, like all megaprojects, will cost more and deliver less than promised. A CEEW report warns of the risks of the NRLP: 'Such large-scale interlinking may result in environmental degradation, evaporation losses, losses in the aquatic ecosystem, waterlogging, salinity, and the submergence of vast areas of land in reservoirs and the huge network of unlined, open canals [and] massive displacement of people.' It also finds that the likely final cost of the water would be around 6 cents per m^3 – 33 per cent more than the 4 cents per m^3 for wastewater reuse, which could be done more easily everywhere, near to source, with far less environmental cost. Currently, out of 27B m^3 of wastewater generated per year in India, only 8B m^3 is treated. But, suggests Neog, it's easier for politicians to sign off on one megaproject than to work with a thousand local authorities to build wastewater-treatment plants, or a million farmers on water-efficiency measures.

This mentality still abounds. On 20 February 2022, Ethiopian president Abiy Ahmed (a Nobel Peace Prize winner since accused of war crimes[8]) turned on the Grand Ethiopian Renaissance Dam, now Africa's largest hydropower project, taller than the Great Pyramid in Giza and six times the size of Ghana's Akosombo Dam. The reservoir will take 4–7 years to fill up. Egypt claim the dam will cause it lose 110 billion m^3 of its own water, calling it an existential threat.[9] Iran and Iraq are locked in regional arms race of dam building, with Iran having constructed at least 600 dams nationwide in the past three decades, including turning rivers away from crossing national borders.[10] It is a race which all sides will eventually lose to siltation and evaporation. The UN Report of the International Resource Panel Working Group on Sustainable Water Management in 2015 found that, 'On average, national policy responses to the growing water scarcity have largely focused on expanding supply through substantive investments in water engineering infrastructure such as building large dams, canals, aqueducts, pipelines and water reservoirs. With a few exceptions . . . The amount of water lost through evaporation from water reservoirs is higher than the total amount of water consumed.'[11]

Annually, due to its high desert location and large surface area, Lake Powell, behind the Glen Canyon Dam on the Colorado River, loses an average of 1 km^3 (860,000 af) of water to evaporation and bank seepage – that amounts to 6 per cent of the Colorado River's annual flow, more than three times Nevada's annual allocation, and enough to supply the entire city of Los Angeles for a year.[12] 'Dams not only impound the water in rivers, they also interrupt the downstream transport of sediment, leading to its accumulation in reservoirs', causing accelerated deterioration, says the UN Working Group report. Overall, the dam-building mindset must be replaced by 'emerging ideas and practices on the alternative types of water storage [and] nature-based solutions.'

To recharge with rainwater, aquifers instead require a connected, functioning river system. Consider what California looked like just a hundred years ago, before the Bureau of Rec got carried away with concrete. A 2015 report from the Sonoma Resource Conservation District paints an inviting picture:

> the diverse collection of habitats including redwood forests, oak woodlands, native grasslands, riparian areas, coastal dunes, and wetlands [left] virtually undisturbed. Rivers and streams, capturing and conveying rainwater, flowed from upland areas though rivers and creeks to the Pacific Ocean and San Pablo Bay along sinuous unchannelized corridors. Intact wetlands functioned as natural filters and buffers from major storms. Under these pre-development conditions, as much as 50% of rainwater infiltrated . . . the soil replenishing groundwater supplies, contributing to year-round stream flows, and sustaining ecosystem function.[13]

Water storage, groundwater replenishment, and flood control, therefore, was better provided for by nature prior to all the grey engineering of the twentieth century.

Joe Poland's pole

In talks and lectures, Jay Famiglietti likes to show a slide titled 'Cumulative Groundwater Depletion in California's Central Valley from 1962 to 2014'. It shows a constant downhill trajectory. There are upticks in wet periods, such as 1978–85, and 1993–8, but they're never as high as where the previous downtick started. From 1962 to 2014 there was a cumulative loss of 90 km^3 (73M af) of groundwater in California. The 2011–15 'drought' wasn't an anomaly. It was part of a century-long trend.

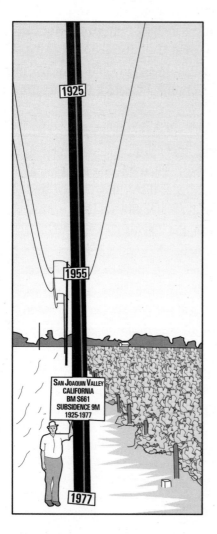

As a faculty member of the University of California, Irvine, at the time, Jay found that period very difficult to watch unfold. 'I'd been looking at these data since 2002. I wrote a paper on the first phase of the drought from 2006 to 2010 and knew exactly what was happening, that this groundwater was being poorly managed, was disappearing, that people weren't understanding that there's a finite amount of groundwater there.' This propelled Jay into science communication – just writing academic papers 'was no longer going to do it'. Everyone in media and politics at that time was 'on message . . . people got it'. Jay says one of the biggest changes he saw in his then hometown of Pasadena, Southern California, was that people stopped watering their lawns. However, when the drought was declared over, and it rained again, 'everybody was like, "Yay, I'm gonna water my grass and wash my car!"' With the rain, his op-eds, TV appearances* and public talks all stopped too.

In fact, the drought had no neat beginning or end: 2015 was merely another slight uptick amid a trend line of decline. The US Geological Survey hydrologist Joseph Poland recognized this back in the 1940s, when land in the fertile San Joaquin Valley was sinking. Underneath the land, groundwater was being pumped too rapidly for crop irrigation, creating subsidence. To demonstrate this, he erected a telegraph pole showing where the ground level would have been in 1925 compared to 1955. He returned in 1977 as an old man to repeat the exercise, posing for a photograph next to a pole towering 9 m above him, every metre indicating ground level lost. Jay likes to show Poland's photo in his talks because 1977 is early on in Jay's graph, when around 30 km³ (24M af) of groundwater storage had been lost – 73 km³ (59M af) would be lost by 2015. You'd need another couple of telegraph poles to show what's been lost today. 'That storage capacity is lost forever,' says Jay. 'The pore water pressure is keeping the aquifer pumped up, keeping the minerals apart from each other: when you take the water out, those minerals actually fuse together, so it becomes rock.'

California has seen an average groundwater depletion of 1.7 km³ (1.5M af) a year since 1962 (see Graph 2). Between 2003 and 2010, Jay's team found that the loss increased to 4.4 km³ (36M af)

---

* The pinnacle of which was appearing on *Real Time with Bill Maher*, when a visibly bemused Jay is cheered on by a whooping audience. He even got a chance to show a slide from his 2015 paper, showing groundwater decline in California. 'It was green and now it's red,' says Bill, sardonically. 'Now, Doc, I'm not a scientist, but red is bad.' (Available to view at: https://www.youtube.com/watch?v=qztALfK4OGs)

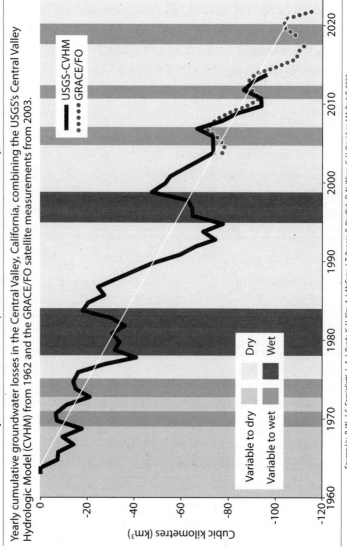

Graph 2 Groundwater depletion in California's Central Valley, 1962–2021

Yearly cumulative groundwater losses in the Central Valley, California, combining the USGS's Central Valley Hydrologic Model (CVHM) from 1962 and the GRACE/FO satellite measurements from 2003.

Source: Liu, P.-W., J. S. Famiglietti, J. A. J. Purdy, K. H. Kim, A. L McEvoy, J. T. Reager, R. Bindlish, D. N. Wiese, C. H. David and M. Rodell, 2022. Groundwater Depletion in California's Central Valley Accelerates During Megadrought, in review, Nature Comm.

annually – equivalent to two Loch Lomonds of water lost every year.[14] 'People just don't understand the scale of it', says Jay. 'If you take Interstate 5 out of LA and look out over the valley, you realize right away why we're using all that water . . . It's monstrous. Just a vast expanse of agriculture, as far as the eye can see. And there's no rivers in sight.' Across the US as a whole, 1,000 km^3 of groundwater were depleted between 1900 and 2008, about twice the volume of Lake Erie, the thirteenth largest lake in the world.[15]

Traditionally, the use of groundwater was limited by its relative inaccessibility: we could only reach it via hand-dug wells, natural springs and, later, windmills to pump it out of the ground. But deep drilling and machine pumping introduced in the 1970s enabled rapid expansion in its use. Suddenly groundwater was being attacked from both fronts: its supply line – the riparian rivers and streams and floodplains – were being cut by dams, while drills came from above to abstract from the aquifers directly. University of California, Santa Barbara, groundwater guru Debra Perrone explains that, where declining water levels are not managed, increasing numbers of groundwater wells will run dry – which is, to her, 'one of the most pronounced manifestations of water scarcity'.[16] By analysing data from nearly 12 million wells across the United States, in records stretching back decades, Perrone and her colleague Scott Jasechko found that groundwater wells provide drinking water to about 120 million Americans, and that 70 per cent of those wells have been deepened to compensate for groundwater depletion. In California's Tulare Lake region, 89 per cent of wells demonstrate 'declining water levels'.[17] They repeated the study globally in 2021, looking at 39 million wells across forty countries, and found literally millions of wells at risk of running dry if groundwater levels decline by only a few more metres.[18]

The Memphis Sands aquifer, a crucial water supply for Mississippi, Tennessee, Arkansas and Louisiana, is overdrawn by hundreds of millions of gallons a day. Much of the Ogallala aquifer – which accounts for roughly a third of America's irrigation groundwater – could be gone, fused into rock, by the end of this century, if pumping trends continue. Many parts of the world are seeing similar rapid ground-water decline. More than 2 billion people and 40 per cent of global agricultural production relies on unsustainable levels of ground-water extraction.[19] Gurugram district in northern India is 77 per cent rural and agriculture-based, with no perennial river. According to local water charity Gurujal, only 10 per cent of the agricultural area is rain-fed, while 98.7 per cent of the irrigated agriculture comes from boreholes. This led to a near collapse of the ground-water, with over-abstraction reaching 308 per cent and groundwater levels falling by up to 5 m per year. The number of water bodies in the district reduced from 644 to 124 between 1994 and 2019, with ponds and temple pools disappearing one by one as the water table dropped.[20]

Water lost in La Mancha

The Spanish capital Madrid, Nuria Hernández-Mora tells me, has just had a freak winter blizzard. The heaviest snowfall in fifty years brought the city, totally unused to snow, to a standstill. Bad in the short-term, says Nuria, but good for aquifer recharge. Hernández-Mora is a senior water-policy specialist at the New Water Culture Foundation (FNCA) and water-policy adviser to the Spain's envir-onment secretariat. Her paper 'Sticks and carrots: managing groundwater over-abstraction in La Mancha, Spain' tells the fascin-ating tale of over-abstraction in a former UNESCO biosphere reserve wetland. I say 'former' because it was, to all intents and purposes, sucked bone dry.

In 1981, the Mancha ecosystem – a 400,000-ha wetland region – was designated a UNESCO biosphere reserve. The reserve includes Las Tablas de Daimiel, deemed an internationally important wet-land and given national park status in 1973, fed with water from the Western Mancha aquifer. Las Tablas are 'shallow wetlands formed with the connection between the groundwater and the surface water', explains Nuria. 'It is about 5,500 km^2 of underlying aquifer across grassland. But the aquifer was severely exploited; there had been a disconnection between the aquifer and water. I first went in 1993 – there had already been fifteen years of intense agricultural irrigation development and groundwater tables had dropped. The national park itself had dried up.' The authorities initially responded by dredging the river supplying the wetlands, effectively turning it into a subservient canal. But the dredging, Hernández-Mora says, 'actually destroyed and disconnected the wetland ecosystem. The water table continued to drop.' Other places were then dredged, until very little remained of the extremely rich area. According to Hernández-Mora, in 2010, Spanish environmental groups wrote to UNESCO asking it to de-catalogue it from the biosphere-reserve list because it was no longer a functional wetland. She cites the US Corps of Engineers definition that a wetland must have three characteristics: water, soils and vegetation. Degrade one, and you no longer have a wetland.

The Mancha ecosystem should be a network of lakes, riverine wetlands and the Tablas' flat, shallow pools dotting hundreds of square kilometres. In thirty years, Hernández-Mora has seen it in its true form only once – around 2011, following two years of above-average rainfall. Her research finds that, between 1989 and 1995, an uncontrolled expansion of agriculture surrounding La Mancha Wetlands (Humedales de la Mancha, in Spanish) pumped out 650 megalitres (mL) annually against a natural recharge rate of just 230 mL.[21] This 'gold rush', as she puts it, occurred due to a combination of factors. First, in the 1970s, the submersible well pump became widely available, so any farmer could, with very little money, pump water from the ground. Second, hydrogeological understanding developed, with detailed maps of groundwater being produced in Spain for the first time. Third, due to fears over mosquito-borne diseases, the dredging and drying of wetlands was a state policy with public support. 'Farmers pumping and converting wetland areas into irrigation was seen as something positive . . . and irrigation farming is more profitable than rain-fed farming.' When Spain joined the European Economic Community in 1986, farmers' subsidies for irrigated crops tripled or even quadrupled. To make matters worse, archaic water laws dating back to 1879, long before electric pumps, made groundwater free to use.

In 2001, already well into the 'gold rush', the Western Mancha aquifer had 40,000 wells, irrigating 133,000 ha of farmland. By 2009, hundreds of thousands of hectares of UNESCO biosphere-reserve wetland had reduced to just 15 ha. The peat was so dry, it caught fire. The ecosystem had collapsed.

The devastation of La Mancha Wetlands is a reminder that groundwater and surface water are not separate, but intimately, intricately connected. Conor Linstead, freshwater specialist with WWF-UK, also speaks of the 'very direct connection between groundwater levels and surface water . . . Rivers are, to a greater or lesser degree, supported by groundwater inflow. So, if you over-exploit your groundwater, then you're depriving rivers and therefore ecosystems of flow at critical times of year, in the dry seasons . . . If you pump the water out of the aquifer and it's not being replenished, then it collapses.' Hence the subsidence seen in California's Central Valley, or Mexico City. In October 2021, the charity International Rivers released a letter signed by more than 570 experts, scientists, engineers, researchers and practitioners, from ninety-seven countries, urging the United Nations and national delegates at COP26 in Glasgow to act: 'wetlands are being lost three times faster than forests, with 64 per cent of them lost globally since 1900 . . . breaking the connectivity that supports ecosystem functions.' The damage, they said, 'is accelerating'.[22]

Fossil hunters

Dr Jennifer McIntosh, professor of hydrology and atmospheric sciences at the University of Arizona, is an internationally renowned groundwater expert, best known for her work on 'fossil water' – groundwater that remains from different geological eras, unconnected to the current water cycle. There are numerous methods for dating water. Any groundwater younger than the 1950s is easy to tell through the presence of tritium. 'Nuclear-weapons test-ing post-1950 released tritium . . . [which] got into rainwater and then precipitated all over the globe.' Similar modern tracers leave dateable fingerprints too, such as CFCs. For older water, 'the most common technique is radiocarbon, or carbon-14. That's the same way that you date ancient bones, or material in archaeology.' That works for anything up to around 50,000 years old. Older than that, and you measure the radioactivity from rocks that release helium-4 into the groundwater. The oldest water that's ever been discovered is around a billion years old, McIntosh tells me. Typically, such ancient water is highly saline – either because salt rocks have dis-solved in it or the water came from earlier oceans – highly mineralized, and quite likely radioactive; you wouldn't want to drink it. Water as far down as 10 km, beyond any commercial drilling cap-ability, would be millions to billions of years old.

Around 12,000 to 14,000 years ago, however, is a sweet spot: the last Ice Age melted away, leaving behind huge stores of relatively pure glacial water below ground. Getting to groundwater laid down over 12,000 years ago, though, is synonymous with drilling for oil – once it's gone, it's gone. This is fossil water. Famous examples include the Disi aquifer shared between Jordan and Saudi Arabia (see page 17). Saudi Arabia has also already used up the fossil water in the 20,000-year-old Wadi As-Sirhan Basin, thanks to a short-term boom in centre-pivot irrigation farming, which came to an abrupt end in 2016 (see page 168). The Nubian Sandstone Aquifer System under the Sahara Desert, meanwhile, is both the world's largest known fossil water aquifer, and the world's largest reservoir of fresh groundwater, covering nearly 2M km². This fell as rain during the African Humid Period, between 15,000 and 5,000 years ago. Des-pite it being totally non-renewable, former Libyan leader Colonel Gaddafi made it central to his government's water policy, at one time providing 70 per cent of all the country's fresh water via the so-called Great Man-Made River Project.

McIntosh also stresses that fossil water being non-renewable doesn't make the younger, shallower, rain-replenished aquifers fair game to overexploitation either. 'We're extracting *all* groundwater at rates so much faster than any natural recharge, that the natural recharge rate almost doesn't matter,' she says. This, in many ways, sums up the England water story (as we'll see in Chapter 6). In reality, says McIntosh, 'it's very challenging to think of any aquifer system that *is* renewable, that *is* sustainable, where the recharge rates *are* fast enough to keep up with the amount that we're pumping.'

Groundwater pumping, including fossil groundwater, is being done at such a rate that it's even contributing to sea-level rise, McIntosh informs me: 'we pump groundwater, we use it, it goes down our toilets to the wastewater-treatment plant and eventually goes out to the ocean.' Overall, around 2 per cent of man-made sea-level rise comes from our pumping of groundwater, says McIntosh.

The dying Dead Sea

The Dead Sea, the saltwater body shared between Israel and Jordan, has shrunk by a third since the 1960s. Slow-motion environmental decline like this suffers from generational memory. With each new generation, we hit reset as we look out at nature: what we see being all we've ever known, and so a new, lower bar is set for what we deem acceptable. This came up later, too, when I visited the River Windrush in England, excited to see a hatch of maybe twenty or thirty mayflies. One landed on my shoulder. When I later told Ash Smith, a local river campaigner, he looked at me incredulously: a mayfly hatch should be a significant annual event, he told me; there were stories in the nineteenth century of the air being so thick with them you could barely see – from a distance, the river would appear as if covered in a fog of them. This still happens in more remote parts of rural France. Seeing a hatch of twenty or thirty, which I had taken as a sign of nature ticking along nicely, would have horrified my great-grandparents' generation.

On the banks of the Dead Sea, as the water declines each year, the tourist hotels unwittingly offer a visual reference to what has been lost. Each year, they must build more steps for their guests to get down to the water. Most, when built, would have been level with the lake. Now, there are hundreds of steps to get down. Some have even filled their swimming pools with Dead Sea water instead, to save their guests the walk (the same feeling of opulence, but protected from the reality). At the southern end of the sea, where the water is harvested in giant evaporation beds for potash and minerals,* the banks are dotted with sinkholes. On the Israeli side, sections of road have disappeared and whole villages have been abandoned. As the Dead Sea recedes, the previously inundated land dries out, including a thick layer of underground salt. When rain falls in winter, the fresh water seeps in and dissolves the salt, causing the ground to collapse inward. Over 7,000 such sinkholes have been recorded, some several metres deep.[23]

The lake itself is reducing by over a metre every year. The River Jordan used to refill it, but now arrives with a whimper, over-allocated by each state taking their pre-agreed share of a rain-fed river that effectively no longer exists. The rainfall of the 1960s and 1990s was plentiful and predictable compared to the climate-changed 2020s. The flow of the River Jordan fifty years ago was 1.4 km^3, all received by the Dead Sea; now, it's lucky to get a tenth of that.[24]

I arrive at the Dead Sea in the late evening, with Nasser and Yana, my hosts from EcoPeace Middle East. The banks are coloured a deep red by the setting sun. With only minutes before the end of the day, and no hotel to retire to, we park up in a near-deserted roadside car park. The water is far down below us, but too tempting to resist. 'I'm going for it,' I say, and start the descent down the muddy banks.

---

* On their respective sides of the sea, the Israel-based Dead Sea Works and the Jordan-based Arab Potash Company extract potash and other chemicals from the Dead Sea via evaporation ponds pumped with approximately 600M m^3 per year from the northern part of the Dead Sea, with an evaporation loss of approximately 330M m^3 per year, accounting for around 40 per cent of the total annual decline of lake levels.

Every 10 m or so there is a mini-bank of 60 to 90 cm, testament to a former waterline. Closer to the water is a row of makeshift shelters, metal frames covered with rags, set up by an entrepreneurial family to attract the lower class of tourist, like myself, who don't want to pay the hotel 'beach' prices. A boy aged around eight tracks me and tries to sell me bottled water. Discarded plastic bottles are strewn along the famous mud and rocks for as far as the eye can see. As I approach the shoreline, the mud is so squelchy I'm in danger of losing my shoes, but I'm not giving up until I can put my hands in the water. When I do, it feels extraordinary – the salt and minerals make it swirl thick, like translucent petrol. The mud seems pure, almost otherworldly – I'm reminded of old episodes of *The Outer Limits*, where travellers arrive at the seas of distant planets. And the silence at the water level is eerie. Everything is flat, calm and peaceful. It's easy to forget the region's problems down here, facing the hills of the West Bank on the opposite side, my back to the plastic pollution. It's easy to forget the Dead Seas problems, too, at water level. It still looks too wide, too vast to be in any real danger. And yet, as I turn to walk back up, ascending each previous shoreline of a formerly mightier lake, it's clear that a reckoning will one day come.

I scrabble back up the bank, the sun now very low. Nasser and Yana suggest we go to a nearby hotel bar to enjoy the last few minutes of the sunset. We sit in the beer garden, the sunset just visible from one far corner table. I look down at the menu and see that the name of the hotel bar is 'The Rovers Return' – the hotel bar is an English theme pub based on the TV show *Coronation Street*. I sit in Jordan, looking across to the Israeli shore, sipping a pint and thinking – unexpectedly, and a bit disappointingly – of Manchester. As we watch the red sun pass behind the hills, Nasser comments, sardonically, 'Of course – it is still sunny in Israel!'

CHAPTER 5

The New Pollution

Mr McCarthy and the New Legon Residents' Association

New Legon, Ghana, east from central Accra, is no slum – far from it. It may be lacking infrastructure, but it's also a spot for people with a pioneering spirit to seek a plot of land to build their own home. Alongside makeshift shacks fashioned out of shipping containers and half-finished houses of grey breeze-block are occasional grand modern houses hiding behind high security fences. Somewhere in between the two is the home of trendsetter Mr Charles Tetteh McCarthy. Seven years ago, when he built his home – a handsome single-storey whitewashed building with a blue tin roof – he was one of the first to move here; now, he's president of a 200-strong residents' association. And top of the agenda in the residents' meetings is water. Access and quality of water here is poor. In particular, they are petitioning for an official water connection from Ghana Water Company.

I'd imagined Mr McCarthy as perhaps an officious sixty-something in a grey suit. I was close on the age, but he's dressed in baggy leopard-print shorts and a bright Ghanaian-print shirt. And rather than an office he shows us into his small music studio, where two heavily-used Casio keyboards are pushed up against a computer and speakers, with guitars hanging on the wall. Before Accra spread out this far, and before Mr McCarthy (pronounced Mr 'MaCatty' in Ghanaian) arrived, New Legon was a wetland.

Drainage ditches were dug and road heights raised, but Mr McCarthy says they are positioned in the direct path of the water as it journeys from the mountains to the sea. When it rains, it floods. Stagnant standing water is everywhere. When we arrive, Mr McCarthy sits down and finishes a large bowl of thick green 'tea' – a herbal mixture, he tells me, to ward off malaria.

Without water supply, Mr McCarthy harvests rainwater from his roof guttering into a 1,000-L (450-gallon) polytank of the kind that stands beside most houses in Accra. This is the first I've seen that actually collects rainwater, though, instead of being expensively refilled from private water trucks. 'If it rains twice, it fills up,' he tells me, smiling. When the rainwater runs out, he buys from the tankers, but he trusts the quality of that water less than the rainwater from his roof: 'The source of that water we don't know. So, we don't drink that water, only to bath, maybe to flush,' he says. Mr McCarthy is relatively unusual in having a flushing toilet, in an area without any sewage pipes. He tells me he attempted to dig his own borehole to reach groundwater, but gave up after 21 m* and turned it into a pit latrine instead, building a concrete toilet block above it. The pit latrine isn't emptied, but rather will slowly decompose the solid waste – with the water leaching out into the soil and, eventually, the groundwater.

There are some boreholes connected to diesel pumps in the neighbourhood, but the groundwater is both deep, making it expensive to drill and pump, and saline; clothes washed in it are left with white salt-residue marks. When full, his rainwater tank lasts at least two weeks, covering all household uses. Mr McCarthy also prefers to drink from his rainwater tank, though his family – which includes his adult children and his grandchildren who live there – prefer to buy 'sachets' from water sellers. Sachets are sealed plastic 500-ml bags, sold either individually on the street or in sacks of thirty. Mr McCarthy shows me one. The label reads *Voltic* in lettering clearly

---

* When I ask a water official later how deep the boreholes in New Legon would need to be to hit water, I'm told at least 100 m.

reminiscent of the more famous Volvic brand, but reassuringly gives a nod to the water's provenance: 'Treated drinking water. Produced in Ghana under the franchise of Voltic (GH) Ltd.'*

All water except rainwater is expensive. A 500-ml sachet of drinking water costs between GH₵3 and GH₵6 (between 29p and 58p),† depending on the seller – comparable to the cost of mineral water in the US or UK, where salaries are many times higher. A top-up of non-potable water from a tanker costs GH₵500 (£60/$71) for 250 L, working out at GH₵2 (24p) per litre. Some years ago, Mr McCarthy was quoted GH₵7,000 (£850/$1,007) to drill and fit a borehole, which led to him manually trying to dig the one that's now a toilet. He says even those with boreholes still find that 'in the wet season they get water, but in the dry season they don't'.

Mr McCarthy and the New Legon Residents' Association are petitioning GWC to supply piped water to the area, but residents would have to pay for the installation and additionally for each individual household connected to it, plus the subsequent monthly water bills. 'The water companies have been to the area to see the challenges we face. But still we are waiting for the decision,' Mr McCarthy says. He expects the cost will be 'very huge': more expensive than boreholes, but better-quality water.

The McCarthy family used to live in East Legon, a more established area with a connected water supply. But still the taps ran only 'twice a week', meaning you had to store water in buckets, yellow gallon jerrycans or – if you could afford one – a polytank. Announcements are made on local TV and radio stating which areas get water at which times, on which days – or sometimes supply can be cut off unannounced (the same for electricity). When the water supply is as irregular as that, the reality of living without it isn't so different – it still comes down to how much you can personally store. Mr

---

* Later, when I buy a larger bottle of Voltic, I see that it is owned by, and is a registered trademark of, Coca-Cola.
† Prices and exchange rate of 8.3 Cedi to the pound at the time of visiting (September 2021).

McCarthy wants to upgrade to a 2,222-L (1,000-gallon) polytank. He tells me that, in the rural village where he grew up, fifty miles inland from Accra, the villagers shared a concrete rainwater tank large enough to last for everyone through the whole of the dry season. Many villages now simply waste rainwater, he laments; capturing rainwater is no longer seen as 'modern'. 'They are just letting it wash away – why? In my time, you were proud to harvest rainwater. When visitors came, you would offer them to drink it, because the alternative was to take water from the stream.' That said, Mr McCarthy is not alone in still doing so. On a walk around the area – as I struggle not to slip in the mud, suspiciously eyeing each puddle for mosquitoes – we see several houses with guttering leading to tanks, small or large. Most impressive of all is George 'Mack' Ameyaki's partially submerged concrete tank, roughly 2 m², topped with metal sheeting. Clothes are laid out to dry on top of it. George, a thirty-something with small children, tells me this can last his family two months when full. He agrees it's strange that not everyone invests in this way, given the cost or dubious quality of the alternatives. His next-door neighbour has a borehole and diesel pump, he says – far more expensive, with less usable water. The more 'modern' middle-class homes seem to uniformly have polytanks above their roofs, positioned solely for water tankers to fill.

The area's main problem is pollution. New Legon, built on 'the lowlands', as Mr McCarthy describes it, floods regularly. There used to be a law, he says – perhaps still is – that no one could build within 15 m of a stream or lagoon. 'If you saw a strip of green trees fifty feet wide, you knew there was a river,' he says. But if the law still exists, it is widely ignored now, the trees long felled and concrete foundations poured to take their place, even in a wetland area like New Legon. George and Mr McCarthy point out several near neighbours who have built too close to the stream. One house with concrete foundations built up against the muddy banks is already sagging and buckled by damp. The narrow stream itself has in effect became an open sewer, but at least one that runs fast. Nearby, George's young son is catching small tilapia fish from it.

When it rains at night, people close to the lagoon will sometimes sleep on their roofs, says Mr McCarthy, as they expect their ground floors to flood. Not Mr McCarthy, though, who built his home with due observance of the 15-m law. 'I did some geography at school,' he grins. 'We are cutting all our trees down to export for timber, and then wondering why the rivers dry up? A tree feeds the clouds; it is all connected. But they say it is all global warming and can't be helped . . .' The remaining rivers are then diverted by the illegal mining, known locally as *galamsey*, he says. 'I or you can't stop it. It is the authorities.'

When we get back to Mr McCarthy's house, being first and foremost a musician, he reaches for his guitar. We have hit upon the topic of one of his songs, he says. With his permission, I film him on my phone. The tune is fast, upbeat, hard to stay still to. His smile is consistently infectious. But the lyrics hit hard:

'Everybody is crying for climate change, but,
They are the same people who are causing it.
World leaders are calling for solutions, but,
They are lying to us and causing it.
European leaders are crying for climate change, but,
They dig up our forests for construction.
The [Ghana] politicians are debating for solutions, but,
They fund the equipment for the *galamsey* boys.

I say, everybody is crying for climate change, but,
They are the same people who are causing it.

They are crying, Booo-hooo-ho!
They are crying, Booo-hooo-ho!

They are crying and at the same time,
Polluting our river bodies.
They are crying for drinking water, but,
Dumping mining waste into our river bodies,

And crying that our river bodies are drying up.
Building indiscriminately on our waterways, but then,
Crying for flooding.

I say, everybody is crying for climate change, but,
They are the same people who are causing it.

They are crying, Booo-hooo-ho!
They are crying, Booo-hooo-ho!'

He ends with a laugh and a big smile, fixing me in the eye, and I can't help but beam back at his infectious energy and the brilliance of his song. But, while smiling widely, I notice my eyes have welled up, too. Booo-hooo-ho.

Muddy waters

As Mr McCarthy's story shows, an abundance of water is one thing; an abundance of fresh, potable water is quite another. Water pollution is the other side of the water-scarcity coin. Maggie White, of SIWI, argues that it's an even 'bigger issue than the quantity of water. We've always had the same amount of water on our planet since the beginning. It has just shifted where it exists. But once we've polluted that water, it's really the quality that becomes the problem. The more water we pollute in our natural environment, the more we're actually dooming ourselves.' She points to arsenic found in many groundwater bodies, industrial pollution, agricultural pollution, and perhaps the most widespread pollution problem of all: raw sewage. 'Still, 80 per cent of our wastewater is being thrown back into nature without being treated,' she informs me. 'So, we are just throwing all that waste into nature, polluting our groundwater and killing our ecosystems, [which are] our buffers when it comes to producing water and ensuring water quality.'

Jay Famiglietti admits that often the water-quality academics and water-scarcity academics are 'two different communities'. But, when you bring them together 'and start overlaying the maps of ground-water contamination with the maps of water scarcity, it is just an incredibly dire picture.' In short, an aquifer full of contaminated water is as useless as an empty one.

Some pollutants occur naturally. Arsenic, for example, is wide-spread in arid and semi-arid basins, and is especially problematic in Bangladesh and West Bengal, India, where millions of people are exposed to arsenic poisoning through groundwater wells. Some-times, as seen in parts of Argentina and northern Chile, mining activities have further dislodged arsenic deposits into water bodies. Fluoride, too, often appears alongside arsenic, with high concentra-tions found throughout Central Asia, the Middle East and Africa, with hotspots in Argentina, Australia, Mexico and Russia. It's esti-mated that fluoride exposure from drinking water impacts 200 million people in twenty-five countries, with 66 million people affected in India alone.[1] Fluoride overexposure can cause dental and skeletal fluorosis in humans, causing damage to joints and bones. Heavy metals, such as lead, aluminium, copper, cadmium and zinc, also make their way into drinking water, typically through aging and corroding pipes. One of the most infamous cases was in Flint, Michigan, where, in 2014, the city's drinking-water source was changed from Lake Huron to the Flint River without consideration of the inevitable corrosion a change in pH would cause: the more acidic river water corroded the ageing lead pipes and exposed resi-dents to potentially deadly levels of lead contamination.

Sewage is an even bigger problem. Open defecation, leaking latrines and raw wastewater are major polluters of groundwater and surface water the world over. A 2019 World Bank report found that 80 per cent of the world's wastewater – and more than 95 per cent in some developing countries – is still released into the environ-ment without treatment.[2] Even sewage passing through pipes rather than pits isn't necessarily being treated, leading straight into streams and rivers. Along the Ganges (Ganga) river in India, some thirty cities, seventy towns and thousands of villages release 6 billion litres (6B L) of untreated sewage into the river every day, making it the

single biggest cause of pollution in a region with a long list of con-
tenders. An Indian government audit in 2017 found that biological
oxygen demand (BOD: a standard water-quality test, representing
the amount of oxygen consumed by bacteria in the water) was
higher than the permissible limit in all cities of the Upper Ganga
Basin, and coliform levels (an indicator of disease-causing pathogens) were hundreds of times higher.[3] The World Bank finds that,
when the BOD of surface water is at a level at which rivers are considered heavily polluted (exceeding 8 milligrams per litre), GDP
growth in downstream regions is lowered by a third due to the
spread of disease. Cholera alone, an acute diarrhoeal disease linked
to water polluted with faeces, infects up to 4 million people each
year, killing an estimated 21,000 to 143,000.[4]

We've known for centuries that human sewage poses a health
risk. But what's in that sewage has changed over time. Duke University Marine Lab describes 'a cocktail of pollutants, including
endocrine disruptors, heavy metals, pharmaceuticals, and pathogens'.[5] According to the FAO (Food and Agriculture Organization
of the UN), of the 3,928 km^3 (3.18B af) of fresh water withdrawn
globally every year (a cool stat in itself, by the way), it is estimated
that over half (2,212 km^3) is released back into the environment as
urban wastewater, industrial wastewater – including power-station
cooling water – or agricultural waste: 'The composition and level of
treatment of these "wastewaters", and therefore the risks for human
and environmental health, vary.'[6] This is far from confined to low- and
medium-income countries. In New York Harbor, around 10,200 m^3
(27 billion gallons) of untreated sewage is released annually.

*

Dr Hans Sanderson, senior scientist at Aarhus University in Denmark, has two ageing illustrated posters on the wall of his office:
'Birds of the Garden' and 'Fishes of the Sea'. He's a keen
environmentalist – 'that's what it's all about', he says – and he used
to keep a diary of when the migratory birds arrived each year: 'they
arrive earlier and earlier, now.' He has a shock of recently blond,

now white, hair and thick-rimmed, Scandi-designer glasses. The ornithology is a hobby; professionally, he is one of the world's pre-eminent environmental chemists and toxicologists. If the WHO want a water-quality expert, they call Hans (and regularly do, including for the seminal report *Pharmaceuticals in Drinking-Water*[7]). When the US government put together a presidential task force on risk assessment of pharmaceuticals, they called Hans – the only non-American on the panel. 'So, let me give you a brief history of the regulation of chemicals in water,' he tells me, and I happily sit back and listen.

Before 1981, any chemical could be marketed in Europe, without the requirement to provide evidence of any problematic environmental impacts. In September 1981, that changed, with the introduction of the first chemical legislation. But the 100,000-plus chemicals already on the market were allowed to remain (known in pharma as 'grandfathering'); only new chemicals released after the regulations were subject to regulation. 'So, you could take any of those [existing ones] and market it however you liked, no questions asked, in any product that you wanted to.' According to Sanderson, this in effect meant that European innovation in the chemical industry was stifled after 1981 – 'why would anybody make a new chemical, and then have to submit data on that?' Producers got lazy and just made use of the old, unregulated chemicals. Only in 2001, after a full twenty years of this, did the EU Registration, Evaluation, Authorization and Restriction of Chemicals (REACH) legislation come in to close this loophole: data was now required for *all* compounds, including the old ones, proving they weren't causing a negative environmental impact. Today, the EU Green Deal ushers in a sustainability strategy for chemicals in Europe that 'starts to ask, What do we need to know about chemicals so that we can protect the environment?' says Sanderson. He believes previous 'very crude assessments' have been motivated by market regulation rather than environmental protection. Only now are we starting to ask what a water-quality test should look like 'to see how it interacts in an ecosystem.'

Over the space of my lifetime, then, we have gone from a chemical free-for-all for all industrial and domestic chemicals and water pollutants, to mild regulations, to only now – and only in the EU, not the UK – starting to look in depth at the environmental impact of everyday chemicals released into our waterways. Little wonder, then, that not a single river in England is currently classified as being in chemically good health.

'This is actually a process going on globally,' continues Sanderson. 'In Canada, it's called the Domestic Substances List. In the US, they're revising the Toxic Substances Classification Act (TSCA). In Korea and China, they have a REACH-type approach. And it basically all springs from a consumer's right to know.' This is paradoxically both reassuring and worrying. Reassuring that (some of) the world is getting its act together, but worrying in terms of the amount of chemicals already released into the environment over decades, much of which will still be slowly leaching down into groundwater and will be stored up in reservoir sediments. The lack of government interest to date is a big problem. All living things, including humans, have receptor systems that respond to chemicals in the environment and act accordingly. Drugs developed for particular chemical responses in the human body will, therefore, likely have an unpredictable impact on plants and animals once released into waterways. Sanderson cites the example of a cholesterol-lowering drug whose working pathway inside the liver is the same as that within plants when they develop chloroplasts and chlorophyll during photosynthesis. 'But, of course, these materials will be able to do something that we might not know in fish, in daphnia,* algae and other non-target organisms that have similar receptor systems and pathways [to us].'

With the help of students, Sanderson took the entire Martindale

---

* I had to look up daphnia: they are tiny crustaceans that look like a microscopic shrimp – and are especially popular in the water-quality community because they are very sensitive to changes in water chemistry and grow to maturity in only a couple of days.

Drug Reference Index – the 'drugs bible', published by Pharmaceutical Press, listing all 6,000-plus drugs and medicines available throughout the world – and began testing their toxicity, one by one. 'At this point in time, nobody really knew anything about the general picture of toxicity of these compounds.' The definition of toxicity, Sanderson says, 'is whether you are able to permeate the cell membrane. Toxicity happens inside the cell. If you are outside the cell, you are, per definition, not toxic.' While working through the Index, Sanderson found that the toxicity of a compound relative to how well it disturbs the cell membrane had not always been picked up. Some pharmaceuticals, such as pesticides, insecticides and antiparasitic drugs, are even actively designed to be toxic. These substances are designed with the exact opposite intentions we'd want for our freshwater systems, so their emergence in our rivers and groundwater inevitably causes environmental problems. In environmental risk-assessment toxicology, says Sanderson, the concern is not necessarily individual organisms, but the function of the entire ecosystem. 'Whereas, with humans, we're concerned only about individual organisms or individual people.' That, in effect, was all that was being tested for, and our water bodies were degrading accordingly.

It's not a problem for the medical community and Big Pharma that 90–92 per cent of a common drug like paracetamol passes into the urine.[8] But all these drugs have an afterlife in our sewers and waterways. Some are readily degradable or very absorptive and will be easily dealt with by ordinary wastewater-treatment plants, says Sanderson. Others, however, including 'thousands of different compounds' – such as some epilepsy and cancer drugs – are not. 'The vast majority of our wastewater-treatment systems today are not designed to capture these materials,' warns Sanderson. 'They are designed to remove nutrients and, to some extent, pathogens, through sand filtration, flocculation, etc. But these micro-pollutants are not really a target.' Plenty of common drugs have long afterlives too. For instance, diclofenac (Voltaren), a non-steroidal anti-inflammatory used globally, is known for having low biodegradability and high persistence in wastewater treatment, leading to bioaccumulation in rivers and sediment.[9] Sanderson adds that many cities, even entire countries, have no wastewater-treatment plants at all, with no 'net' for chemicals to slip through. And systems that do exist may be run down and inefficient. A 2016 study in the Puget Sound, Washington, found twenty-five contaminants in effluent from sewage-treatment plants detected in nearby waters. Interestingly, many of these contaminants were detected in the fish tissue, indicating a high potential for bioaccumulation up the food chain – of which we are at the top.[10] The researchers summarized, 'of the hundreds of chemicals that are likely present in the Puget Sound ecosystem, only a small percentage are currently monitored or regulated and there is little or no environmental toxicity information for the vast majority of these compounds . . . it is possible that a substantial load of potentially harmful chemicals are introduced into streams and nearshore marine waters daily.'[11] And that's the *treated* stuff.

Pakistan's water scarcity is made far worse by its water pollution. 'There's no water treatment,' Maheen Malik of AWS informs me. 'For example, in the entire city of Lahore, we don't have a single municipal treatment plant.' Wastewater is either directly discharged and ends up in the Ravi River, or it's treated to varying levels by effluent treatment plants within industry facilities, then discharged into untreated canals, benefiting from 'a dilution factor and nothing else'. The sewage canals were initially designed as storm-water drains, Malik tells me: 'they've been converted, but are unlined, so sewage water is seeping into the groundwater.' In Lahore, that groundwater is extracted for drinking water. 'The quality of water has changed tremendously over the last couple of years,' she says, 'because of this seepage issue.' Like large parts of India, there is a lot of arsenic in Lahore's groundwater too, and there's much debate about whether this comes from industry or the area's geology.

The textiles and garments industry is a major contributor to Pakistan's GDP, employing approximately 40 per cent of the country's total workforce, but it's also a major water polluter. In Punjab and Sindh, says Malik, it's not uncommon to see whole stretches of river run a vibrant red, yellow or purple: 'A lot of the dyes will go straight out into streams.' Progressive factories may treat it, but there'll still be traces of dye that are then discharged; while other sites discharge the water without any treatment at all. According to WWF, only 1 per cent of the generated wastewater in Pakistan is treated, with no regulations on waste disposal and treatment.[12]

In New Zealand, meanwhile, water pollution largely comes from the white-settler agriculture that has entirely reshaped its natural landscape. 'Our waterways are in crisis,' say New Zealand conservation organization Forest & Bird: a shocking 90 per cent of the country's wetlands have been drained for farming and development, while 74 per cent of freshwater fish are threatened with extinction. 'The number of dairy cows in New Zealand has risen by 69 per cent nationally in the past two decades and by up to 500 per cent in some regions . . . Our rivers, lakes, and streams are being polluted and too much water is being taken from them.' In particular, this means vast amounts of animal manure causing rivers and 'favourite swimming holes' to be left increasingly unsafe 'due to high bacterial levels, or toxic algal blooms.'[13]

*

Much of our knowledge of algal blooms comes from the research of one woman: Professor Nancy Rabalais. In the early 1980s, she was the only other scientist working at the newly founded Louisiana Universities Marine Consortium (LUMCON) alongside its director, Don Boesch. There, she began her research on the growing 'dead zone' of hypoxic, oxygen-free water in the Gulf of Mexico. Given a team of two researchers and a small grant from the National Oceanic and Atmospheric Administration, she began to map oxygen levels in the Gulf. Taking monthly samples in a small boat, they did a mapping cruise in the summer of 1985 to determine the area and conditions. They found a large and continuous area of low oxygen many miles wide. The next year, they found two distinctive areas of hypoxia in the waterways flowing from the Mississippi

River into the Gulf. In 1993, the size of the Gulf's hypoxic 'dead zone' had doubled. And by the 2010s, few pleasure cruisers would dock at the red, dead stretch of Louisiana coastline.

As huge volumes of fresh water from the Mississippi meet the sea, nutrients from agricultural run-off in the river feed giant algal blooms that turn the water red. Dead algae pile up at the bottom and are broken down by bacteria, consuming all the oxygen. This dead zone of oxygen-free water has been measured by Rabalais and her team, who have found it reaches over 20,000 km^2 (the Dead Sea in Jordan, by way of contrast, is just 605 km^2). This is not, however, a natural phenomenon. The nutrients feeding the algal blooms are caused by nitrogen and phosphorus pollution from agricultural fertilizer and animal manure.

Almost every year since beginning the project, Rabalais has taken the LUMCON research boat *Pelican* out of the mouth of the Mississippi. From the ship at night, 'you see crabs, shrimp and eels that usually live on the bottom, but there's not enough oxygen, so they are forced up to the surface,' and are picked off by gulls.

Nitrogen levels have tripled since the 1950s, says Rabalais. Taking sediment cores to discover when low oxygen levels have been present, she found that the indicators only started showing up in the 1970s. The turning point was cheap and accessible fertilizer, liberally applied, which led to dissolved nitrogen and phosphorus leaching into the groundwater and rivers as run-off. The cause and effect is so direct that Rabalais can now use her measurement of the nutrient load coming out of the river in May to predict the size of each year's Gulf of Mexico dead zone, to within 80 per cent accuracy. The largest hypoxic zone she has measured was 22,720 km^2, in 2017. Each year, her findings are duly reported by the local press, but she sees little substantial change in farming practices. Indeed, in smaller years, such as 2020, when it was 'just' 5,480 km^2 of dead sea, it is somehow reported as a victory ('the third-smallest ever measured!' exclaimed Associated Press).[14] At first, Rabalais says, it was hard to convince the public that the problem was real and was related to agricultural practice – a story as old as time, in environmental circles. Now that it's accepted, the hope is for change. Much like a carbon footprint, this is a nitrogen footprint: 'It's agriculture, it's meat consumption – people can do things,' says Rabalais. A global study of ocean dead zones worldwide (already over a decade old, and therefore now a likely underestimate) found 405, cumulatively covering 245,000 km², a larger area than the UK,[15] while the World Bank finds that 'more than half of nitrogen fertilizer leaches into water.'[16]

Artificially fertilized algal blooms increasingly happen in freshwater bodies too. In 2011, a massive cyanobacteria-related algal bloom on Lake Erie, the fourth largest of North America's Great Lakes, was caused by phosphorus and nitrogen run-off from nearby farms. The cyanobacteria, called microcystis, a form of blue-green algae, were particularly nasty, emitting neurotoxins and hepatotoxins causing nausea, vomiting and potential liver damage, as well as being deadly for other animals and aquatic life. They're also difficult to treat using conventional methods. Local residents couldn't even boil the water – that only increased its toxicity.[17] Lake Erie is the primary source of water for more than 12 million people across Canada and the United States. The microcystis bloom is now an annual event, visible from space.[18] In 2014, the water supply to 400,000 people in Toledo, Ohio, was shut down because of it. In October 2021, Lake Erie's algal bloom was reported to cover 1,373 km² (530 square miles), with the Michigan Environmental Council criticizing the state government for 'failed policies that don't result in real reductions to agricultural runoff pollution. We need an effective strategy to curb the damage that unsustainable agricultural practices inflict on our fresh water.'[19, 20]

Agriculture is arguably the biggest pollution threat to fresh water globally, even more so than human sewage, and attacks the water system from many fronts. Although mineral fertilizers have been used since the nineteenth century to supplement natural nutrients, their use has increased dramatically in recent decades. According to the FAO, the world now consumes ten times more mineral fertilizer than it did in the 1960s, when the Green Revolution was already

in full swing. A 2018 FAO report found that, 'the mobilization of
nutrients may already have exceeded thresholds that will trigger
abrupt environmental change in continental-to-planetary-scale
systems, including the pollution of ground and surface waters'. On
the North China Plain, for example, the use of nitrogen and phosphorus fertilizer is reportedly 66–135 per cent more than crops can
absorb: 'This excessive use of fertilizer directly [causes] severe environmental degradation already evident in many of China's rivers and
lakes.'[21]

The World Bank agrees, saying that, 'the world may have already
surpassed the safe planetary boundaries for nitrogen' and even suggests that 'it is the world's greatest externality, exceeding even
carbon.' One particularly shocking paragraph in this 2019 report
finds that 'fertilizer runoff and release of nitrates into the water
poses a risk large enough to increase childhood stunting by 11–19
percent and decrease later-life earnings by 1–2 percent.' This, they
say, means that 'fertilizers likely generate damage to human health
that is as great as, or even greater than, the benefits that they bring
to agriculture.'[22] A 2022 investigation alleged that agrochemical
nitrate contamination was killing children in Nebraska, with nearly
2,000 paediatric cancer cases in Nebraska between 1999 and 2018 –
more than anywhere else in the country.[23] Researchers at the
University of Nebraska Medical Center had noted an alignment
between these high numbers and watersheds containing higher
levels of nitrate fertilizer or weedkiller.

Animal manure is another big source of nitrate and phosphorus.
According to the FAO, 'The total number of livestock has more
than tripled from 7.3 billion units in 1970 to 24.2 billion units in
2011'.[24] A 'unit' appears to be FAO speak for a sentient animal – and,
given that there's circa 8 billion human units, you can see the relative scale of the poo-nami. Livestock production accounts for 70
per cent of all agricultural land use and 30 per cent of the land surface of the planet. Since 1981, global milk production has doubled.
Intensive aquaculture and fish farms also result in faeces (plus
uneaten feed and drugs) being released directly into water bodies,

and aquaculture output rose from 47M tons to 74M tons between 2006 and 2014.[25] In the UK, intensive chicken farms are leaching waste into rivers. In Denmark, intensive pig farming 'is a significant source of release,' says Hans Sanderson, 'both at the farms and also into drainage canals into streams and rivers.' Waste from livestock is a greater cause of eutrophication than pesticides, he says, at least in the EU.

Other hotspots for pig production include eastern China, Brazil and north-eastern USA, especially Iowa and Minnesota. In 2017, *Environmental Health News* ran an in-depth series of reports titled 'Peak Pig'. They found that Iowa had more than 6,300 'hog farms', most with over 1,000 pigs (sorry, units), with more than 10 billion gallons of liquid manure – 'high in fecal coliform, nitrogen and phosphates' – applied annually to Iowa fields.[26] About 300,000 Iowa residents get their water from backyard wells; a University of Iowa study of such private wells found that half tested high in nitrates and 43 per cent for coliform bacteria, an indicator of faecal contamination. Another study of just one Iowa hog farm found MRSA, a bacterial infection known for its antibiotic resistance, in half of the 200 pigs tested and nine of the twenty workers.

Antibiotic resistance is growing globally. For years, antibiotics have been given daily to healthy livestock – not to treat sickness, but because they encourage rapid growth. The EU banned avoparcin in 1997 and another four antibiotics in 1999, following similar bans in Sweden as far back as 1986. However, this brings us to the ever-present danger of unintended consequences: according to Sanderson, prohibiting the use of those antibiotics led to the use of alternatives, namely 'antibiotics that are used for humans, such as tetracyclines.' Pumping animals full of the same antibiotics that we take – and on a far greater scale – and then letting their waste manure and urine leach into watersheds, creates the perfect breeding ground for bacteria to develop immunity. 'This is a huge risk that we run every day,' says Sanderson, bleakly. 'We've been talking about this for decades, and there's always been another important problem overshadowing it. But, at some point, we might run out of antibiotics, and then we will be back to a world pre-antibiotics. And we already have a lot of people dying. In Europe, around 25,000 people die from preventable infections because we just don't have effective antibiotics [any more]. And this number is going to increase.'

Plastic surgery

In south-west Uganda, Nirere Sadrach struggles to be heard over the sound of the weir behind him as his friend hits *record* on the camera phone. 'We are mapping the extent of plastic waste in the River Rwizi,' he shouts, gesturing behind him, where hundreds, if not thousands, of plastic bottles float on the fast-moving water. 'As you can see, our rivers are flowing with plastic, and there is not that much action from companies and our government.' He signs off saying, 'We would like to see the companies producing these plastics take greater action to end plastic pollution now.'

On the phone, a few days later, Nirere tells me that, since his childhood, 'I have been exposed to the effects of a climate breaking down and an environment that is being degraded, every day.' He studied a diploma in sustainable business, volunteered with a local climate NGO, and in 2018 joined the Fridays for Future movement started by Greta Thunberg. In 2019, Nirere began the #EndPlastic-Pollution campaign to tackle what he saw as one of Uganda's biggest environmental problems. Rivers flowing with plastic is a depressingly long-term problem. Nirere has recorded similar videos in many locations, including Lake Victoria, Africa's biggest inland water body. The River Rwizi itself flows through ten districts and two large cities, as well as protected national parks and agricultural land: 'important places for our country, for the ecosystem of our country . . . So, this plastic pollution problem is really devastating for us.'

Nirere's aim in founding End Plastic Pollution Uganda was to demand greater corporate responsibility from those generating the plastic. On endplasticpollutionnow.blogspot.com, one of his more shocking posts relates to Lake Victoria: 'Plastic in Lake Victoria is degrading to form microplastics that have already been confirmed in fish like the Nile Perch (Empuuta) and the Tilapia fish (Engege) . . . the waters are becoming a green plastic soup near the shores.' Nirere tells me that the water is heavy and unusable. In the city, he says, 'water quality is really very, very poor . . . you can't drink it, you can't use it for anything' because of waste being pushed through the national water channel and ending up in the lake.

Over 250 million people across seven countries draw their water from Lake Victoria. However, the problem of plastic waste is 'out of control'. Each year, Nirere and over 150 plastic-waste pickers conduct a 'plastic audit' of the manufacturers of the bottles they find. Their (unofficial) top five are: 1. Coca-Cola; 2. Pepsi; 3. Mukwano Industries; 4. Unilever; 5. Fanta. The End Plastic Pollution message is simple: 'Companies should stop blaming people and consumers – it requires the company to change its product design or model.' Failing that, the Ugandan government should legislate accordingly, as neighbours Kenya have by banning single-use plastics in protected areas from June 2020.* 'In Uganda particularly there is a double standard from these companies,' Nirere tells me. 'The way they behave in your country is not the way they behave in our country.'

Much of the problem lies in the lack of municipal water supply. As I saw in Ghana, plastic bottled water is often the only reliable source of clean drinking water. Its supply chains reach far further than utility company pipes. 'It is very effective; these guys are producing a bottle for as little as 500 Ugandan shillings,' says Nirere (at the time of writing, that's 11p (14¢).

---

* Just two months later, in August 2020, a lobby group representing oil and chemical companies, including Shell, Exxon, Total, DuPont and Dow, was exposed for pushing the Trump administration to use a US–Kenya trade deal to expand the plastic and chemical industry across Africa, raising fears of undermining the ban and instead using Kenya as 'a dump site for plastic waste'. (Emma Howard, 'Oil-backed trade group is lobbying the Trump administration to push plastics across Africa', *Unearthed*, 30 August 2020)

Wherever there is plastic in water, there are microplastics, typic-ally defined as any plastic particle measuring less than 5 mm. Most of this comes as larger plastic debris in water degrades into smaller and smaller pieces. A significant amount comes, already in micro-form, from clothes-washing too: a recent study found that machine-washing 6 kg of clothing could release more than 700,000 microfibres, most of which goes out into wastewater.[27] Not only are microplastics now everywhere in the environment – they've been found at both the North and South Pole[28] – but they are also in our drinking water. A global study of tap water in 2017 found microplas-tics in 83 per cent of 500-ml samples taken; the average number of microplastic particles found in each ranged from 4.8 in the US to 1.9 in Europe. The study was repeated a year later, with bottled water, in nine countries. Despite often being sold as a healthier, more expensive 'premium' product, the bottled water contained an average of 161.5 microplastic particles for every 500 ml – 3,000–8,000 per cent more than ordinary tap water.[29]

Hans Sanderson has just completed a study of microplastics and polymers for Cefic, the European Chemical Industry Council, and P&G in the US. He's less concerned about microplastics than many other environmental risks, however, because microplastics are too large to puncture a human cell membrane and are therefore not, by definition, toxic. Nanoplastics, defined as anything below 100 nano-metres (nm, or 0.0001 mm), *could* puncture a cell, but haven't yet been studied in enough depth; the ratio of microplastics that break down into nanoplastics is unknown. Sanderson likens the mechan-ics to paper-folding: you can only fold paper seven times before it becomes impossible to fold any more; the material quality of some plastics may similarly only break down a certain number of times, possibly never reaching the nano scale. Microplastics can, however, carry other pollutants, such as flame retardants or dyes, on their surface and harm us that way. In that sense, says Sanderson, it remains the compounds that we should worry about, not the plastic itself, which is inert. That said, in high enough quantity, inert or not, they can 'clog up gills, clog up digestive tracts . . . And this is also what we saw in the polymers that we worked on: that they could have a mechanistic effect on the organisms.'

Sanderson nonetheless argues that there's a lot of 'fear mongering' around plastics: 'we badly need plastics, otherwise we cannot conserve our food and our food production would implode.' As with sewage, it's how we deal with the waste that matters. 'As a person, you probably just excrete all of it. And it's not causing any problem. And, for most other organisms, it's just passing straight through them without causing any problem.' Although, Sanderson admits the jury is still out. While microplastics may in themselves be inert and not toxic, all those little shards passing through us may well be causing inflammation in our organs – and inflammation leads to diseases and cancers. A 2021 Chinese study found that people with inflammatory bowel disease have 50 per cent more microplastics in their faeces. The most common type of plastic found in the stool samples? PET – used, among many other things, to package bottled water.[30]

Sanderson's main point is that microplastics, as troublesome as they are, are just the tip of the water-pollution iceberg. A mass closure of gold mines in South Africa in the late 1990s, for example, saw many mines fill up with water, causing the major environmental problem known as 'acid-mine drainage'. Mark Dent of the AWS in Pietermaritzburg, South Africa, explains: 'The geology here is dolomite, which is honeycombed, full of water. When they mined the granites, they exposed salt and rock, and salt exposed to oxygen, sulphur dioxide and water is sulfuric acid. Once the mines filled up, hundreds of kilometres of tunnels of acid is going back up and leaking out through springs and rivers all over the place . . . bringing a whole load of metals and other things into the sediments.' According to the Department of Agriculture and Rural Development in Gauteng province, toxic and radioactive mine residue covers an area of 321 km^2.[31] Just one closed mine, near the town of Krugersdorp, reportedly released 15M L of acid-mine drainage per day.[32] In 2019, some 6,000 derelict and ownerless mines were documented across the country.[33] Acid-mine drainage is a problem globally, from

Patagonia to Papua New Guinea,[34] while pollution of rivers from artisanal and illegal mining was also one the biggest problems I found in Ghana – the *galamsey* – where the dangerous work of hand-digging small open-mining pits requires a constant flow of water to wash through mud to find gold particles, and the muddy waste, often mixed with diesel from the generators, floods back out into the rivers.

A less discussed form of water pollution is heat. 'You need a lot of water to cool down power plants and nuclear-energy plants,' says Maggie White at SIWI. That water is then put back into nature. She cites the example of French energy supplier EDF being forced to shut down their nuclear plants during a huge heatwave in the early 2010s 'because they were not allowed to put it back into the river – the water was denaturalizing the river ecosystem.' This happened again in 2018 and 2019, and is now seemingly an annual occurrence in the French summertime. The low flow rate on the Rhône River due to lack of rainfall also restricted output at EDF's Saint-Alban nuclear plant in September 2020.*

But perhaps the most infamous modern water pollutant is per- and polyfluoroalkyl substances, or PFAS, which include PFOS (perfluorooctane sulfonate) and PFOA (perfluorooctanoic acid). Known as 'forever chemicals', they're sufficiently notorious to have their own film starring Mark Ruffalo (*Dark Waters*, 2019). Manufactured by 3M and DuPont, PFAS are synthetic polymers so water resistant and indestructible that they became a key ingredient in Teflon, Scotchgard, fire-extinguishing foam and stain-resistant furniture fabrics. You'll likely have something coated with PFAS in your house. The problem (much as with CFCs, another in the DuPont back catalogue of misery) is that they don't break down in the environment, and in high doses cause birth deformities, thyroid disease and cancers. Documents show that 3M knew as early as the 1970s that PFAS accumulated in human blood, and its own

---

* Some plants, however, are air-cooled – which surely needs to happen more in a water-efficient future.

experiments on rats and monkeys concluded that PFAS compounds should be regarded as toxic.[35] Even so, 3M still didn't pull it
from the market until May 2000. DuPont, meanwhile, just kept
going. As its own website admits: 'In 2002, DuPont began manufacturing PFOA for its own use in order to maintain continuity for
customers with critical applications.' It didn't come to an agreement
with the US Environmental Protection Agency (EPA) to phase out
its use of PFOA until 2015.[36]

The longevity of PFAS in water became evident, says Hans
Sanderson, thanks in part to a large fire near the Pearson Airport,
Toronto. A quantity of firefighting foam containing PFOS ran into
the nearby Etobicoke Creek, and Professor Scott Mayberry at the
University of Toronto developed the analytical methods to detect
them in the water. Working with Mayberry, Sanderson wrote his
PhD on PFOS while 3M were still producing them. Much of the
interest died down when 3M, and belatedly DuPont, agreed to cease
manufacture, but it piqued again when PFOS began turning up in
all drinking water, everywhere. 'Also in Denmark, in my local
drinking-water supply,' confirms Sanderson. 'I know I've been drinking PFOS and PFOA . . . If I took a blood sample of your and my
blood, we will find these compounds.' The question, Sanderson says,
is: 'Is this a problem or not? What does it mean? It means that you
have an elevated risk. And this is what environmental toxicology
and human toxicology is really about: assessing risks and accepting
risk.' It is impossible to remove all risks, so environmental and
human toxicologists set a health-based threshold for the acceptable
level of pollutants in the water. In December 2018, Health Canada
published drinking-water guidelines for PFOA of 200 nanograms
per litre (ng/L) and for PFOS of 600 ng/L. The US EPA, meanwhile, set its health-advisory limit much lower, at 70 ng/L for both
PFOS and PFOA (although this is advice, not a federal limit, and
several states choose to ignore it).[37] In England and Wales, the
Drinking Water Inspectorate (DWI) prefers a limit of 10 ng/L for
PFOS and PFOA. The decision-making process isn't purely
scientific – otherwise all countries would pick exactly the same

number. Ultimately, it's 'an ethical, political and economic discussion,' Sanderson acknowledges. For his part, when he fills up his glass, 'I know it's super low concentrations, so I'm not really concerned about it.'

PFOS and PFOA are now regulated simply because they are best known (thanks to Mark Ruffalo, or more specifically the real-life campaigning lawyer he played, Rob Bilott). Unfortunately, there are around 4,000 other PFA compounds – typically made simply by adding or subtracting another fluorine atom. In England, none of these other PFAS are tested for or have a health-based limit. Denmark, perhaps the best in class, regulates twelve of them. In fact, it's not just PFAS. The United States receives notices for more than 1,000 new chemicals every year. I ask Sanderson if water-quality professionals and health regulators are constantly playing catch-up. Is it likely that something as harmful as a PFOA has already been released on the market, and we just don't yet know about it? 'We hope that the systems that we have would prevent new compounds like PFOS and PFOA being emitted,' he says. 'But it's difficult to say. We have many millions of compounds that can be used; we have an exponential growth in the development of new chemicals. We consume annually the equivalent of half a kilo of raw chemicals per person on the planet.' One of his concerns is that a lot of production is being moved into developing countries, where there's much less oversight and control over what's produced, released and potentially causing health problems. Indeed, the World Bank finds that, 'keeping up with such a growing range of risks is difficult even in countries with significant resources and nearly impossible in developing countries.'[38]

Given that Sanderson knows more than anyone what is really in the water we drink, and how best to remove it, I finish our call by asking him what water he drinks. Does he have a bottled brand of choice, a favourite filter jug, an under-the-sink filter system, or even a domestic reverse-osmosis machine? 'Straight tap water,' he says, without hesitation. Although, he admits, 'the tap water is great in Denmark.'

CHAPTER 6

England's Problems

It's 9 a.m. on a miserable grey day in early October, in a large, near-empty car park. There's nobody around except one overeager boater and a man climbing out of his car, donning a hi-vis jacket. This must be Duncan Gibbons, from the Thames Water press office, so I get out of my car too. I'm here to see Farmoor Reservoir, South Oxfordshire, one of Thames Water's largest: 8 km in circumference and holding some 14,000M L of water (14M m^3). This is where the tap water in my house, some 42 km away, comes from. I want to understand how water gets from source to tap, and back out again via the sewage system. And, in so doing, tell the story of England's water problems. Because, oddly, little old England manages to encompass many global water problems – scarcity, over-abstraction, pollution, underinvestment, government and regulatory failings, environmental degradation and corporate misconduct – all within the confines of one small country in the far west of Europe, where I happen to live.

Will Lawrence, Farmoor's young site manager, strides out to meet us like a rugby player coming onto the pitch. We exchange pleasantries about the weather (it's grim, we all agree) and then crack on with a walking tour. Farmoor is an unassuming concrete bowl of a reservoir, grey engineering at its most brutalist, but the years since its construction in 1976 have softened it, with now-mature trees at its edges. The reservoir's intake is direct from the adjacent River Thames, but were that to dry up (it never has) Farmoor's 14,000M L is the equivalent of three weeks' supply. That is both a lot, and rather precarious.

You might think that a reservoir next to a river would be great for storing storm water and excess flow caused by heavy rain. But, Will tells me, that's actually when they close the intake: 'heavy rainfall's not great for us, because it causes turbidity – we don't want to be pumping in silty, stirred-up water, full of pesticides washing off the fields, with high nitrate and high ammonia levels.' A monitoring station by the intake sends a data update every ten seconds. Turbidity, explains Will, is a measure of particulate matter in water: how many little solids are floating around in it, counted by shining light through the water. Nutrients in the reservoir, coming in from the river, sometimes cause algal blooms in summer as the concrete warms up. Will calls this 'crypto season', due to the levels of cryptosporidium, a diarrhoea-causing parasite that must be removed by the treatment works. The quality of the water coming in dictates the level and expense of water treatment needed.

The concrete intake tower, overgrown with moss and lichen, looks like something from a 1970s dystopian film set. The intake is around 60M L a day, at a constant rate, making it eerily silent – there are no air bubbles to create a rushing sound. Outside, Will points to a nearby hill with a suspiciously flat top: 'that's Beacon Hill, another reservoir, with about 14M L of water. Imagine a big concrete box, completely sealed. We pump potable water up from the water-treatment works to that reservoir with our high-lift pumps. Then we send that out by gravity to as wide a supply area as possible.' Beacon Hill therefore acts as one very large water tower, using gravity for the water pressure, and pumping stations to hurry it along. And that, towards the very end of its reach, includes my house.

A few steps down the sloping grass banks of the reservoir is the water-treatment works itself, an industrial sprawl of red brick and magnolia paint, tall chemical tanks and large-diameter pipes. A visitor centre includes a hollowed-out tree trunk used as a water pipe in the 1600s ('That's why they are still called "trunk" mains', Will explains). He takes me through the water-treatment processes, building by building. I peer down into noisy, gunky, green clarification tanks, dissolved air-flotation tanks and sand-filtration beds, and fondle beads of activated carbon (for the full step-by-step process, see Figure 1, pp. 130–31). In the control room at the top, overviewing the lake, I meet water engineer Dave, who tells me that they typically take 150M L a day from the River Thames, ranging from 300M L to zero – although zero days are rare. Even in the height of summer, the Environment Agency can't stop Thames Water from taking water from the river, though they can reduce it to a maximum of 60M L – the last time Dave can remember that happening was for a couple of weeks in 2012.* Turbidity is his main headache, not drought. The turbidity reading on his screen is currently 11.7 NTU (Nephelometric Turbidity Units), and if it went above 20 NTU he would likely turn off the inflow from the river. The nitrates in the river are currently 8.1 parts per million (ppm). 'It becomes alarming for us at 10 ppm,' he says. Ammonia is more variable, and endangers the disinfection process; the water intake is turned off at only 0.3 ppm of ammonia. On Dave's screen, it's currently 0.14 ppm. This isn't just from local farms, but also from antifreeze used on the local Royal Air Force base runway – illustrating how any outdoor chemical use will, eventually, find its way back into our water system. Dave says the other worry is pesticides – organophosphates, the 'banned nasties' – though the ozone treatment process should still break them down. I ask if they monitor for and remove drugs and pharmaceuticals. Will steps in: 'We don't actively monitor for that sort of thing. That's more a question you should ask our wastewater colleagues.'[†]

What really strikes me is how we – England, Thames Water's

---

* I'm told more than once during my research that London came perilously close to Day Zero in 2012, with one industry insider saying it was 'days away from standpipes' before a rainy summer saved the day just before the Olympics.

† Spoiler, I did, and they don't.

customers and its shareholders – are the beneficiaries of a legacy of infrastructure largely built in the 1960s and 1970s. We get cheap water thanks to the investment of previous generations and governments. Will agrees: 'It really is about maintaining the assets that we've been handed down and trying to get the best out of them.' But this old infrastructure is now struggling to meet modern demand.

England's north–south inequality is usually heavily tilted in one direction, with the London-dominated south sucking in all the money and talent. With water, however, the balance tips the other way. The north not only has far more rainfall, but also more reservoir storage. Dr Peter Melville-Shreeve at Exeter University informs me that Kielder Water, in Northumberland – the largest artificial lake in Europe, holding 200 km^3 of water – 'was built with the assumption of future industrial use increasing.' In the 1960s, the northern industrial heartland saw water demand outstrip supply, with British Steel in particular requiring greater volumes, while domestic water usage was also rising due to growing employment. Margaret Thatcher's industrial reforms in the 1980s sold off British Steel, killing both the industry and the water demand. Kielder Water may now be 'just an expensive place to go hiking,' says Peter, 'but it was bought and paid for by a previous generation of water engineers and water bills.'

The state of (foul) play

The Waterwise Annual Conference on 19 March 2019 was a fairly small affair for water-industry types. But it was about to make national news. Sir James Bevan, chief executive of the Environment Agency, had heavily altered the pre-prepared speech he'd been given. An expected tame welcoming address instead became known as the 'Jaws of Death' speech. As his audience shifted uncomfortably in their seats and a handful of trade journalists suddenly woke up, he said, 'The jaws of death is the point at which, unless we take action to change things, we will not have enough water to supply our needs . . . many parts of our country will face significant water deficits by 2050, particularly in the south-east, where much of the UK population lives.' The water-abstraction system his agency oversees was 'designed more than fifty years ago, for a world with less environmental protection, less demand for water and no awareness of climate change.' Just as important as new infrastructure was changing human behaviour: 'We need water wastage to be as socially unacceptable as blowing smoke in the face of a baby.' The industry audience didn't know whether to whoop or weep.*

On 9 July 2020, the Public Accounts Committee said that all the bodies responsible for the UK's water supply – the Department for Environment, Food and Rural Affairs (Defra), the Water Services Regulation Authority (Ofwat) and Bevan's Environment Agency (EA) – had 'taken their eye off the ball' and must take urgent action to ensure a reliable water supply. The committee's chair, Meg Hillier MP, said, 'It is very hard to imagine, in this country, turning on the tap and not having enough clean, drinkable water come out – but that is exactly what we now face . . . Defra has failed to lead and water companies have failed to act.'[1]

By June 2021, the EA's chair, Emma Howard-Boyd, warned that summer rainfall could decrease by 47 per cent by 2070. The EA's National Framework for Water Resources now showed that, by 2050, water availability in England could be reduced by 10–15 per cent, leaving rivers with 50–80 per cent less water during the summer, and 'unable to meet the demands of people, industry and agriculture.'[2] By 2050, an extra 4B L of water *per day* will be needed for England, up from a current demand of 14B L – a 29 per cent increase, with less water and rain coming in. This extra demand would be the equivalent of 11.6 more Farmoors – yet no major reservoir had been built in England since Farmoor, in 1976.

Without significant action, The National Audit Office (NAO)

---

* 'I was the next speaker at that event,' Andrew Tucker at Thames Water tells me, a year later. 'I was on the front row as he did his talk, saying "jaws of death . . . blowing cigarette smoke in the face of babies." And we're sitting there thinking, Jeez, I bet that didn't go through the EA press team!'

Figure 1 How my water is treated – Farmoor Water Treatment Works (WTW)

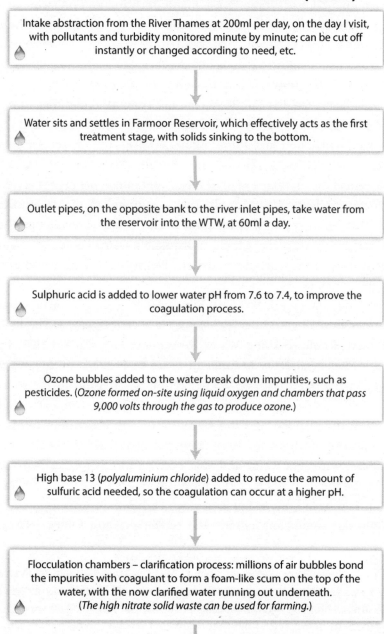

How my water is treated
– Farmoor Water Treatment Works (WTW)

Intake abstraction from the River Thames at 200ml per day, on the day I visit, with pollutants and turbidity monitored minute by minute; can be cut off instantly or changed according to need, etc.

Water sits and settles in Farmoor Reservoir, which effectively acts as the first treatment stage, with solids sinking to the bottom.

Outlet pipes, on the opposite bank to the river inlet pipes, take water from the reservoir into the WTW, at 60ml a day.

Sulphuric acid is added to lower water pH from 7.6 to 7.4, to improve the coagulation process.

Ozone bubbles added to the water break down impurities, such as pesticides. (*Ozone formed on-site using liquid oxygen and chambers that pass 9,000 volts through the gas to produce ozone.*)

High base 13 (*polyaluminium chloride*) added to reduce the amount of sulfuric acid needed, so the coagulation can occur at a higher pH.

Flocculation chambers – clarification process: millions of air bubbles bond the impurities with coagulant to form a foam-like scum on the top of the water, with the now clarified water running out underneath.
(*The high nitrate solid waste can be used for farming.*)

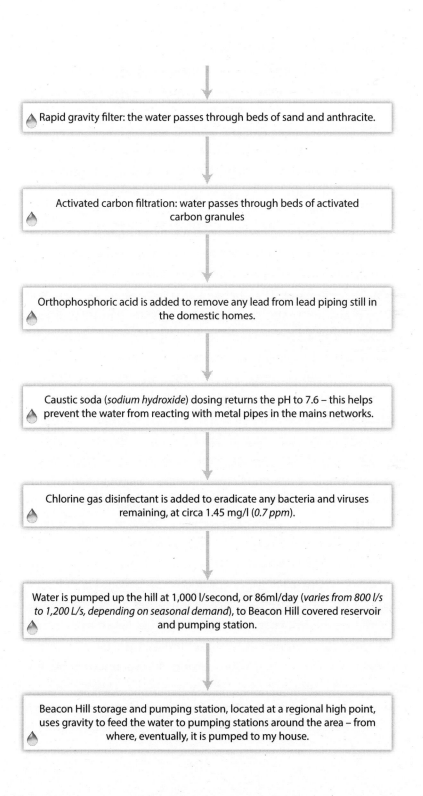

Rapid gravity filter: the water passes through beds of sand and anthracite.

Activated carbon filtration: water passes through beds of activated carbon granules

Orthophosphoric acid is added to remove any lead from lead piping still in the domestic homes.

Caustic soda (*sodium hydroxide*) dosing returns the pH to 7.6 – this helps prevent the water from reacting with metal pipes in the mains networks.

Chlorine gas disinfectant is added to eradicate any bacteria and viruses remaining, at circa 1.45 mg/l (*0.7 ppm*).

Water is pumped up the hill at 1,000 l/second, or 86ml/day (*varies from 800 l/s to 1,200 L/s, depending on seasonal demand*), to Beacon Hill covered reservoir and pumping station.

Beacon Hill storage and pumping station, located at a regional high point, uses gravity to feed the water to pumping stations around the area – from where, eventually, it is pumped to my house.

forecasts that the total water demand will start to exceed supply in England no later than 2034.[3] However, water companies have already been abstracting too much water, leading to environmental degradation and the disappearance of rivers, including the internationally unique chalk streams of the south-east. A *reduction* of 480M L per day is needed by 2045 just to lower existing groundwater abstraction to sustainable levels. Something, or someone, has to give, or Day Zero will be arriving on UK shores within a decade of this book's release.

Part of England's problem goes back to its unique system of private water companies being handed state monopolies. Prior to 1989, water supply was publicly owned, like it is everywhere else in the world. But Margaret Thatcher put a stop to that. In 1989, the ten water authorities spanning England and Wales had their assets and personnel transferred into limited companies and floated on the London Stock Exchange. Today, almost everyone in England and Wales receives their water and sewerage services from those same ten water and sewerage companies and thirteen water-only companies. Each has its own fiefdom on the map, with no competition, run for private profit; you're obliged to sign up to your regional supplier. The Water Act 1989 also removed previous restrictions on the statutory financial amounts water companies could borrow or pay as dividends.[4] To protect the interests of customers and the environment, however, privatization was coupled with regulatory oversight, most notably from the EA, the Drinking Water Inspectorate (DWI) and Ofwat. Because of the lack of competition, Ofwat sets limits on the price water companies in England and Wales can charge. And, to ensure that those companies don't just snaffle the cash and let the infrastructure they inherited fall to ruin, every five years the water companies must submit statutory water-resources management plans (WRMPs) setting out their intended investment approach for the next twenty-five years. Despite this, a fair amount of cash-snaffling still goes on. A 2017 study by the University of Greenwich found that water-company shareholders had received a total of £56 billion since privatization, with some water CEOs on £2-million annual salaries.[5]

By 2018, the then leader of the Opposition, Jeremy Corbyn, was calling for the water companies to be renationalized. Even the incumbent secretary of state for Defra, Conservative MP Michael Gove, attacked the water companies (to their faces, at the Water UK conference) for 'playing the system for the benefit of wealthy managers and owners, at the expense of consumers and the environment', and suggested that they had 'shielded themselves from scrutiny, hidden behind complex financial structures, avoided paying taxes, have rewarded the already well-off, kept charges higher than they needed to be and allowed leaks, pollution and other failures to persist for far too long'.[*] In cash terms, over £18.1 billion was paid out to shareholders of the nine largest water companies between 2007 and 2016, accounting for 95 per cent of profits. Gove added, in his conference speech, that the water companies 'appear to be as intent on financial engineering, just as much as real engineering.'[6]

Even the *Financial Times*, traditionally a champion of privatization, found that total capital expenditure by the ten big water monopolies had declined by 15 per cent since privatization, from £5.7 billion to £4.8 billion a year: 'Over the same time the companies – which were sold off with no debt and handed £1.5 billion – have borrowed £53bn, the equivalent of around £2,000 per household. Much of that has been used not for new investment but to pay £72bn in dividends.'[7] Australian infrastructure firm Macquarie owned Thames Water between 2007 and 2017, leaving it with £2 billion of debt, while paying its investors, according to one analysis, on average

---

*   On this point, Thames Water's Andrew Tucker fights back: 'We succeed every single day. Every single day, people can turn the tap on and they've got perfect drinking-quality water. You don't get thanks for that; you only get crucified in the media when a pipe fails somewhere, even if it's the first time the pipe has failed in 120 years.' In 2020, Thames Water lost on average 22,200 L/km of pipe, every day – more than twice as much as the next-worst offender.

between 15.5 per cent and 19 per cent in dividends a year.[8] UK news-papers accused it of aggressive asset stripping.*

Thames Water makes for an interesting case study because of its sheer size. Its Beckton sewage-treatment works is the largest in Europe, and it has the largest abstraction borehole site too. In total, it supplies 2.6B L of drinking water a day to 9 million customers, and provides wastewater services to 14 million. Steve Tuck, abstraction manager for the company, admitted to me, when I was reporting for the BBC in 2019, that the current system 'is a little hand-to-mouth'. Overuse by customers, coupled with a growing population and less rain, had led to 'an exacerbation of low flow problems', as he mildly put it. Another company insider tells me that London holds only ninety days of water – 'after that, she's dry'. About 75 per cent of that supply comes from rivers, mostly the Thames, and 25 per cent from groundwater aquifers. In Thames Water's patch, 25 per cent of the water is used by business and agri-culture, and 75 per cent by households. And in one of those households is me.

Before I began researching this book, I considered myself a water-conscious consumer. I use rainwater to water the garden; I shower rather than run a bath; I wouldn't dream of leaving the tap on while brushing my teeth. But my bill in February 2019 revealed an uncom-fortable truth. My family of four was consuming the same as the average British household of six: circa 233 m^3 of water a year. That means we were using around 160 L per person, per day – above the already high UK average of 143 L per person, per day (143 L was the average in February 2019; it's since risen to 153 L). I hadn't even realized I had a water meter, because I pay a fixed monthly direct debit; and, despite my high use, it was pretty cheap – easily my low-est utility bill. But that little infographic at the bottom of my bill,

---

* You'd think that would be a lesson learned. Instead, in August 2021, Ofwat approved a new £1bn equity takeover of Southern Water. The new owner? Macquarie. (Jillian Ambrose, 'Sink or swim: Macquarie plunges back into crisis-hit UK water industry', *The Guardian*, 30 January 2022)

comparing my water use to others', was a nudge in the right direction.

Andrew Tucker, water-efficiency manager at Thames Water, tells me bluntly that, in London and the south-east, 'we basically don't have enough product, going forward.' They don't have any new water resources, yet '100,000 people a year are moving to our patch alone . . . And that increasing demand on water just means that the maths doesn't work out.' He admits that it's 'the most perverse business model in the world.' Thames Water has a monopoly: 100 per cent of people in their designated region are their customers and use their 'product' (water) every day. 'And yet we reinvest and spend a huge amount of effort and time to get our customer base to use less of our product.' They do this because, increasingly, they can only achieve their statutory requirement to supply product to everyone that moves into the area, or any new house built, by persuading existing customers to use less. 'Most of our customers are pretty efficient,' he says, making me feel sheepish. But the per-capita consumption (PCC) figure that wavers around 'roughly 150 L per person, per day' is only giving us the mean average; 'when you look at the mode, the majority of customers are around the 100 L per person, per day mark. That graph has a long tail of high users that's skewing the average.' What is also clear is that everyone, uniformly, underestimates their water use. Research by Water UK in 2020 found that 46 per cent of Brits believed their household use was under 20 L of water a day, compared to the 153 L per person reality.

After the call, Tucker arranges a home water audit for me. A nice Scottish guy called Andy appears at my door later that summer; after turning off all my appliances, he takes one look at my water meter and says, 'See that wee thingy going round and round? That means you've got water leaking in your house.' A search through our appliances identifies a leaky loo. A third of all water is lost through leakage in customers' homes, Andy tells me, not out in the pipe network itself. He also fits a low-flow showerhead for good measure, much to my wife and children's displeasure later. Andy's audit concludes that over half of my household's water use comes from the shower alone, the annual carbon footprint of which – including heating the water from our boiler – is, he says, higher than that of our car.

Reducing demand is one thing, but 'ultimately we will need to bring new water supply into the system. It's going to be a twin-track approach to maintain water supply going forward,' says Tucker, back on our call. 'All water companies in the UK rely on winter rainfall to recharge these systems . . . If we don't get that winter recharge, it just drops and keeps on dropping, because our raw water storage [reservoirs] is actually quite small.' Tucker is Australian and says mates back home find it funny that England has a water problem, given its wet reputation. 'We do get a lot of grey days. But grey doesn't mean rain. Even drizzle doesn't mean rain. "Oh, today we got 0.25 mm, it was wet all day." Well, 0.25 mm is going to do fuck all.' I laugh and he gives me a quiz question: 'Which Australian state capital city gets more rain on average every year than London?' I guess Sydney. Nope. 'They all do.'

There are a several reasons for England's comparative lack of water storage, Tucker says. First, 'every square inch of land has been used pretty intensively for the last thousand years and there's not much room to play with.' Second, the south-east is pretty flat, with no valleys to dam. Third, we have a population poorly educated in the need for water saving or living with drought. And water is too cheap – or at least not valued. When we speak, Thames Water's combined water supply and wastewater charge is around £2.20 per 1,000 L. 'You pay the same for *one* litre of water at WHSmith at the train station.' Presumably the price they can charge is dictated by the regulator, Ofwat, not by the Thames Water management board? 'Very much so,' agrees Tucker. But old infrastructure needs investment, and that's 'constrained because the overriding priority of the regulatory system is to keep customer bills down.' That's all well and good, he says, but there's a balance to be struck. In December 2019, Ofwat announced that it expected water companies to *reduce* household bills by 12 per cent over the five years to 2025.[9]

Tucker argues that education and cultural awareness are even greater drivers of water efficiency than price, though. He says they're trying to advocate the introduction of water-use restrictions, like in Australia, so 'it becomes a cultural thing'. He notes that, in England, companies are 'crucified in the media' for suggesting ways the public can reduce seasonal demand. When I suggest that, after months of heavy winter rain, the public perception is, 'Well, why wasn't that banked somewhere?' he quickly interjects: 'There is no bank. That's the number-one thing that you can tell the public: there is no bank. We want a bank. We've been prevented from getting a bank for twenty years.'

RAPID and the Abingdon Reservoir

Thames Water estimates that, by 2045, it will need to find an extra 350M L of water supply *per day*. The 'bank' that they've wanted for twenty years is the long-planned but never-built Abingdon Reservoir. First proposed by Thames Water in 2006, it would be the first major reservoir built in the south of England since Farmoor in 1976, and capable of supplying their entire shortfall with one big project. The problem is, as Andrew Tucker identified, the region has no valleys to flood. The site near Abingdon, Oxfordshire, is perfectly flat, fertile farmland. The only way to build is upwards – what's known as a 'bunded reservoir' – and Thames Water wants 150M m^3 capacity (twelve times larger than Farmoor), which would take nine years to build, covering more than 6 km^2 of land, with bunded walls 30 m high – making it the largest bunded reservoir in the world. This wouldn't just be a blot on the landscape, it would *be* the landscape. Unsurprisingly, local villagers don't want it. A campaign group promptly sprang up: the Group Against Reservoir Development (GARD). In 2010, Brigadier Nick Thompson, then chairman of GARD, said, 'The noise, pollution and disruption of the project would destroy the quality of life for many people . . . Together, we can defeat Thames Water's proposals.'[10] And they did. A public inquiry in 2010 produced a 329-page report concluding that the Abingdon Reservoir proposal did not meet the statutory requirements and that a reservoir of that size was not justified by the evidence. That didn't stop Thames Water from putting it – with minor tweaks and changes – into every five-year management plan since. Now rebranded the 'South East Strategic Reservoir', as a joint proposal between Thames Water and Affinity Water, it looks likely to finally go ahead.

I meet Derek Stork and John Broadbent, GARD's current chair and vice chair, on the roadside behind the village of Steventon, near Abingdon. They show me the large, flat expanse of fields that would be lost: golden wheat ripples to the horizon, dotted with solitary ancient oak trees standing like totems, having seen centuries of crops and communities come and go. All that we can see before us, says Derek, a retired Yorkshireman who had hoped to see out his days here in peaceful tranquillity, will be under water, should the reservoir go ahead. The proposal is to dig just 6 m down and build 24 m up – so, from where we stand, we wouldn't be looking out over water, but looking up at a steep embankment. It would take three years just to fill. 'It's got to be pumped upwards from the Thames,' says Derek, incredulously. That three-year fill time is just the minimum: the 'hands-off flow' river level, below which you're not allowed to abstract during periods of drought, says Derek – like those in 1976 or 2012 – would mean no pumping would be allowed. That's also why it needs to be so big: once full, 'it would last for something like 500 days of supply' – compared to the region's current storage total of ninety days. Although building and pumping upwards seems crazy, this part of Oxfordshire is known for its clay soil, and the clay cap beneath could be used to make a waterproof lining. Back in the 1990s, there was talk of building a reservoir about a tenth of this size, says Derek, but plans quickly escalated. 'Now I'm sure that they want to sell water to everybody else in the south-east.' By 'everybody', he means the other water companies. Currently, water companies operate almost entirely within their own territories, only using and selling the water within their borders. They can sell to others, in what are known as water transfers, but they don't – in 2019, just 4 per cent of water supplies were transferred. But this, everyone agrees, will have to change if England is to survive the droughts to come.

Derek and John show me another part of the site, 1.5 km away, towards the village of Hanney. We pull into a farmer's side road, separating a field of corn from a field of solar panels. John keeps bees here, proudly pointing out two hives buzzing away. We're now in the middle of the reservoir, says Derek. We'd need to drive another kilometre and a half to reach the end of it. Nearby is a coppice of tall mature trees that Derek says would be below the 24-m watermark. 'It's not a Site of Special Scientific Interest,' says John, referring to a term that would protect it under environmental legislation. 'It is very historic, but there's nothing unique about it, so we can't play the "Roman ruins" card.' Derek hopefully suggests that Alfred the Great wasn't based too far away, so there 'might be some Saxon stuff'. The A34 and the railway run strategically past, but not through the site, making it ideal for a major building works. The area is also prone to flooding, being marshland up to the Middle Ages. The surrounding villages were referred to as 'the island villages' until the twentieth century, and still are by some older locals. It isn't Derek and John's intention, but it's starting to sound like a pretty good place for a reservoir.

Our final stop is Derek's house, in a picturesque part of the village of Steventon. In Derek's garden, amid a choreographed tangle of climbing plants, we sit around a table and talk at length about what living next to the biggest building site in Europe would mean to them and the community. The fight raging between GARD and Thames Water has been going on for so long that there's battle-weariness in Derek and John. They remain passionate, staunch defenders, but admit they're tired of making the same arguments, winning both in court and parliament, only for Thames Water to return with a rehashed proposal of the same idea. I'd come here

expecting to meet two Nimbys – retired men with time on their
hands and too used to getting their own way, to borrow one of campaigner Feargal Sharkey's phrases – but in truth we share many
concerns about water shortage in the south-east. They have seen
their own little village almost double in size with a new-build housing boom, and similar expansions of nearby villages and towns.
They just don't want Thames Water to get away with building one
big, brash, sledgehammer of a solution, when there are several less-
damaging options available. 'In a normal year, the reservoir won't be
feeding water for more than forty to fifty days,' argues Derek. 'This
is a back-up solution; most of the time it won't be used. Because
most of the time the base of the system – the Thames and its tributaries, and the London storage reservoirs – will handle things.'
There's 'Not in my backyard', and there's 'Please don't build a pyramid to England's water mismanagement.' I still feel, on balance, that
a reservoir should be built here. Just not one this big. Surprisingly,
Derek and John agree: they would accept a smaller reservoir on the
same site, perhaps the size of Farmoor (a twelfth of the proposed
size). But they would prefer a water transfer from the River Severn
to the River Thames, wastewater reuse, fixing leakages, and
desalination – all of which are also on the cards.

The fate of Derek and John, GARD and the South East Strategic
Reservoir now lies in the hands of a newly formed body called
RAPID. In response to dire report after dire report, and James
Bevan's 'Jaws of Death' speech, the UK government set up the
Regulators' Alliance for Progressing Infrastructure Development
(RAPID) in 2019. Recognizing that new water supplies equivalent
to that consumed by over 9 million people would be needed by the
mid-2030s, and that private water companies weren't investing in
this, RAPID was formed to assess and then green-light around
seventeen major water-infrastructure schemes and help fast-track
them with public money.* RAPID was a recognition that
privatization hadn't worked; the original theory that it would bring

* While the light is still on amber for the Abingdon Reservoir, it looks

more investment into an ageing water sector was disproven. Thirty years on, the water sector simply continued to age while investors creamed the profits – someone had to step in before taps ran dry. And that someone turned out to be Paul Hickey, RAPID's managing director.

All the recent doom-laden water reports point to the same conclusion, Hickey tells me, in 2021: 'We need to be able to move water about the country better than we do at the moment, and we need to bolster and create new strategic sources of supply.' Candidly, he says it was 'a disappointment' that the water companies didn't do that themselves. So, now, finally, Ofwat and RAPID are stepping in with £469 million of ring-fenced money to make it happen.[11] 'Some schemes will drop out as we go through this process and understand which is the best mix of schemes,' he says. But, if this was the water infrastructure version of *The X Factor*, the frontrunners include:

1. A new South Lincolnshire Reservoir for Anglian Water and Affinity Water: a 50,000M L reservoir, with an abstraction from the River Witham, supported by a transfer from the River Trent.[12]

2. A new Fens Reservoir in East Anglia for Anglian Water and Cambridge Water: a 50,000M L reservoir with an abstraction from the River Ely-Ouse, possibly restoring local wetlands.[13]

3. Using the Grand Union Canal to transfer 50M–100M L per day of treated wastewater from Minworth in the Midlands to Affinity Water in north-west London.[14]

4. London Effluent Reuse: upgrading four sewage works that discharge into the River Thames and the River Lea, including a

---

very likely to start construction in 2025. Thames Water already own most of the land.

pipe from Beckton transferring up to 300M L per day of recycled water to King George V Reservoir.

5. An interconnector transfer of raw water from the River Severn
    to the River Thames (joint activity with Severn Trent Water
    and Thames Water), amounting to 250M–400M L per day.

6. And, finally, giving itself one last shot for the title, the Abing-
    don South East Strategic Reservoir. Sizes under consideration
    range from a slightly more locals-friendly 75,000M L up to the
    150,000M-L behemoth that GARD dreads the most.[15]

Southern Water's plan for a £600-million, 75M-L-per-day desalination plant near Southampton was rejected in the early rounds for
its prohibitive cost and likely environmental damage.[16] What the
shortlisted proposals have in common* is their inclusion of long-
distance transfers of water or water recycling – trying to break the
water companies out of their 'always done it this way' comfort
zones. The old (and current) model of abstracting everything on
your land and hoping for winter rain is no longer tenable. 'That's
what RAPID is about,' says Hickey. He tells me that RAPID is also
considering how schemes might work together, in combination:
'Abingdon may well fit with a Severn–Thames transfer,' he suggests,
solving the issue GARD raised about whether the River Thames
actually has enough water to fill the reservoir. 'A new Fenland Reservoir could potentially serve Cambridge Water,' which currently
'doesn't have any reservoirs and is solely reliant on abstraction from
iconic chalk streams . . . That might be a way of us getting ourselves
out of that issue.' And that's quite an issue, to put it mildly. Because
the plight of England's chalk streams is a national scandal caused
purely by over-abstraction by water companies.

---

*   It's not really a competition – all the 'finalists' can, and most likely will,
be given the go-ahead.

Feargal and the chalk streams

Feargal Sharkey calls me on my birthday in July 2020. It's quite the birthday present. He introduces himself, in his characteristically emphatic Northern Irish tones, as 'chairman of a thing called the Amwell Magna Fishery, which is the oldest fly-fishing club in the United Kingdom.' But he's much more than that. He was the lead singer of The Undertones, the punk balladeer of legendary DJ John Peel's favourite song. And he's somehow now England's foremost river defender. Not that the 4 km of the Upper River Lea knew it at the time, but its future was saved the day that Feargal Sharkey first cast his fly line over it. 'It's what brought me into contact with the EA and water companies and abstraction and sewage, and sent me off on this whole journey!' he laughs. In the late 1990s, his fellow anglers began noticing that the Lea's water levels were falling rapidly, without any obvious link to weather or drought. 'The EA instantly pointed the finger at Thames Water,' he says, 'who had a licence to abstract about 100M L of water per day directly out of the river, about 1.5 km upstream from the fishery.' After a decade of legal wrangling and diminishing river flow, Sharkey became the club's chair and personally took on the fight. Sharkey's club spent almost two years preparing a case to take the EA and Thames Water to the High Court. By this point, he says, the river was stagnant and no longer flowing. The case was settled out of court and abstraction from the river catchment was greatly reduced to avoid further action. Sharkey says the river is now 'looking in fantastically mighty shape'. But, for him, that was just the beginning: 'If a little fishing club had to take the bloody Environment Agency to the High Court, just to get them to do the right thing – what the hell else might be going on?'

Sharkey discovered that water companies were being allowed to over-pump groundwater to the extent that many other rivers in the south-east were running dry. This was happening under the nose of the EA – the very regulator that was supposed to stop it. And these weren't just any old rivers, but chalk streams.

Chalk exists elsewhere in the world, but nowhere is there such a mass of it – from the famous White Cliffs of Dover to the Needles of the Isle of White, and hundreds of kilometres of underground aquifer in between – than in south-east England. A softer, whiter form of limestone, chalk is made up of countless billions of tiny marine fossils that settled on the seabed during the Cretaceous period. Water percolates easily through chalk, with organic matter filtered out on the way, leaving crystal-clear, almost immediately drinkable spring water. Chalk streams are nearly entirely fed by these springs, with beds of glittering golden gravel and fish and invertebrates that thrive in the exceptionally clean water. Botanically, they are the most diverse of all England's rivers. The estimated 246 chalk streams are described in the draft report as the 'equivalent to the Great Barrier Reef . . . a truly special natural heritage and responsibility'.[17] To suck their source dry, therefore, seems unconscionable.

Sharkey's years of researching and campaigning have led him to conclude that this is happening because of 'a political decision to artificially suppress the price of water. That idea and that goal, coupled with pretty abysmal oversight and regulation, has created the situation.' He says lack of investment has led to the south and south-east of England coming close to running out of drinking water. 'The water from the chalk aquifer is as good as you're going to get anywhere in the world. And, for a water company, that's incredibly seductive stuff. Because it's so pure, you have to do little to treat it before you can send it on to your customers.'

Water companies had, for years, denied that over-abstraction was happening. When I interviewed a Thames Water abstraction manager in 2018 for the BBC, he told me, 'Chalk streams will dry up anyway; it's part of their natural process.' And he's far from alone among water-company managers in thinking that way. But, says Feargal, 'they are talking through their arse. If you go back to the Domesday Book . . . the River Ver had something like thirteen water mills on it. So, where's all that water gone?' The River Beane once had a corn mill so large on its banks that it was widened four times by a nineteenth-century owner to maximize production. 'Four times!' exclaims Sharkey. 'Now, there's not a fucking drop of water within miles.'

The stretch of the River Lea hosting Feargal's fishing club had got so bad that 'there was on average thirty-six days per year where there was actually no flow at all.' After the water company reduced abstraction, it magically reappeared. A website of EA data, called gaugemap.co.uk, provides a map of all river flow and groundwater levels. Feargal tells me to look at a groundwater borehole called (don't laugh – remember, this is England) Lilly Bottom, in the Chilterns. In August 2019, it was at the lowest level ever recorded, thanks to a combination of over-extraction and three relatively dry winters. Affinity Water, he says, nonetheless continued abstracting water; they didn't implement their Drought Management Plan and the EA didn't make them do so. No abstraction restrictions were put in place. 'And yet, at that point, every single chalk stream in Hertfordshire had lost mile upon mile [of river].'

Another example Feargal gives is the hands-off flow requirement for the Lower River Test: 'Clearly, as it says on the packet, the notion of hands-off flow is that it is the minimum volume of flow needed simply to maintain the ecological health of that river.' The threshold is set at 355M L – any lower, and 'the river starts failing'. But, in 2018, Southern Water successfully applied to the EA to reduce the volume requirement of the hands-off flow by a third. The details are there in black and white on the permit: plans to 'abstract from the Test Surface Water source below the Agency's proposed hands-off flow of 355M L/d down to 265M L/d'.[18]

'The Environment Agency approved that permit,' says a still angry Sharkey. 'Here is the premier fucking chalk stream on the planet, which should be the finest river ecosystem known to man, with a legal minimum-flow requirement, and the EA still grant that Southern Water can reduce it by a third.'

Feargal says he doesn't blame the water companies, however. Rather, he believes the original sin was their privatization, leading to profit-maximization, and the government's subsequent weak regulatory and environmental oversight. 'You don't have an open competitive market. These guys are operating state-granted monopolies.' Collectively, UK water companies are 'carrying something like £48-billion worth of debt on their balance sheet . . . Yet, at the same time, they've also paid out something like £57 billion in dividends.[19] And, by the way, with the full knowledge, consent and cooperation of your regulator Ofwat.'

One power that Ofwat *does* effectively wield, however, is the price that the companies can charge their customers for water. They keep this very low, which keeps politicians happy, but – as Andrew Tucker noted – hamstrings the water companies' ability to invest in infrastructure. 'The price of water in England is stupefyingly low compared to pretty much any other Westernized nation on Earth,' claims Feargal. 'I don't even have a water meter. I could just turn on my tap and let thousands of litres of water run down the drain and not be charged extra for it. And, as an Affinity Water customer, I know that water came from a chalk stream.'*

*

I meet Sharon Moat in late September, in another empty car park, this time near some allotments in semi-rural Cambridgeshire. I'm here to see the River Ivel chalk stream, or what's left of it. Sharon founded both the RevIvel Association and the Chalk Aquifer Alliance in 2020 – campaigning against the over-abstraction of the aquifer by the water companies, which is causing the collapse of the chalk streams. She walks and talks at pace. Her knowledge of, and enthusiasm for, the river and the plight of the chalk streams flows out of her like a spring. Which is just as well, because there's no longer a spring at the source of the Ivel. 'We want the abstraction linked to the flow rate,' says Sharon, which seems like a very humble

---

* Just 51 per cent of customers in England had a water meter in 2017 (albeit up from 17 per cent in 1999); according to Defra, those with water meters use 12–22 per cent less water than those who pay a flat rate.

request. 'So, if your flow rates drop, they stop abstracting. And, so far, they've not agreed to do that.' 'They', in this region, being Affinity Water. She complains that 'the Environment Agency' is also a misnomer: 'They really haven't done their job.' She accuses them of superficial fixes rather than addressing the root cause of the problem: over-abstraction.

A visitor sign at the 'Ivel Springs', the Ivel's source, stands impotently to tell us of the river's importance and the freshwater wildlife it supports. But there is no water, only weeds. We walk for a while before we see standing water, but even that, says Sharon, is just a puddle from last night's rain; a couple of days ago, there was nothing there. We pass a sluice gate that once served a flour mill, but the water here now wouldn't even turn a hamster wheel. 'It's just so sad to see it as it is,' she sighs. 'This is a perennial river; it used to be famous for its trout, for its mills, for watercress farms . . . Now, this only has water in it in February, but by June, July, it's dry again.' Along the way, we meet an elderly woman walking her dog, who reminisces about courting with her husband here in the early 1960s. 'It used to be such a pleasant, clear stream,' she says. 'It was so beautiful.' Her equally elderly Jack Russell terrier stands forlornly in a green puddle, as if acknowledging the loss. 'It was crystal clear. You know, like chalk streams are supposed to be.'

All the water supply across East Anglia, Sharon tells me, comes from boreholes drilled into the chalk aquifer. The nearest borehole to our walk is just a kilometre or so away, in Baldock. Since 1970, according to RevIvel, 47 per cent of species in Hertfordshire's rivers and wetlands have 'noticeably declined', with 7 per cent now extinct.[20] But, even if you don't care about the silver-hook moth or the white-clawed crayfish, such a decline is not great news from a water-supply perspective either. By Affinity Water's own figures, there will be a gap between customer demand and the company's ability to supply water of 250M to 410M L per day by 2050, equivalent to between 2 million and 3 million customers.[21] 'They know that's not sustainable,' says Sharon, 'that their business model won't work moving forward, because they are going to be called to account for the damage . . . and London will eventually run out of water.'

The immediate difference on my next visit – to Bury St Edmunds – is that the River Lark actually looks like a river. When I arrive, there's water in it, it's reasonably wide, and there are signs of river vegetation and bird life. But it isn't a chalk stream. The water is murky, slow, barely moving. Its flow comes mostly from the town's storm drains and gutters. The expansion of this Suffolk town and its surroundings has sucked up the once-pristine chalk stream with its rich flora and fauna. The world-renowned 'gin-clear' water is now a run-off ditch. Libby Ranzetta, who chairs the Bury Water Meadows Group, shows me around and introduces me to local river volunteer Glenn Smithson. Glenn has lived here all his life. He used to swim in the river after school with his mates. You wouldn't – and couldn't – swim in its shallow, stagnant waters now.

The patch we're now on is Anglian Water's, the largest water and sewerage company in England geographically, covering 20 per cent of England's land area. It also hosts the largest number of chalk streams. Bury St Edmunds is home to two of them, strategically built in the eleventh century on the confluence of the Lark and the Linnet. Glenn and Libby walk me down to a stretch of river near the ancient abbey grounds – it's pretty, but made so only by their volunteers, covering concrete walls with bales of coconut husk and hazel to allow nature to literally take root again. 'This was absolute rubbish,' says Glenn. 'There was no habitat here whatsoever.' All river volunteers face the same dilemma: their dedication allows nature to retake a small foothold, but in the process covers up the worst of the environmental crimes. And the main crime here is lack of flow caused by the over-abstraction of the aquifers by Anglian Water. 'The aquifers are just not recharging,' says Glenn. He's volunteered on local rivers for nearly two decades and says you can easily spot true chalk-aquifer water because it comes out a glacial blue, 'just a completely different colour to this run-off,' which he describes as just a 'vegetated ditch.' Here lies the River Lark: once one of the world's pristine chalk streams, now a vegetated ditch.

They take me to a local Tesco supermarket. When it was built in 1996, what remained of the passing chalk stream was channelled into a concrete culvert that looks like an American storm drain – the only biodiversity in it now is the variety of empty crisp packets. This was, they tell me, a navigable stretch of the Lark for river traffic in the 1800s, with a dock for coal barges. I doubt any passing shoppers would know it's even supposed to be a river, now, let alone chalk-fed. 'The only reason this hasn't dried out completely is they dredged the river out and dropped the river down, like, six feet,' says Glenn, as we gaze down at the concrete ditch. 'All of the riverbed was just dumped in the fields.' A short drive away, Glenn shows me the Lark's source. It should be a bubbling mineral-rich spring filtered through the chalk at a constant temperature of 10°C. But instead, it's the effluent outflow pipe from the local sewage works.

Maybe sensing that I'm now thoroughly depressed, Glenn and Libby finish by taking me to a restored stretch of the upper river that runs through Breckland – a strange, sandy savannah unique to East Anglia. Here, the river runs fast, crisp and clear. Gravel twinkles beneath the surface, and water-crowfoot flutters in the flow like Ophelia's long hair. Butterflies and birds emerge from bushes and riverbank bulrushes. Glenn has been working on this stretch for over a decade and is proud to show it to me. It's what a chalk stream should look like. But there is a final sting in the tale. Like a war-bombed building rebuilt to an estimation of its former glory, or a famous bridge moved brick by brick to a new site, this is a loving reconstruction. The path of the river should be hundreds of metres away; it was moved long ago by a farmer who found it inconvenient. The glittering gravel was brought in by truck and added by Glenn and his fellow volunteers. And the water, most likely, although it is hard to know for sure, is the treated effluent from the sewage works.

Later that year, in an online Chalk Aquifer Alliance talk, I see a plan that could save the chalk streams and solve the abstraction problem. Charles Rangeley-Wilson is a fisherman and journalist who became a river campaigner. In the 2010s, he focused his long-held interest in river restoration on the plight of the chalk streams, and, with the help of the WWF, began beating a drum that few others were at the time. By 2020, together with engineering consultant John Lawson, he introduced the concept of Chalk Streams First – which became the Chalk Stream Restoration Strategy – that began to gain traction with water companies and RAPID. It's as simple as this: stop all abstraction from the chalk aquifer boreholes upstream, let the chalk streams fully recover and flow again, and abstract at the bottom of the catchment, downstream, instead. I can and will elaborate, but that's it in a nutshell.

It is, says Rangeley-Wilson, in his Chalk Aquifer Alliance Zoom presentation, 'the golden opportunity to realign the way we abstract water from our environment.' He describes a quarter of a century of zero progress, despite infinite numbers of reports. 'The way a chalk stream works is rain falls on the hills, it sinks into the ground and down into the aquifer,' he explains. 'There is a very direct relationship between the height of the groundwater, the saturated zone and the power of the springs. So, the higher the groundwater gets, the longer the river gets, and the more water comes down the river.' There is even a direct mathematical formula for this, he says: a 10 per cent increase in the height of the groundwater above the valley bottom amounts to a 25 per cent increase in river flow. He presents graphs from various rivers, showing groundwater levels and flow rates tracking each other precisely, even when the boreholes are far away from rivers. The chalk aquifer should be viewed as a regional entity, he says, not as a series of discrete valley-by-valley entities. It's more like an interconnected underground lake. Bad news for water companies, in that they can't pick and choose which chalk stream they want to save – abstracting from the aquifer anywhere has a negative effect on all. But good news because a significant overall reduction in abstraction benefits all too.

The suggestion in the 1990s that there was no real evidence that abstraction caused low flows was, Rangeley-Wilson says, 'a whole load of old guff.' The UK Technical Advisory Group on Chalk Streams, in response to the EU Water Framework Directive, set a minimum flow rate of 7.5–20 per cent to achieve good ecological status. On the Ver, for example, that would leave an acceptable abstraction rate of between 3M and 6M L a day; in reality, 25M L a day is what was abstracted in 2020. The story is similar across other chalk streams. Rangeley-Wilson argues for something radically different: for groundwater abstraction to be replaced by water taken from outside of or downstream from these chalk valleys. He namechecks the Abingdon Reservoir and the Severn–Thames transfer as other ways of alleviating groundwater abstraction. The solution, he says, lies within the problem: stopping the abstraction would increase the amount of water coming down the river. Likewise, getting the water back up to the towns that currently have local boreholes can be addressed by a new grid of pipework and pumping stations, called Supply 2040, already planned by Affinity Water. This is a vital component to various new supply options, including its ability to get water from a future Abingdon Reservoir or Severn–Thames transfer. In short, Rangeley-Wilson says, 'Supply 2040 is the thing we've been looking for all these years.' He thinks it would allow for 'total cessation in the groundwater abstraction in the chalk streams of the Chilterns.'

In September 2020, Affinity Water issued a press release committing to 'ending environmentally unsustainable abstraction from these precious river catchments and to work in collaboration with other water companies, industries, universities and NGOs to develop alternative, sustainable water supplies away from chalk river catchments.'[22] This was undoubtedly a positive step, and a big win for campaigners. But it still wasn't a time-bound, concrete commitment. In early 2021, a member of Affinity Water's management board, Jake Rigg, tells me, 'Yeah, absolutely right, we need to change what we're doing. The Chalk Streams First proposal is a really important piece of the jigsaw. And we're exploring that.' The sticking point, he says, is that 'no one really knows' what will happen if they turn the boreholes off. Paul Hickey at RAPID tells me that he's also 'very keen' that Rangeley-Wilson's plan 'is properly investigated.' If such a simple idea works, he says, 'it would be absolutely brilliant.'

In March 2022, I call Charles Rangeley-Wilson for an update. Chalk Streams First's idea is gaining currency. Paul Hickey has apparently advised Affinity Water that RAPID want them to investigate it seriously. It is no longer a dream among NGOs, but a coalition of stakeholders, including the water companies, Ofwat and the EA, now under the auspices of the Catchment Based Approach (CaBA) initiative. The water companies are now 'talking about abstraction reduction,' he tells me. 'But the idea that it would allow you to take that water from lower down the catchment is still not alongside the strategic options such as the Abingdon Reservoir or the larger ideas.' He thinks Affinity Water are 'underplaying the potential' of ceasing all abstraction from the chalk aquifer. However, he has got the water companies to sign up to his definition of 'sustainable groundwater abstraction' in chalk-stream catchments, namely: 'that which causes a maximum reduction from natural flows of circa 10 per cent.' Rangeley-Wilson tells me that the negotiation to get this 'took up half the discussion of CaBA . . . There was quite a lot of resistance to it. They don't instinctively like it.' Some water companies were better than others, though: 'Thames Water were actually really good, and they helped broker at least a consensus around that definition.' He believes that putting that definition into practice would mean 'enormous numbers' in terms of abstraction reductions.

Reducing the abstraction on the Colne catchment boreholes alone to reduce the river flow by no less than 10 per cent would, by CaBa's calculations, mean a reduction of 100 mL (megalitres) a day. If Rangeley-Wilson and his cross-sector group of collaborators are correct, at least 80 per cent of that would re-emerge downstream. Even if they're wrong,* the Grand Union Canal could deliver an

---

* 'I'm as certain as the sun will come up tomorrow that it will be proven,' Rangeley-Wilson tells me. 'It's just physics: you stop taking the water out of the ground and it doesn't disappear into the centre of the Earth. You will get a very large percentage of the water back at some point in the catchment.'

extra 100 mL of treated water a day from Birmingham (whose largest wastewater-treatment plant produces 400 mL a day). 'So, it's eminently deliverable,' he insists. 'And it's eminently deliverable in quite a short timescale.'

Algal green and unpleasant land

In the height of summer, I visit one of the jewels of the Cotswolds: Bourton-on-the-Water. Here, the River Windrush runs through the town under picturesque low stone footbridges and past riverside pubs, before opening up in a wide flirty flourish for the delight of tourists and paddlers. I meet former police officer Ash Smith on the outskirts, the sun beating down and the lushness of the greenery almost iridescent. His dog, Archie, an immaculately trained Labradoodle, joins us for the walk. We're far outside of chalk territory here – although, as Ash says, 'There's no such thing as a "bad river".' There isn't much of a problem with flow rate here, either: there's always something for the day-trippers to paddle in. The issue – to my surprise, given the regional importance of this tourist destination – is river pollution. As we've seen, the other side of the water-scarcity coin is water quality. And boy, does England have a sorry tale to tell here too.

Ash moved here for his retirement, but spends most of his time talking to people like me for his campaign group Windrush Against Sewage Pollution (WASP). In 2015, he discovered that the local sewage works, run by Thames Water, was illegally sending raw sewage out into the river. Being a good citizen and a lover of rivers,* he reported it to the EA, who duly investigated. 'Although, I use that term quite loosely,' he says. 'Thames Water self-assessed that the environmental impact of them having dumped untreated sewage over many years was "low". And the Environment Agency accepted

---

* Ash tells me he used to be a super-keen fisherman. But he once went fishing for eleven days in a row, and by day eleven thought, 'Actually, this is boring, isn't it?' and never fished again.

that without question and took no further action.' That really rattled Ash's cage. Which was probably a mistake, because, since then, Ash, along with fellow campaigner Peter Hammond and journalists like Rachel Salvidge and Sandra Laville, has elevated river pollution to the level of national scandal.

The story really broke in 2020: during the previous year alone, water companies in England had discharged raw sewage into rivers on more than 200,000 occasions, for more than 1.5 million hours.[23] This was a shock to many, but also a massive underestimate. Before 2015, the waste sources – known as storm outflows or combined sewer overflows (CSOs) – weren't even monitored.* Since then, the EA required that, by 2023, all overflows be fitted with 'event-duration monitors' to track how often and for how long they're being used. Nearly 18,000 CSOs overflow with raw sewage when the plants are overwhelmed during (in the EA's wording) 'exceptional circumstances' – meaning heavy rainfall. By 2021, when event-duration monitors covered 80 per cent of the network, the true scale of the scandal began to emerge. It was twice as bad as thought. The water companies had discharged raw sewage into rivers and coastal waters in England more than 400,000 times that year, for 3 million hours.[24] Exceptional circumstances were happening exceptionally often. By mining the data now readily available on the EA website, Ash even found numerous examples of sewage works pumping out raw sewage into rivers when there wasn't any rain at all. This was happening across the board, by every

* We have Sir Joseph Bazalgette to thank for CSOs. His sewer system was a wonder of Victorian engineering, and saved London from the 'Great Stink' by essentially dumping London's sewage further downstream. But it left us with the unfortunate legacy of collecting surface-water drainage and sewage into a single combined sewer. This flaw was recognized as early as the 1915 Royal Commission on Sewage Treatment and Disposal, which observed that the two waste streams were better kept separate, but that it was too expensive to rip it all up and start again. Unfortunately, the CSO system was still copied throughout the UK and distributed throughout the then British Empire.

wastewater company. Thames Water were bad, spewing raw sewage into rivers for 237,000 hours, but Severn Trent were worse (559,000 hours) and United Utilities worse still (726,000 hours).*

In July 2021, Southern Water was taken to court for breaching its environmental permits fifty-one times over a five-year period, with 8,400 unlawful and non-compliant incidents of sewage 'escape' – 4,000 of which lasted for over an hour.[25] They were issued a record fine of £90 million (albeit Southern Water's profits of £139.6m that year had increased from £118.8m in 2019).[26] One of the worst individual cases recorded by Ash and Peter was Thames Water's Mogden Sewage Works in south-west London, which, in 2020, spilled 7B L of untreated sewage– 2B L (or sixteen Olympic-sized swimming pools) of which poured during one forty-eight-hour period alone. All of this was illegal, yet it was happening, so claims Ash, with the tacit approval of the regulating agency.

What *should* happen, he says, is that the water industry treats sewage to a required standard before putting it out as effluent. 'This is the whole basis of environmental protection: you must treat sewage to a standard that the EA, by some calculation, decided would be non-damaging to that water course.' This calculation, written on the wastewater company's permit to operate, will cover total suspended solids, biochemical oxygen demand, ammonium and – but only for plants servicing a population of 10,000 or more – phosphate (note that nitrogen is not on the list). A company's environmental-performance assessment is then based on how well they performed in achieving those standards. 'Thames Water were over 99 per cent successful – all of the companies are – and they've done that by dumping huge quantities of untreated sewage, unmeasured and unmonitored,' outside of the permit process. Ash describes

---

* To complete the set, from bad to worst: Anglian Water – 170,547 hours; Northumbrian – 178,229 hours; Southern Water – 197,213 hours; Wessex Water – 237,035 hours; South West Water – 375,370 hours; Yorkshire Water – 420,419 hours; Dŵr Cymru Welsh Water – (in England) 21,300 hours; (in Wales) 872,976 hours.

photographs they have of one sewage outfall fitted with an event-duration monitor alongside one without; treated, permitted effluent pours out of the first, while untreated sewage pours out of the second. 'One gets measured and passes the standard, while the other is dumping the excess.'* He still sounds astonished just retelling the story. 'They're being allowed to break the law on an epic scale,' alleges Ash, 'because of Defra and its principle of not wanting to interfere with the profit of industry.' When Defra was formerly called the Department of the Environment, he says, the joke went that 'DOE' stood for the 'Destroyer of Everything'. Ash describes it as 'a total collapse of regulation.'

Walking along the riverbank of the Windrush, we duck under brambles and low branches. Dangling from a cable in a tree is a submerged water-quality monitor Ash keeps, which gauges dissolved oxygen and temperature. 'Dissolved oxygen is a good surrogate for general water quality,' he explains: big spikes indicate eutrophication from too much algae. We arrive at a modest Victorian-looking earthenware pipe poking out of the bank just above the water level. An unusually large clump of water iris has sprouted there, enthusiastically feeding off the artificial phosphorus and nitrogen injection that sewage water delivers. The pipe isn't flowing today, but Ash shows me films of it 'chucking out raw sewage, really stinking up the river', with sewage fungus growing underwater – a nasty, brown, woolly bacterial growth that feeds off raw sewage and smothers plant life (sewage fungus typically needs a full week of sewage to form, Ash says). This pipe is technically a storm overflow, but Ash claims it was called a CSO until the requirement to monitor them came in, and a large number of pipes were hurriedly reclassified.[†] Either way, it's a pipe that pours raw

---

* You can see this picture, plus a whole talk by Ash, on the Chalk Aquifer Alliance talks archive, at chalkaquiferalliance.wordpress.com
† Such quibbling was also recognized as merely semantic by the Storm Overflow Evidence Project, prepared for Water UK, the EA, Defra and Ofwat in 2021: 'the term "storm overflow" is synonymous with the terms

untreated sewage into a river. 'You could call it corrupted out of all sense of moral and environmental justice, twisted in favour of industry,' says Ash. 'And it didn't have to be this way.'

RAPID's Paul Hickey was the EA's deputy director for water resources from 2017 until 2020, and Anglian Water's environmental-standards manager for fifteen years before that. He makes the case for sewage overflows. 'It's important to remember why they're there,' he told me. 'Sewer systems will be beaten by storms, and if there isn't an outlet, then you end up with flooding at people's houses. So, that's why they're open to rain- and surface-water ingress . . . the issue is that they're located in appropriate places and their operation is minimized to the most extreme events.' Which, as we all now know, clearly isn't the case. 'No, and that's what needs to be addressed,' agrees Hickey. We all agree. Hickey probably agreed while he was at the EA. And yet, the shit still hits the river fans. Professor Kevan Martin, a civil engineer whose master's thesis was on sewage sludge, tells me later that 'in the UK, the sewage treatment is medieval, it's staggeringly poor. I worked in Switzerland for twenty years, and the water that comes out of the sewage works into the rivers, you could drink it. You would never put something from [a UK] sewage works in your mouth.'

I ask Ash if our rivers *are* our sewage system, but that no one wants to admit it? 'Rivers are being used to provide tertiary treatment,' he says, of the supposedly treated effluent. And untreated sewage? 'Absolutely: we can show it, we've got the numbers.' Nearby Witney spilled for about 1,300 hours in 2019 and 1,600 hours in 2020. Didcot doubled year-on-year. 'These are places where new housing is going in.' He claims the EA wouldn't have even investigated such matters 'had it not been for us, spreading word of this around the country.' Some water companies claim that spilling via CSOs is necessary due to groundwater infiltrating cracks in the pipes and needing to come out somewhere. But 'spilling due to

Combined Sewer Overflow (CSO), Storm Tank Overflow and Intermittent Discharge.'

groundwater infiltration is not a permitted activity,' says Ash. 'We had that written down, right up to the secretary of state. And they really regret ever mentioning that,' he alleges. But 'Defra forced the EA to only prosecute a tiny fraction of offences and to ignore most of them.' Corroborating that, in January 2022, a leaked document reported by *The Guardian* and the *ENDS Report* showed that the EA has instructed its staff to 'shut down' and ignore reports of low-impact pollution events because it has insufficient money to investigate them. The EA's own leadership team 'made it clear to government that you get the environment you pay for.'[27] James Bevan said as much again in a speech in March 2022, admitting that 'we are still seeing too much pollution from sewage', but saying that, while it was the agency's job to 'regulate robustly',* they need to 'have the powers and resources to do so.'[28]

We leave the river so Ash can take me to a Land Treatment Area. Being kind, you could call this a nature-based solution, using reed beds to treat sewage before it flows out, naturally filtered, into the adjacent Windrush. Ash, however, calls it 'an optimistic phrase for dumping sewage on a field and hoping everything will be all right.' Here, raw sewage pours through a concrete channel with metal grates above it, which we stand on to watch. Plastic mesh sacks, the frontline of our sewage system, are placed on the end of four outflow pipes feeding into a ditch that flows through this 'managed wetland' or 'damp sewage field' (depending on your point of view). Old plastic mesh sacks, full of gunk, hair, wet wipes and tampons, are strewn to one side. It surprises me just how crude it all is. And if it's happening here, on the edge of a famous Cotswolds tourist hotspot, how much worse are they getting away with in more secluded areas?

'It shows you just how much bacteria, hormones, drugs and chemicals pass through,' says Ash. Even if things pass through a water-treatment plant or sewage works, he says, those 'treatments' are often inadequate. Ash believes that people don't realize their own government has 'sacrificed the environment for business', as he

---

* I can almost hear Ash snort, 'Or at all,' in response.

puts it. 'Anywhere else, it would be called proceeds from crime.' (Remember, he used to be a police officer.) It's cheaper to pollute than to do the job properly, he says. 'It's a deliberate choice, created and facilitated by Defra, the EA and Ofwat, just to lower water bills.'

This issue is no longer simply the purview of campaigners such as Ash and Feargal Sharkey. The House of Commons Environmental Audit Committee's *Water Quality in Rivers, Fourth Report, 2021–22* doesn't pull punches either: 'rivers in England are in a mess', it says bluntly. 'A "chemical cocktail" of sewage, agricultural waste, and plastic is polluting the waters of many of the country's rivers. Water companies appear to be dumping untreated or partially treated sewage in rivers on a regular basis, often breaching the terms of permits that on paper only allow them to do this in exceptional circumstances.'[29]

The EU Water Framework Directive in 2003 required that every single river should be in good overall environmental health by 2015. It became obvious in 2014 that England was going to fall woefully short, with only 17 per cent of rivers meeting the criteria; by 2015, this dropped further to 15 per cent. The government applied for a 2021 extension, and then another for 2027. In 2020, Brexit meant that the UK left the EU anyway. By 2021, what would have been the first extended deadline, that 15 per cent had dropped to zero: not a single river anywhere in England meets good overall health.[30] As for the 2027 deadline, even if it did still exist, 'You can forget that too,' Feargal tells me on the phone. The government's new twenty-five-year environment plan, he says, requires 75 per cent of rivers to be in 'a natural state' – whatever that means – 'as soon as is practicable'. Firm, time-bound environmental regulation, then, replaced by meaningless platitude.*

Despite the increased popularity of wild swimming in the UK,

---

* A timebound commitment did eventually materialise. In late December 2022, no doubt hoping to bury the news amid the Christmas holidays, the UK government announced a revised extension of the 2027 deadline: they would now give themselves until 2063.

Feargal would 'genuinely advise anybody to really think carefully about wild swimming in any river in England.' A favoured recreational river site in Ilkley, Yorkshire, on the River Wharfe, successfully applied for 'bathing status' in 2020 in order to get the river water regularly tested. Unlike the thousands of blissfully ignorant bathers each year, the campaigners from the Ilkley Clean River Group know precisely what the reports will uncover: unfit and unclean. According to Feargal, in 2019, Yorkshire Water owned up to dumping sewage into the Wharfe on 196 separate occasions, 'right in the town centre, in Ilkley'; when E. coli levels were tested, they were found to be 42,000 times above that required for safe bathing water.

In a stunning press release in January 2021, Defra admitted that 'infrastructure has not kept pace with development growth over decades.'[31] The chair of the Environmental Audit Committee in 2022 was a Conservative (the ruling party), as were half of its sixteen-strong committee. Yet their wording could have come from Greenpeace: 'The sewerage system is overloaded and unable to cope with the increasing pressures of housing development, the impact of heavier rainfall, and a profusion of plastic and other non-biodegradable waste clogging up the system. Successive governments, water companies and regulators have grown complacent and seem resigned to maintaining pre-Victorian practices of dumping sewage in rivers.'[32]

Until a detailed map of sewage outflows went up on the Rivers Trust website (which everyone living in England and Wales should look up immediately*), no one – possibly not even the wastewater companies themselves – knew the nationwide scale of the problem. But, Feargal explains, rather than immediately clean up their act, the EA created a new loophole. Now, they'd 'categorize a continuous spillage, right up to fifty days, as one spill event.' In agreement with the water companies, the EA also won't investigate unless

---

* Trigger warning – you'll never want to go for a river swim again. Seemingly no river or stream in England is exempt: 'Is My River Fit to Play In?', Rivers Trust, arcg.is/19LiCa.

there's at least sixty spill events per year. As Feargal points out: 'There's not enough fucking days in the year, if they count fifty days as one event; you'll never get to sixty.' Not just that, but 'exceptional circumstances' was also amended to 'due to rainfall or snowmelt' – not even heavy rain.

Plenty of muck makes its way into our rivers from farmland too. In the summer of 2021, campaigner and journalist George Monbiot streamed a live one-hour film called *Rivercide*, charting the pollution of one of the great rivers of England and Wales: the Wye.[33] The Wye is a tourist destination, the closest thing we have to France's Dordogne, where pleasure kayakers journey slowly along the water, stopping occasionally to lie on the riverbank or swim. I had visited with my father in 2020 and even had a swim in its waters. Despite the high tourist numbers, there still feels something secluded and undiscovered about it, a million miles (yet barely fifty) away from the south-east's urban sprawl. As a rather sunburned-looking George walks along a stone shingle beach, he says to camera, 'You can see that something is very wrong here. These stones, a few years ago, were completely clean, and now –' he points to the ones submerged below the shallow water – 'they're covered in green slime . . . There should be far more fish, far more insects.' (Hearing him say it, I remember that slime underfoot from my swim.) A local conservationist tells him that 95 per cent of the water-crowfoot – a freshwater plant essential to this river ecosystem – disappeared in just two years, between 2019 and 2021. 'There's nothing left.' Stimulated by an excess of nutrients – nitrates and phosphates – constant algal blooms have taken over the river. 'It's like watching someone you love die of cancer,' a local fishing guide tells Monbiot. Here, farming is an even greater culprit than sewage outflows. There are approximately 20 million chickens held in intensive farms across the Wye catchment. Chicken manure contains five times the amount of phosphate of sheep and cattle manure. Some of this manure is spread on the land by farmers happy to have the fertilizer; but, being far more than the plants can naturally take up, it leaches either into the groundwater or directly into the river.

Agriculture is the dominant source of nitrogen in our waterways (70 per cent) compared to sewage effluent (25–30 per cent). Over half of England is designated as a nitrate vulnerable zone due to elevated nitrogen concentrations in groundwater and rivers. The largest source of phosphorus (P), meanwhile, the most common cause of water-quality failures under the Water Framework Directive in England, is still sewage effluent.[34] Charles Rangeley-Wilson writes: 'if P concentrations keep on rising, the river's ecology will switch over to an algal-dominated plant community. Benthic algae smothers the river bed and . . . the gravel in which many insect species live . . . algae cloaks the leaves and stems of the higher-order plants, reducing their ability to photosynthesize.' There comes a point, he says, 'where, if the P is very concentrated, the river turns into an anoxic soup and nothing much survives.'[35] This is exactly what is happening to the Wye. Meanwhile, in the summer of 2022, England's greatest freshwater lake, Windermere, the jewel of the Lake District, saw an algal bloom so large that it spanned the entire width of the lake.

Sewage works in progress

I wanted to complete the circle. Having visited Farmoor Reservoir, where my water comes from, and the many rivers in between, I wanted to see where my sewage ends up too. During the first hard frost of November, with a bright, low sun on my back, I set off from my house on foot to Thames Water's Banbury Sewage Treatment Works.

It's a large, surprisingly picturesque site, and, on this crisp, frosty morning, the mist is slowly rising from the circular settlement tanks. Large grassy lawns glisten in the morning sun. 'We get deer here too,' Diana, the site manager, tells me. First, she takes me to see the inflow pipe, the 'cess', which sits inside a large concrete tank. All of Banbury's sewage enters through a vertical pipe pointing skywards – producing an effect strikingly similar to (apologies to the squeamish) a large chocolate drinking fountain, with the chocolate-brown water flowing over in a perfectly uniform mush-room shape. A few metres further down is a smaller, milkier mushroom. 'That's from Adderbury,' Diana says. Apparently, it's just a coincidence that they're producing a paler sewage than Banbury today, unless the good people of Adderbury are hiding something. Once mushroomed out and over, the sewage inflow then moves over a slow metal conveyor belt that travels up and out of the water, removing the worst of the solids. As we watch, wet wipes and tam-pons emerge. The most memorable thing Diana has seen come out is a dead cat. 'That upset me – I like animals,' she says. But the one constant, without fail – on every chain of the conveyor belt – is sweetcorn. 'We *really* can't digest it,' she says.

Next, our chocolate milk moves on (sans sweetcorn) to the large circular 'primary settlement tanks' we probably all picture if we think of a sewage works. Here, the water ('Don't call it "water"; this is "flow",' Diana constantly corrects me. 'Remember, it's still sewage') slowly spills over the tops of the tanks onto the next process, while the solids settle as sludge in the bottom of the tank. At this point, sludge and sewage water follow two separate paths. The water moves to an aeration tank, where bubbles separate out any remain-ing solids (largely cotton buds, judging from what's floating on top). Next, it's chemically treated with 'ferric' (ferric chloride – one of the few water-treatment chemicals that can sequester odours), and then it goes to secondary settlement tanks before a final sand filtra-tion and then out into an open channel that leads to the River Cherwell. Becky, the Thames Water PR who set up my visit, tells me that some smaller sites skip the sand-filtration stage and go straight from the secondary settlement tanks to the river.

The final outflow channel (Diana still won't call it 'water,' even at the end of the process; it's 'final effluent') is surprisingly substantial, like a narrow canal. The water is gushing out at 2.09 mL per day, according to a digital display. This also gives readings for phos-phorus, potassium and NO_3, but Diana asks me not to write down the figures. It's all compliant with the EA standards, she assures me. She grabs a sampling cup on the end of a metal pole and scoops out some water for me to look at. I decline to drink it. 'No, don't!' she says, hurriedly – but then adds that, given the choice between drinking this final effluent or the raw water in Farmoor Reservoir, she'd choose this any day.

The sludge, meanwhile, goes from the first settlement tanks into an anaerobic digester in which it creates methane gas, providing 50 per cent of the energy needed to run the site, while the digestion process breaks down the solids – which are then part-dried and compacted into 'cake'. Large, steaming piles of cake. Up close, the piles look surprisingly hygienic: very dark, almost black, palm-sized flakes – like thin cuts of peat, and perhaps organically not all that different. The cake is sold to farmers, who spread it on their fields (not for crops for direct human consumption, but for indirect consumption, such as animal fodder or wheat). It's popular among farmers, says Becky, and there's plenty of it to go round.* I ask how they monitor for the pharmaceuticals and antibiotics that carry through into our urine and faeces. 'We don't,' says Diana, blankly. What about raw-sewage pollution? Before my visit, I checked 'Is My River Fit to Play In?' and found that, in 2019, Banbury Sewage Treatment Works' storm overflow spilled thirty-three times, for a total of 298 hours, into the River Cherwell. When I ask Diana if any untreated sewage goes into the river, she gives me a straight, 'No. Otherwise that completely takes away everything we are trying to do.' Becky tells me that Banbury's storm tanks, regardless of how much it's raining, are big enough to handle it (and I do see them, huge and empty, on my tour). In other Thames Water sites where the storm tanks are fewer, however, such as Witney (which, in 2020, spilled ninety-seven times, for a total of 1,563 hours), it doesn't take

---

* Buckinghamshire farmer Steven Lear later tells me that he's used this 'sewage cake' in the past. 'But I'm not sure I would use it any more. Since doing a bit of research into the amount of microplastics and chemical products that people have flushed down the toilet and whatnot – it worries me a little bit from a soil-health point of view.' But, he added, it's 'very good because it's high in phosphate. A lot of farmers do use it.'

much for them to spill. And, while the wastewater companies get the blame for this, a Thames Water insider I speak to directs the blame at Ofwat. Thames Water's previous business plan had apparently included storm-tank upgrades precisely to prevent such situations happening. Ofwat allegedly put a stop to it, fearing an increase in customer bills (apparently it would have meant a single-figures annual increase a year – only pennies per month – which many customers wouldn't even notice). In England, you get what you pay for. And we pay little for rivers to flush away our sewage.

To see the effluent at this final stage of its journey, I use the Rivers Trust website map to find where the outflow meets the River Cherwell itself. I follow a canal towpath, before taking a turn into a farmer's field. When I eventually reach the river, I can hear the rush of the outflow pipe before I can see it. A heron is sitting on the metal barrier above it – the fish perhaps enjoying the warmth of the effluent. Here, I meet that 2.09 mL again, now some way away from the sewage works. The Cherwell is not a large river, much depleted by industry, canalization and weirs – one of those rivers where you're never too far away from a sunken shopping trolley. From here, though, with all that extra flow, it swells by perhaps a third. If it's good enough for the heron, perhaps it should be good enough for us. But I still have Diana's words ('It's not water, it's effluent') going round in my head, and I can't stop thinking about the lack of screening for modern drugs that go through our systems, as described by Ash and so meticulously by Hans Sanderson. The treatment process got rid of all the solids, but everything else flows back down river, joins the Thames and is recaptured by Farmoor Reservoir. I've completed the circle all right. I will eventually drink, flush and shower in this effluent all over again.

CHAPTER 7

Where Does Asparagus Come From?

At ground level, you can't even make out that it's circular. A long, straight metal arm on wheels runs from where I'm standing on the periphery of the field, back into the centre, 500 m away. Hanging from the arm, which stands perhaps 3 m above the ground, are hundreds of evenly spaced water hoses, with spray nozzles at the ends, dangling just 30 cm above crop height. Like a giant mathematical compass, it performs the Sisyphean task of etching out the same perfect circle every day, spraying water in the process. At this farm, near the Gila River Indian Reservation, Arizona, the crop it's spraying is alfalfa. Humans don't eat alfalfa. But cows do.

Fly over the farmlands of the American agricultural interior and crop circles are the one ever-present feature of the landscape below. Not the alien kind, though; this is an agricultural phenomenon called centre-pivot irrigation. Each field is drawn and watered by a single remote-controlled irrigation arm that radials out from the centre, every hose calculated to provide a perfectly even spread of water. The one I'm looking at makes a field of 1 km in diameter. On Google Maps, I zoom out to see that I'm standing in the middle of seven such circular fields – a modest number, by American standards. The earth outside each lush green circle is bare, dry sand, giving – from the satellite image – the effect of green dinner plates dropped on a beach. We are, after all, in the Sonoran Desert.

According to Valley Irrigation, the company who produce the centre-pivot arm I'm looking at, alfalfa is one of the largest irrigated crops in the state. An article in *American Dairymen* magazine admits that 'alfalfa requires more water than most other crops grown in the U.S.'[1] Between 4.9 and 6.8 km^3 (4M–5.5M af) of water is used for alfalfa irrigation annually in California;[2] just one crop using the equivalent of California's entire Lake Mead allowance.* A centre-pivot field can cover 52–62 ha with a single arm-sweep. Flow rates and volumes differ according to conditions and crop type, but the University of Tennessee suggests most centre-pivots are designed to apply 0.3 inches (0.76 cm) over twenty-four hours, at 27 L per minute, per acre, meaning you could potentially apply just over 2 inches (5 cm) in a week[3] – close to the average *annual* rainfall of Nevada. That fact alone shows that reliance on rain-replenished groundwater simply isn't sustainable.

In the *Smithsonian*, agricultural and environmental historian Joe Anderson, at Mount Royal University, writes: 'By the time irrigators sensed the limits of groundwater irrigation, the region was stuck in what historians call an infrastructure trap: The success of center pivot irrigation has thwarted alternative visions for developing these dry areas.' In recent years, the fragility of the Ogallala aquifer, integral to the Midwest, has become apparent under such intense abstraction. Anderson asserts that most parts of the aquifer have declined faster than natural recharge – sometimes by 30 cm per year – because of centre-pivot agriculture. In 2017, the latest data from NASA's GRACE-FO satellite showed that, 'About 30 per cent of the water once stored beneath Kansas is already gone, and another 40 per cent will be gone within fifty years if current trends continue. Nonetheless, pumping continues.'[4] A 2019 study led by

---

* I focus on the Lake Mead and Colorado River system as a case study in this book, but of course California has many other sources of water. According to a Pacific Institute study, total water use in California averages around 53 km^3 (43M af) annually, of which agriculture accounts for 80 per cent or 43 km^3 (34.6M af), much of it groundwater. (Heather Cooley, 'Urban and Agricultural Water Use in California, 1960–2015 (Oakland, CA: Pacific Institute, June 2020)

Stanford University finds that irrigated crop production is impractical once an aquifer's saturated thickness drops below 9 m: 'Under status quo management and climate, the proportion of the [High Plains] aquifer below this threshold is predicted to grow from 25 per cent to 40 per cent between 2012 and 2100.'[5]

Further west, in Blythe, California, more than half of the crop grown in the irrigation district is alfalfa. Here, they are not centre-pivot groundwater irrigated, but flood irrigated from the Colorado River itself. And much of this alfalfa will be fed to cows in Saudi Arabia, not South Arizona. Fondomonte Farms, a subsidiary of Saudi Arabia-based Almarai – the world's largest vertically integrated dairy company – sends the alfalfa to King Abdullah City, where it becomes cattle fodder and eventually milk and pastry products. Saudi Arabia was big on centre-pivot irrigation itself, until its water ran out. The Wadi As-Sirhan Basin was once pockmarked with thousands of green circles in the same way as the Great Plains are today. The craze hadn't begun until the 1980s, but by the 1990s farmers were pumping an annual average of 18.5 km^3 (15M af)* of mostly fossil water. Before the centre-pivot farming boom, there was an estimated 500 km^3 of water beneath the Saudi desert. Experts estimate that four fifths of that has now been lost.[6] Alarmed by the rate of decline, in 2008 the Saudi government announced the reduction of wheat production; by 2016, it was stopped completely.[7] And so, to meet demand for wheat and dairy, they turned to other countries whose own wasteful water practices were not yet restricted. In 2014, Fondomonte and Almarai bought their first land in Arizona.[8] Professor Cody Friesen at Arizona State

*   If it's hard to visualize that amount of water, 3.5M af is about the same as 4 km^3 (or a cubic mile). So, picture a mile-high cube of water looming over your town, twice the height of the world's tallest building. While 15M af – what Saudi farmers were pumping each year, with effectively no recharge – is closer to 18.5 km^3 – a watery cube that would reach close to commercial aeroplane cruising height. That was literally consumed, gone, eaten by cows and humans, in one year.

University tells me that, since then, 'the Saudis have been buying large tracts of land south of Tucson to grow alfalfa.' When they import alfalfa into Saudi Arabia, he says, they're effectively importing water in compact form. In 2015, Saudi firms expanded across the state border into Blythe – an agricultural town surrounded by vast fields irrigated by canals filled with the Colorado River, fed by the ever-dwindling Lake Mead.

In Blythe, I meet Bart Fisher, a third-generation farmer in the valley and trustee of the Colorado River Board of California and the Palo Verde Irrigation District, of which Bart's grandfather, Wayne H. Fisher, was a founder member. Wayne lost his oil fortune during the Great Depression and came to Blythe to start again, in agriculture. That wasn't so unusual at the time, Bart tells me, as we drive in his pickup truck. 'All the original farmers in our valley were mostly Oklahoma Dust Bowl expatriates, who moved looking for water.' We would call them 'climate refugees' now. Wayne H. Fisher's master stroke was to apply for rescue money for the Palo Verde Irrigation District via the new Reconstruction Act: 'at that point, the federal government is just printing money left and right, trying to keep the economy going and keep the whole country from collapsing,' explains Bart, who, now a senior himself, vividly remembers such tales from his childhood. The Palo Verde Irrigation District ended up with a lot of money, a lot of land and, in time, a lot of water.

The first site Bart drives me to is the Palo Verde Irrigation District Diversion Dam. It looks like any other large reservoir, buttressed by reeds, water spanning off into the horizon. The difference is that it's privately owned. Built by the Federal Bureau of Reclamation in 1957, it was fully paid off and passed into the Irrigation District's ownership in 2007. A plaque stands outside, dedicated 'to the many pioneers that recognized the potential and value of our Valley' – with Bart's grandfather prominent among the names. 'The only intent of this dam is to divert this flow of [Colorado River] water out into our irrigation system,' explains Bart. The system is a tapestry of irrigation canals, starting as wide as a river and tapering down to concrete-lined ditches and dykes that feed into individual farms and fields, to be pumped out and flooded over the crops. That might seem an extravagant private use of the Colorado's precious water, but it's in keeping with the wider federal infrastructure. 'The Colorado River meanders from the Hoover Dam all the way up through Parker,' says Bart. 'This water comes out of Lake Havasu, which itself is a settling basin for the Central Arizona Project and Metropolitan' – by which he means the Metropolitan Water District of Southern California, but everyone here refers to it simply as 'Metropolitan'. Bart says their dam diverts 22.6–28.3 m^3 per second of the Colorado River into their irrigation canals – a modest amount, he says, compared to Imperial Valley users downstream that take 3.8 km^3 (3.1M af) annually. You get a very real sense, here, of how everyone takes their slice of the river. Satellite pictures show the Colorado River bordered by green farmland all the way down, canals and fields radiating out like vines, until it reaches the Mexican border – and then nothing. There's barely anything left. The death of a river by a thousand cuts. Forget reaching the sea; the Colorado barely reaches Mexico.

Bart personally owns over 4,654 of the Irrigation District's 42,290 ha. I ask if he equates large-scale crop exportation with sending Colorado River water elsewhere and he points out that, in food production, water is used multiple times – in his case, around 0.62 km^3 a year is diverted, but 0.37 km^3 is returned to the river. That goes on to grow crops downstream, then water vapour is transpired through the crops into the atmosphere and that in turn irrigates something else further east. 'So, we export grain to Bangladesh. Is that sinful behaviour? Not really. It's just what we do. Does that use precious water from the West? It sure does.' He then adds, 'But it's a strategy that we're not involved in developing. We welcome markets; all we do is exist within the business framework that's been established for us by federal treaties.' Farmers will farm, unless governments stop them – like they did in Saudi Arabia in 2016. But food and fodder must come from somewhere. At some point, serious global discussions are needed about where farming happens, and who has enough water to do it.

There are early signs of those decisions in California. Via a $38-million programme funded by the Bureau of Reclamation, Metropolitan and other water agencies in Arizona and Nevada, Palo Verde Irrigation District farmers are now being paid to fallow their land, receiving $909 per acre of farmland left fallow in 2021 and $925 in 2022. In theory, this could conserve up to 220,000 mL (180,000 af), which stays in the river and can be used by Metropolitan for other – largely metropolitan – purposes. In 2005, Metropolitan signed a thirty-five-year Memorandum of Understanding with Palo Verde Irrigation District, limiting fallowing to a maximum of 29 per cent of irrigated lands in the valley. As soon as the agreement was signed, says Bart, Metropolitan pushed to achieve maximum fallow, and the 29 per cent maximum quickly became the 29 per cent norm.

In 2015, Metropolitan purchased 5,140 ha of its own, making it the largest landowner in the Irrigation District. By mid-2021, it owned 11,736 ha of land with senior water rights. When the MoU runs out in 2040, will water-starved Metropolitan stay in the farming game, or will it use its water rights for other things, like, say, water supply? A 2021 report to Metropolitan's board of directors stated its intention quite plainly: 'acquiring lands with senior priority rights is a potential strategy to obtain reliable and cost-effective water supplies in the face of a changing climate.'[9] 'Metropolitan LA, that's a lot of power over there: 19 million people,' says Bart. 'And industry and factories and aerospace. So, if they decide they want to take our water, what's to prevent that?' Metropolitan are currently playing nice, leasing its farmland back to farmers, albeit with stipulations for crop types and efficient water use. 'But, obviously, they have no interest in owning farmland here, except to take the water,' concedes Bart. At the end of the programme, in 2040, he anticipates that they'll fallow all their land.

To show me what such a post-farming future would look like, Bart takes me to the town of Blythe itself. In the high street, shop after shop is shuttered up. The shopping mall has only one shop unit let. It's not a linear correlation, but, with a third of fields in a farming community already fallowed, then a third of regular jobs have gone too. If you fallowed them all, America would lose agricultural towns like Blythe forever. 'Life is largely boarded up in Blythe, and businesses are pretty much closed . . . And a lot of this has happened since Metropolitan started acquiring their farmland.' It's not only that Metropolitan pays farmers not to farm, but – as the largest landowner in the valley – its own fallowing dollars don't stay in the community: 'It's just a ledger entry on their balance sheet,' says Bart, who estimates that $6 million to $7 million a year is removed from the local economy this way. Perhaps, by 2040, there will be no farming community left to complain.

Back out on the fields, Bart takes a phone call from a fellow Irrigation District member, Dwayne. Dwayne can't believe how bad this year's cotton harvest is. 'Yeah, it's bad. It was just too damned hot,' replies Bart. It's discouraging, says Dwayne. 'It *is* discouraging. *Shit*,' replies Bart.

Dwayne says he's been talking to growers further west: 'And they said, sooner or later, we're not going to have tomatoes in the San Joaquin . . . They have no idea how much water will be available.'

'You know what's gonna happen, Dwayne?' Bart responds. 'The San Joaquin Valley is going to be tumbleweed and it's gonna revert back to desert. And, interspersed in the middle of all those fallowed fields with tumbleweed, you're going to have trees: you'll have pistachios, you'll have almonds; that's where any available water will go.'

'You know,' replies Dwayne, 'when you go up Route 33 to Firebaugh, there's, what, 200,000 acres there that's fallowed? That used to be farmed. Now it's all bare, all weeds. There's a farm shop and a house, and there's nothing else.'

'It's all gonna look like that,' responds Bart. 'It's gonna look like that all the way from Bakersfield, past Paterson, all the way up towards Stockton and to the Delta.'

Bart is no pessimist. He's a pragmatist, a businessman. He loves what he does – he tells me later he turned down a multimillion-dollar offer from Metropolitan to buy his land ('What would I do then?' was his answer to them). In truth, he can foresee the death of his community, his way of life, coming around the corner, but he'll be damned if he's going to contribute to its demise. He'll see out his days, with his friends and family, with Dwayne and other farmers, before Metropolitan fallows the lot. Part of me wants to hug him. He tells me about a new melon variety he's growing, originally from Hami, China, and they sound so good I want to buy them all. But equally I know that this is the end of an era. The era when you could flood your fields with the water from the Colorado. The era of cheap water.

Evan Wiig, director of membership and communications for Community Alliance with Family Farmers (CAFF), California, recognizes the tumbleweed vision of the future too. In fact, he thinks it will be worse. Even the nut trees will struggle. 'For the vast majority of the Central Valley, where we grow almonds,' he says, 'if you look at projections twenty or thirty years out, based on climate change, most of it is not even going to be suitable for almonds. Especially in the southern valley, in five or ten years . . . you're not gonna make a profit. Because it's not just about water allocations; the literal climate is changing.' Even if crops can survive 37°C and longer droughts, they often need a certain amount of 'chill hours' for their production, too. The famous cool layer of fog in the San Joaquin Valley that offered those chill hours has gone now, he says. 'Here I am, in Sonoma County's wine-growing heartland, and we're not gonna be able to grow the same grapes in twenty years' time.' Crop types and trends have always moved and shifted over time. Farmers are used to adapting to climate and market changes. But what does it mean if California, the supplier of 90 per cent of the world's almonds, can no longer grow almonds? Or, in a European context, what if the climate of Champagne can no longer produce Champagne?*

---

* Lelarge-Pugeot, a seventh-generation grower in Champagne, told *Wine Enthusiast* magazine in 2020 that he already harvests a whole

Growing pains

Back in England, I drive to meet an organic farmer, John Bishop, at Rushall Farm, Berkshire. The motorway gives way to an A-road, then a B-road, then just the wonderfully named Scratchface Lane. In the lush fullness of late July, it's an oil painting of English countryside. Gentle rolling hills, fulsome hedges and ancient oaks framing fields of barley, sheep on green pasture, and a babbling, crystal-clear chalk stream – the River Pang – cutting along the bottom of the valley.

All 429 ha of Rushall Farm is organic. The crops John grows are 'fed' only by a clover cover crop. He hasn't sprayed any artificial fertilizer on them since the year 2000. He believes this has a water benefit. 'From a water perspective, because you're not putting on artificial fertilizer, you don't promote artificially rapid growth . . . you've got a longer time when you're drawing from the soil moisture deficit, the water that's actually in the soil.' This helps, he says, during the all-important three weeks in June when the wheat grains fill with water – if they don't get any in those three weeks, your crop fails. And that's crucial, given that England's agriculture is largely rain-fed, not irrigated. Anything that can be done to retain moisture, or slow the withdrawal, is a good thing. He's been farming since 1976, and says it's now 'warmer, and less predictable in terms of how much it will rain and when.' Crops like maize that weren't considered suitable to grow in this country forty years ago are now common. John also writes for the local parish magazine and sends me some of his articles after my visit. In one, from February 2020, he writes: 'It is extraordinary that the value of these essential foods is the same as it was 30 years ago; actually [the] wheat price at one point was higher in 1974! It is not difficult to imagine how all the

---

month earlier than his grandparents did. (Sarah E. Daniels, 'Climate Change Is Rapidly Altering Wine As We Know It', *Wine Enthusiast*, 3 February 2020)

costs, including fuel, tractors and machinery and labour have gone up enormously. The general response has been for farms to get bigger and more industrial.'[10]

As we walk towards the River Pang, down a track that probably hasn't changed significantly in hundreds of years, I ask John what changes are needed in the drier climate of the 2020s. He says that, when he and his wife went to university in the 1970s, they'd been taught by of some of the world's top agricultural researchers. 'They were lecturing us on climate and history and physiology, the lot. There was the Institute for Research into Animal Diseases, the Dairy Research Institute, the Grassland Research Institute; the government had experimental husbandry farms where they were pioneering new systems. Today, all that's gone. So, I see a generation of farmers that aren't so versatile in terms of adapting to changed conditions.' When I later speak with a younger Buckinghamshire farmer, Steven Lear, with a 1,200 ha acre beef and arable farm, he backs this up: 'I've lived on a farm all my life, I went to university and studied agriculture – at no point did anybody tell me about soil health. It was all about the agrichemical model: use lots of fertilizers – that's how you grow a crop.'

Guy Singh-Watson, farmer and founder of the UK organic veg-box company Riverford, wrote, in the dry summer of 2018: 'The only commercially viable option (and the most environmentally favourable) is to build clay-lined winter-fill reservoirs wherever there is a valley bottom wide enough. To invest in an asset that is used so unpredictably . . . is a bold move, but perhaps climate change is shifting the odds.' He wrote again, two weeks later, and still no rain: 'Is this an early manifestation of the predicted resulting climate change? Perhaps it is too soon to say with authority – but by the time we have that authority, it will be too late.'[11]

Globally, agriculture is the biggest user of water by far. Conor Linstead, freshwater specialist with WWF-UK, tells me that around 70 per cent of global water abstraction goes to agriculture, with the vast majority of that to irrigation. 'When you look at the proportion of the withdrawals that are lost to the catchment as evapotranspiration [from crops], then agricultural use is more like 90 per cent . . . at a global level, so you can't really address water-scarcity issues, especially in the arid countries, without looking at agriculture.' In the UK, only 10 per cent of water goes to agricultural irrigation, but, he says, what's often misunderstood is that 'agriculture is a consumptive use, whereas domestic water use is not.' Domestic water use is returned back at a point further downstream. Agriculture, though, is entirely consumptive: it's sucked out of the river or groundwater and then gone.*

In Nevada and Arizona, Colby Pellegrino tells me that about 80 per cent of the water used is for agriculture. 'So, everything that we do, relative to municipal conservation, is only squeezing 20 per cent of the use. And, eventually, you have to get to a point where you're working on more than just 20 per cent.' In the US, water is effectively free to farmers, argues Newsha Ajami, at Stanford University's Water in the West programme: agriculture has a pre-built infrastructure, with aqueducts and dams specifically made to provide irrigation water. Some farmers 'have water rights and just take the water out; they don't need to pay anything . . . if you have land, you can have a well and you can take water out forever.' Farmers used to have a direct relationship with water and drought, says Ajami. 'They used to grow a lot of non-permanent crops that, during drought years, could be fallowed and then, after the drought was over, they would start growing again.' Now, there's a shift from low-value annual crops to high-value perennial crops – but these are 'permanent crops that need water, no matter what,' and by definition can't be fallowed. This has also brought a cultural shift, 'from generational farmers to agribusinesses who are making lots of money and are capable and willing to pay to dig a deeper well . . . people who are in it for the short term, to make a buck and move on – they basically don't really care what's going to happen long term.' Once the water goes, the agribusinesses will go too, moving on to the next place to suck dry.

---

*   Unless there are drainage return canals, as in Blythe.

In Punjab, India, in January 2022, former economics professor Gian Singh stated on his blog that groundwater use for agricultural irrigation is the biggest culprit for groundwater depletion: 'groundwater used for growing crops in Punjab exceeds its available quantity each year and there is a strong correlation between crop combinations and groundwater balance. Two cereal crops – wheat and paddy – account for over three quarters of the total sown area in Punjab and are mainly responsible for the groundwater depletion problem.' In particular, this is down to the boom in available pumping technology. 'In 1961, the number of tubewells* in Punjab was only 7,445. This has risen to around 1.5 million in 2021 . . . The huge increase in the number of tubewells to meet the irrigation requirement of paddy has led to the continuous fall in the groundwater level.'[12]

The Delhi think tank CEEW argues in a recent report, *Reallocating Water for India's Growth*, that India's agricultural sector needs major reform if India is to solve its water crisis: 'India uses two to three times more water to produce a unit of a major food crop in comparison to China, Brazil, and the US. In parallel, the demand for water in other sectors is rapidly escalating, leading to increasing conflicts across regions in India.' An insufficient national reservoir capacity, 70 per cent of which is concentrated in just six states, has led to an over-reliance on groundwater. 'Technological innovations [and] energy subsidies have made groundwater pumping a far more attractive option for farmers. Increased water withdrawal . . . will undoubtedly increase the pressure on India's groundwater resources.'[13]

Kangkanika Neog, an independent water consultant in Assam, tells me that the canal systems built post-independence are increasingly inefficient, thanks to poor investment. Evidence also suggests

---

*   A tubewell is a narrow tube, typically stainless steel, driven into a shallow subsurface aquifer. Boreholes are similar, but wider and deeper, for more difficult-to-tap groundwater. Both then need pumps to bring the water to the surface, either hand-operated pumps, diesel, electric or solar.

a correlation between increased groundwater use and decreased canal-water use over the decades. People began depending on their own water, she says, and groundwater also became more accessible through tubewells and electric pumps. Energy costs were high, which should have restricted use; instead, during the Green Revolution in the 1960s and 1970s, the government subsidized electricity for farmers, causing groundwater pumping to skyrocket. According to Kangkanika, 'Along with pesticide and fertilizer subsidies, this led to land degradation and water levels going down.' Certainly, groundwater in many parts of India is now perilously low. In CEEW's analysis, India will require an additional 41B m^3 by 2030 and 117B m^3 by 2050. The only way to achieve this, they say, is by 'reallocating local irrigation water' – that is, from farms to towns.

According to CEEW, government-set minimum support prices for staple crops in India, such as wheat and rice paddy, have created a 'skewed incentive structure' in favour of water-intensive crops. Punjab receives only 40 per cent of the monsoon rainfall that West Bengal gets; yet – thanks to the state minimum prices – Punjab farmers 'grow paddy, mainly by drawing groundwater during the summer months, when evaporation rates are high.' This is far from unique to India. Most countries have incentivized the wrong water practices. In the EU, the Common Agricultural Policy has long been criticized for incentivizing inefficient water use. A report into exactly that in 2021 by the European Court of Auditors even ran with the sub-title, 'CAP funds more likely to promote greater rather than more efficient water use'. The report found that 'Member States use EU funds to support water-intensive crops in water-stressed areas'. It also finds that the Common Agricultural Policy made the draining of precious peatlands and wetlands eligible for income support.

Perhaps the most misplaced water incentive in the world is reserved for the USA. Here, stories abound of needlessly flooded unsown or fallowed fields, just so the next year's water rights are maintained, under 'use it or lose it' stipulations. According to *Scientific American*, 'use it or lose it' rights are common in state laws throughout the Colorado River basin and give 'the farmers, ranchers and governments holding water rights a powerful incentive to use more water than they need . . . There are few starker examples of how man's missteps and policies are contributing to the water shortage currently afflicting the western United States.'[14] It wasn't until February 2021 that Arizona – arguably soon to become the country's most water-stressed state – allowed water users to apply for a 'temporary reduction' to conserve water without losing their rights.[15]

For drip irrigation, water is run through pipes buried or lying on the ground next to the crops, the water trickling or dripping directly next to the crop roots. But it's not a mainstream option in developing countries because the costs of system installation, let alone maintenance, are prohibitive. CEEW's Vaibhav Chaturvedi tells me that India has been 'talking about drip irrigation for twenty-five years,' but it's not generally viable, even after government subsidy. Drip irrigation isn't necessarily the silver bullet for water efficiency that you might expect, either. 'Very efficient irrigation doesn't always deliver environmentally,' says Conor Linstead, at WWF, 'which is an interesting paradox.' Indeed, it's one with a name: the Jevons paradox. This occurs when technological improvements increase resource-use efficiency, but the consequent decreased cost leads to rising demand and more use of the resource. Linstead elaborates: 'The scientific literature is pretty clear that, if an individual farmer goes from inefficient to efficient irrigation, they tend to consume more water because the water they've "saved" is still available to them – so they will just expand their production area or double the crop. So, it's not really that panacea that people claim.' Spanish water-policy specialist Nuria Hernández-Mora confirms that she's seen this many times in agricultural modernization programmes, for surface-water irrigation: 'It has happened again and again.'

By definition, drip irrigation remains at the root level and is 'lost.' Flood irrigation, however, while seemingly more wasteful, can in fact go down to top-up the groundwater. 'Flood irrigation is very useful for groundwater recharge,' says Dr Helen Dahlke, associate professor in integrated hydrologic sciences at the University of California, Davis. 'One of my colleagues has estimated that, when we started with flood irrigation in the Green Revolution in the 1950s and 1960s, we recharged around two feet [60 cm] to the groundwater.' Drip irrigation is also typically sourced from groundwater as opposed to surface water, because groundwater is being pumped up clean and pressurized, whereas surface water can be more turbid and can clog up nozzles; so, it's typically abstracting groundwater without giving back to it.

Drip irrigation also comes with a high materials-waste footprint. 'The new thing in agriculture is growing a lot of annual crops with subsurface drip irrigation,' says Dahlke. These systems are installed new every year, she says, rather than being left in the ground. It has also allowed agriculture to expand into areas that had not farmed before. 'The big idea of using drip irrigation to conserve water is a complete lie,' summarizes Dahlke. 'Because, what extra water we are saving, we're using it for more growing crops.' For food production, that's fine: more food can be grown for the same amount of water. But for local rivers, watersheds and water supply, it raises important questions. One of which is, should we use up local water resources for export crops?

So, where does asparagus come from?

The British asparagus season traditionally runs from St George's Day, on the 23 April, through to summer solstice, on 21 June. For eight and a half weeks, this miraculous, perennial vegetable reappears from the ground. The perfect growing conditions are a cool, crisp late spring; not too wet, not too dry. Each spear is harvested by hand and is best eaten within two days, before it loses its unique flavour. 'There is always an air of excitement when the first British asparagus comes into season as it signals the beginning of spring,' Chris Ling, M&S vegetable buyer, tells *The Guardian*, among the equally perennial patch of news stories that mark the beginning of asparagus season every year.

Despite this, most British consumers wouldn't be able to tell you when the British asparagus season is. Because asparagus, like most fruit and veg, is available in UK supermarkets all year round. Most of the out-of-season asparagus we eat in Britain comes from Peru, flown over in chilled containers. The stories of how Peru cornered the world asparagus market vary. In 2004 *The New York Times* said the US's 'War On Drugs' subsidies in the 1990s funded Peru's fledgling asparagus industry as a means to encourage alternatives to cocaine farming.[16] A less US-centric telling finds that asparagus production in Peru dates back to the 1950s; since when, production increased thanks to changing agricultural laws, a 280-km irrigation project in 1990, and indeed the 1991 Andean Trade Promotion and Drug Eradication Act with the United States, which lowered trade barriers prompting a flurry of free-trade agreements in the 2000s.[17]

The problem, however, is that much of Peru's asparagus production occurs in the Ica Valley, one of the driest places on Earth, with just 8 mm of annual rainfall. The Ica River was diverted to the asparagus farms, while the groundwater below plummeted. According to the Sustainable Food Trust, in the Ica Valley, 'wells are drying up and water rates are astronomical, forcing farmers to sell their land to the asparagus trade. Supplies are severely rationed, yet one giant asparagus farm can use as much water as the entire city of Ica every day.'[18] The cover of a 2010 report by Progressio, CEPES and Water Witness International showed huge green rectangular fields of irrigated crop standing out against the pure white desert sand. Indigenous communities who made a marginal living here by herding alpaca, sheep and llama were allegedly pushed out as the water disappeared. The report found that, with abstraction 'significantly exceeding the amount of recharge, the water table in the valley has plummeted, typically by rates of between half a metre and two metres a year, and in places by as much as eight metres each year – almost certainly the fastest rates of aquifer depletion anywhere in the world.'[19] It even quoted from anonymous agribusiness people scared about the rate of decline: 'If we don't do something fast the valley will die.'

It's easy to dismiss a report that's over a decade old, however, so what's happened since? In 2015, smallholder farmers in the town of Ocucaje protested against one of the biggest growers in Ica, Agrícola La Venta, who arrived in the region in 2007. Within years, the desert was turned into rows of asparagus. The *Desert Sun* reported in 2015 that 'wells have gone dry, and farmers with small plots complain the newly arrived megafarms are using up their water.' Twenty-one-year-old Joselyn Guzmán told the paper, 'We don't want the water to be taken away because it's the only livelihood we have.'[20] Agrícola La Venta assured those concerned that the few wells it had would do no such thing. Fast-forward to May 2018 and *Ojo Público* reported that, in fact, Ocucaje farmers 'are leaving their land because of a water shortage. The only production that currently prospers is that of Agrícola La Venta, which . . . was able to obtain 23 well licenses in record time in the midst of the declaration of a water crisis in the region.' Ocucaje residents were forced to collect water in buckets because they'd lost the permit for the underground borehole they'd owned since 1984, while nearby agribusiness wells pumped day and night.[21] In 2018, investigative reporter Fabiola Torres used drones to survey the region and found several companies' properties dotted with illegal wells extracting groundwater, some camouflaged beneath green mats. Instead of confronting the companies, Peruvian authorities granted retrospective licences to the illegal wells.[22] All the while, Peru exports of fresh asparagus grew to its main markets in the United States (75 per cent of exports), Spain, the UK and the Netherlands.[23]

*

In November 2020, Carla Toranzo emails to let me know she'll be late for our call. Her Internet has become patchy due to power outages. When we talk over WhatsApp, the reason becomes clear. That very morning, the Peruvian president, Martín Vizcarra, was removed from office in what newspapers were describing as a 'coup'.

Protests on the streets were turning into riots. He was the fourth president to be ousted in just five years.* As Carla goes on to explain, the struggles of politics, bureaucracy and a hand-tied public sector are central to Peru's water problems.

Carla Toranzo, a civil engineer by training, is Latin America coordinator for the Alliance for Water Stewardship (AWS). Based in Lima, Peru – the second-largest desert city after Cairo, Egypt – she is both a student of, and an active consultant in, the agribusiness boom which has brought jobs, wealth and water scarcity to Peru. 'Peru has huge challenges related to water and sanitation in general,' she says. 'Some people don't have water in their houses; they don't have toilets. Their wastewater is not treated.' She says much of the focus is on the amount of water used by agriculture, but there are structural problems in the government too, and the state infrastructure requires modernization. However, it's hard to convince private companies to get involved in public water policy. '"It's the responsibility of the government" – this is the response you receive,' says Toranzo. 'Maybe, when you see things from the outside, you say, "Oh my God, [agribusiness] are terrible, they are not sensitive [to the people's plight]." But here most of the economy in Peru is informal, most people don't pay taxes, so businesses think, I *do* pay taxes and you want me to do *more*?! It's not easy to get companies on board.' Even so, she tells me that the latest documents from the Peruvian National Water Authority show that 'the aquifer [in Ica] is not in balance, it is in negative balance. I'm not saying that agriculture is doing well, because everybody knows here that there are a lot of wells, but there are a lot of structural problems here, too.'

Their southern neighbours, Chile, are suffering from a fifteen-year megadrought, with water problems here, too, exacerbated by agribusiness. Petorca avocados are among the most emblematic and problematic, suggests Andrea Becerra, author of a Natural

* His successor that day, Manuel Merino, formed a far-right government. Just five days after taking office, he too resigned following the deaths of two protesters.

Resources Defense Council (NRDC) report on drought in the San-tiago Metropolitan Region.[24] 'There are huge avocado plantations and avocado is extremely water intensive, leaving rural residents with no running water and dependent on water trucks,' she tells me. 'This has brought the industry tons of bad publicity and led to parts of Europe ceasing to import from Petorca.' Becerra has heard anec-dotal links between large plantations and the demise of the Aculeo Lagoon, a 12 km² lake that completely dried out in 2018. In such instances, 'climate change' becomes an easy scapegoat for agribusi-ness. But 'it's not climate change. It's the agriculture business. It's a combination, for sure . . . but now there is this race to the bottom.' Chile produced about 245,000 tons of avocados in 2018, most of which went to Europe, where demand for avocados is growing by 4.6 per cent each year. The avocado industry puts its average annual growth rate even higher, at 15 per cent.[25] It has become a race to extract what's left while there's still something to extract. Becerra notes that this is no accident – General Pinochet's 1981 privatization of water is a policy the country is now paying for dearly.

Avocados grown in the Petorca region require some 1,200–2,000 litres of irrigation to produce one kilogram of fruit. This gives each avocado an individual water footprint of around 273 L each. Throw-ing away that half an avocado turning brown in your fridge, therefore, wastes 136 L of water. A Petorca villager, meanwhile, is given an allowance of only 50 L of water a day, transported in by truck, to cover all their needs.

The global water footprint

The concept of a water footprint, or 'virtual water', is simple. Every-thing you buy took some water to make or grow. If you buy coffee from Kenya, you're consuming the water from that region of Kenya; if you're wearing a T-shirt made from cotton grown in Egypt, you are, in a sense, wearing water from Egypt. Everything, from elec-tronics to furniture to the pages of this book, has a water footprint. And, if that water originated from a water-scarce region, then the supply chain, the retailer, and you, the consumer, are complicit in the import and export of that water.

Even during Day Zero, some of Cape Town's key exports continued to be among its most water intensive. The region is the heart of South Africa's wine country, which exported 428.5M L of wine, in 2016, to Europe and the US. It takes approximately 1,000 L of water, fed to the vines, to make just 1 L of wine. On top of that, the Western Cape exported an estimated 231,000 tons of citrus fruit, mostly oranges, in 2017. The water footprint of just one orange – the net amount of water used to grow it – averages 80 L. Arguably, a Capetonian drinking a glass of local wine or eating a local orange would have exceeded their daily 50-L water allowance; in practice, of course, such water footprints are roundly ignored. And they are ignored at the expense of local water tables.

Academics Biswas and Tortajada point the blame for the world's 'unsustainable path' of water management directly at our water footprints: 'Some 2,500 litres of water are needed to produce one kilo of rice. However, 10 times this amount is needed to produce one kilo of beef. Thus, when people's dietary preferences move to animal protein, water requirements go up very significantly, a fact mostly missed by water planners . . . As numbers of middle-class households increase, the water requirements to produce their preferred diet will go up very significantly.' India, in particular, the authors say, has been following this unsustainable path for at least half a century.[26]

The water footprint also obscures an uncomfortable truth: many countries have *already* run out of water. If you define meeting domestic water needs as having enough water to grow sufficient food for a population, then that ship has sailed.

Tony Allan, at King's College, London, first worked in the Middle East in the 1960s. He noticed that the region had effectively run out of water to meet its food needs. Why didn't they starve? Why didn't 'water wars' break out with neighbouring countries? It was, Allan deduced, because the global food trade allows for food grown in wetter countries to be exported to drier countries. Speaking at the

American University of Beirut in 2016, he recalled, 'when I started to look at the region as a whole, I found that Libya ran out of water in the 1950s, but . . . certainly by '72, every country in the region had run out of water in the sense that they didn't have enough water to grow their own food . . . in 1972, they were starting to import food, import wheats and grain and flour at a very increasing rate, and then I realized – ah, all you need to do if you run out of water is to import food!'[27] This led Allan to invent the terms 'embedded water' and 'virtual water' to explain the invisible water traded around the world. He also found that trading this invisible water allowed politicians to hide the extent of domestic water shortages: farms may dry up and go out of business, rivers can even dry up and disappear, but, if supermarket shelves are full, then nobody cares – except for the farmers and the fish, both of which struggle to find a voice in metropolitan capitals. 'So, the political leaders found that the solution was to import food . . . economically easy and, above all, no politics,' said Allan.

In terms of supermarket shelves, people have never had it so good. Food that used to appear for a short time annually is now stocked all year round. In 2008, WWF published the water footprint of the United Kingdom, listing it as the sixth-largest importer of virtual water globally. Of the total water needed to grow the food and produce the goods consumed in the UK, only 38 per cent was actually from the UK: 63.6B m^3 of the UK's annual water use is sourced from overseas, over thirty times the flow of the River Thames. So, while the average UK citizen directly uses about 153 L of tap water per day, their indirect consumption of embedded water through imported products is around 4,645 L a day.[28] Arguably, like Libya, the UK ran out of the water needed to sate its consumption many years ago.

The concept of the 'water footprint' followed on from 'virtual water', introduced by a close collaborator of Allan's, Arjen Hoekstra, in the Netherlands. In 2002, Hoekstra refined the theory by dividing water into three categories: green water (rain), blue water (surface rivers and lakes) and grey water (polluted water that needs treating). Every commodity has a green-, blue- and grey-water footprint, and we can calculate its value accordingly – Hoekstra made it his life's work to do just that.*

In *The Water Footprint of Modern Consumer Society*, published in 2013, Hoekstra wrote: 'Until today, water is still mostly considered a local resource, to be managed preferably at catchment or river basin level. This approach obscures the fact that many water problems are related to remote consumption elsewhere.' In fact, he says, the very idea 'that water demand is something local and to be met locally is a misconception. Most of the water consumed in this world is for making agricultural and industrial goods that are traded regionally or internationally.' For example, you might look at a bottle of cola and assume that the total amount of water used is right there within the bottle, at a ratio of 1:1. Hoekstra calculated that, in fact, the total water footprint of a bottle of cola is 168–309 L, depending on the source of the sugar. If it was grown in Andalucía, Spain, for instance, 'this has contributed to lowering water levels in the Guadalquivir River and [the] drying of important wetlands'. Worse still, what if you add up everything you drink in a day? Hoekstra's example included one glass of milk (water footprint of 200 L), two black coffees (260 L), one orange juice (200 L), one black tea (30 L) and one glass of wine (200 L) – giving an overall water footprint of approximately 900 L just for one day's drinks. As Hoekstra noted, a typical bathtub contains 90 L of water – so we're drinking about ten bathtubs per day. Imagine filling your bath ten times over and watching it all just drain away.

The water-footprint flame is now carried by Dr Rick Hogeboom, assistant professor at the University of Twente and director of the Water Footprint Network founded by Arjen Hoekstra. Rick brought his water-footprint message, plus a loud shirt with orange buttons, to a new audience via a Ted Talk in 2018.[29] He surprised the Dutch

---

* Both Hoekstra and Allan were high on my list of people to interview for this book, but tragically Hoekstra died while cycling in 2019, and Allan passed away in 2021.

crowd with the fact that a conventional carnivore's meat-based diet*
has a water footprint of 4,300 L per day, whereas, for a vegetarian
diet, the water footprint is only 2,700 L per day – 40 per cent
smaller. This is because 'the animals that we eat are fed enormous
quantities of soy and maize, which require a ton of water to grow. If
we struggle to feed growing populations in an ever-scarcer world, is
it fair to use 2,000 L for one steak? In total, he says, all the tap water
you use at home – for baths, showers, cooking and cleaning – is just
1 per cent of your actual water footprint.'[†]

In conversation, Rick is more nuanced than in his Ted Talk. He
admits that it's a constant struggle for academics to simplify their
message for a general audience, while also conveying the complex-
ities of the science. In truth, whether a water footprint is 'good or
bad' almost always elicits the answer 'it depends'. For example,
what's the beef with beef? 'It depends,' he smiles. 'If you swap your
beef burger for a dish made with pistachios from Iran, then those
nuts probably have a larger water footprint compared to, say, Irish
grass-fed beef.' Or take cotton. I'm wearing a cotton T-shirt labelled
organic; I try to buy organic so I know it's not polluting the soil or
killing pollinating insects – but is it 'better' from a water perspective? 'It depends,' laughs Rick. 'We did studies with some major
cotton-clothing producers, and it really varies by the farm and how
the farmer goes about their water management. In general, you can
say that there will be a lower grey-water footprint component to
organic production, but often it comes at a cost of a somewhat
lower yield, because you cannot apply all the fertilizers.' In summary, organic is 'a good movement,' says Rick, 'but I'm not sure it
goes far enough to fix the problems.'

Rick argues that it's fair to question whether it's smart to use

---

* Fun fact: the traditional Dutch snack frikandel – made from 'several
kinds of meat' mixed together in a long sausage – is so popular that 600
million frikandellen are made annually in the Netherlands, an average of
thirty-seven eaten per person, per year.

† Or equivalent in frikandellen.

finite resources to 'grow asparagus in the desert to export to rich nations,' for example. But he says you can quickly get tangled in 'philosophical discussions on the sovereignty of state. I guess this is analogous to deforestation in the Amazon. Why should we care what the Brazilian government does by decree of its democratic mandate, right?' And yet we do, and the moral – and environmental – implications of that seem clear. He says the Netherlands is a water-rich country, globally the second-largest exporter of agricultural products by volume – mostly meat, eggs and dairy – 'which is crazy for such a small country . . . we could feed so many other countries that therefore would not have to strain their own water resources . . . we could definitely sacrifice a few fields of maize to grow some asparagus, to relieve the pressure in Peru.' But that is a purely environmental argument, he says. 'If you look at the social or economic argument, employment – then it bites. But, if it's an unsustainable system, the water runs out in Peru and those people will go out of business anyway. So, we'll see.'

PART II:
TURNING THE TAPS

CHAPTER 8

Repaying Our Debt

It's early May, and farmer James Alexander stands in the middle of a wheat field in rural Oxfordshire, holding up a pair of white underpants as I take photos. Tall, young and bearded, James looks slightly sheepish at the absurdity of it all. Then it's my turn. I hand him my camera phone, he hands me the Y-fronts, and I pose, trying – and failing – to decide on the right facial expression. Then he starts digging. We bury the underwear, taking a GPS location and hammering in a red pole to mark the spot. In a plastic bag in his four-by-four truck are two more pairs of white XXL cotton undies, waiting to be buried in two more fields. We drive on, avoiding eye contact.

It all began in 2017, when photos started emerging on social media and in the farming press of farmers in their fields holding up baggy white cotton underpants, with headlines such as 'Farmers in Cornwall are being encouraged to soil their underpants.'[1] There'd be a 'before' photo, in which the pants were pristine, and an 'after' photo, taken two months later, now reduced to sorry brown tatters. This was deemed good, and proudly shared with the hashtag #SoilYourUndies. As the Natural Resources Conservation Service, Oregon, put it at the time, 'The undies in Oregon's farm country are falling to pieces. And that's just the way farmers like it.'[2] The challenge, issued by the Californian Farmers Guild in 2017, was to bury a pair of 100 per cent cotton underpants in your field and dig it up two months later; if you have healthy soil, there'll be nothing left but the elastic waistband, the microbial life having devoured the rest. 'Cotton is an organic material and breaks down naturally just like anything else you'd put in your compost pile,' Evan Wiig, the guild's then executive director and originator of the craze, told *The Daily Telegraph*. 'So, if you bury cotton in soil teeming with life, all those creatures will begin to feast.' Pesticide-intensive mono-crops, meanwhile, produce disappointingly pristine pants.

I sought out James to repeat this challenge in the Oxfordshire clay because he's an unusual farmer in more ways than one. He employs all three farming methods – conventional farming, organic farming and 'no-till' farming. I wanted to repeat the 'soil my undies' challenge in all three field types. As a contract farmer, he doesn't own the land, but rather owns the machinery and runs other people's farms for them. He therefore intimately knows and understands the processes, pros and cons, of all three systems. With the exact same soil type, river basin and geography, the couple of hundred hectares he farms, over a radius of barely 6 km, offer something of a natural experiment: a direct comparison of modern farming methods. And his personal preference, after years of experience, is the no-till method, both for crop yield and water efficiency.

No-till means 'no ploughing'. The earth is never disturbed and is never left bare – there's always a crop growing, either to be harvested – the stubble remaining with the next crop growing green in between – or a 'cover crop', such as clover, to 'fix' nitrogen naturally back into the soil. Fertilizers aren't needed. More importantly, the soil, the water table, and the groundwater aquifers beneath all reconnect, like the natural system is supposed to.

You need very little equipment for no-till farming. Not requiring a plough makes a big saving for a farmer. You do need what James calls a 'drill', which gets pulled behind a tractor with thin sharp discs, like circular saws, to make a surgical cut into the soil to plant the seed – which is spat out of a narrow tube – and then the soil closes up again as soon as it passes through. James uses the term 'direct drilling' synonymously with 'no-till'. The process is this: grow a cover crop over winter, then flatten it with a crimper (like a light steamroller), then direct-drill your seeds straight in. Harvest in summer or autumn, rinse and repeat. The soils, nitrogen and organic matter just build and build. This is called 'regenerative agriculture': instead of mining the soil for nutrients and leaving the earth in an increasingly poor state, it gives back to the soil, which improves year on year. And, from a water point of view, it's potentially a miracle cure.

When I first rang James up with the pants idea, it was still winter. On a recent rainy drive through rural Oxfordshire, I'd seen roads turning into muddy streams as the topsoil of clay-red fields ran off into the road. James told me he'd been out on his field that morning and didn't get muddy boots: 'you can walk some of my fields in your best shoes, you wouldn't get muddy – because our infiltration is so much better, the water just goes straight through.' I had to come and experience this for myself.

*

If our water systems were a bank account, then we are all so far into our overdraft that the bank is at the point of collapse. But it's not too late to start repaying our debt. This isn't about 'giving back to mother nature'; it's about ensuring our own survival. And it begins with refilling our groundwater aquifers.

Based in Santa Rosa, about an hour north of San Francisco, Evan Wiig is a former rancher who now promotes regenerative farming for the Community Alliance with Family Farmers (CAFF). Since he first headed out west to work on a cattle ranch in 2012, drought is all he's known. On a 486-ha grass-fed cattle ranch in the hills of Marin County, 'you're entirely dependent on rainfall,' he says. 'It's not like you can go out and irrigate hillsides.' As the drought began to bite, they experimented with 'biomimicry', moving the herd to intensively graze in small pastures and then quickly moving on. 'Look at the way a natural herd of herbivores moves,' he explains. 'They're not crammed into a feedlot, that's for sure. But they're also not spread out evenly; they're in a tightly knit herd that moves across a large expanse of land.' Wiig calls it a symbiotic evolution that has taken place between the grasslands and ruminants: 'They graze it down, aerate it with their hooves, fertilize it, and, before they entirely destroy that land, they move on . . . By the same time the following year, that grass is healthy and rejuvenated, and has everything it needs.' On his cattle ranch, they did everything they could to operate in the same way, so they weren't relying on irrigation, excessive use of reservoirs or imported feed.

No-tilling is biomimicry too, he says. The idea of no-till, regenerative agriculture is to prevent soil disturbance at all costs. Tilling is 'like a tornado coming through an ecosystem,'* destroying the connections between 'the fungi, the nematodes, the earthworms, all of that subterranean ecosystem. The more you can keep that intact, the better water-holding capacity you have.' Healthy soil is a sponge of criss-crossing roots, wormholes and mycorrhizal fungi, all of which retains moisture, maintains nutrients and captures carbon. 'When you go into a forest or the prairie, you'll see a minimal amount of ground disturbance . . . no one's going through the forest with a giant tractor turning over two feet of soil. What you find is intact ecosystems and long roots.' You also find healthy, recharged aquifers beneath. Without that soft, spongy root structure above, topsoil becomes crumbly, dead earth, leading to soil erosion and water run-off. All this makes sense in theory, but how to show it in practice? By getting farmers to bury cotton underpants into their soils to see which ones degrade the quickest. 'So, in 2017, we did a silly little video, put it on YouTube and Facebook and whatnot to encourage other farmers to do it as well.' The video took off overnight, #SoilMyUndies went viral, and before he knew it Evan was getting calls from international media. 'There were farmers in England doing it – I was invited on the BBC!' he exclaims, still surprised at his fifteen minutes of fame.

No-till, regenerative agriculture is now becoming part of 'the mainstream conversation', argues Wiig. During the 2019 Democratic

---

* We call tilling 'ploughing' in the UK. But I'm largely going to stick with the US term 'tilling', because 'no-till agriculture' has become an internationally recognized term.

primaries, Beto O'Rourke name-dropped 'cover cropping', 'no-till' and 'regenerative agriculture' in a live TV debate – Wiig says no one was talking about this stuff even four years ago. But the semantics is less important than the end goal. Regenerative farming can also reduce inputs, both fertilizer and irrigation, which reduces farmers' overheads. When Wiig talks to Republican types, or the likes of the San Joaquin Farm Bureau and conventional almond growers, he avoids talking about climate change or regenerative agriculture and instead talks about it in terms of money-saving.

Back on James Alexander's farm in England, the challenge is slightly different. Farming in the UK is almost entirely rain-fed, but the rains are becoming less and less reliable. James's family has been farming here for generations. 'There's no seasons any more,' he complains. 'For the last three years, we've just had wet and dry. It does get a little colder in the winter, but not like it used to . . . the last two winters have been two of the wettest on record, but actually that rain's only fallen in about six weeks.' It's May when I visit, and he asks, rhetorically, 'Remember April showers? We only had two millimetres of rain last month.' That's why he now prefers no-till farming. He plunges a spade into the conventional field for me, describing the crumbly red-brown block that emerges as 'sad soil', just a growing medium to hold the plants upright; many litres of pesticide and fertilizer will need to be sprayed to grow anything in it. The topsoil compacts under his machinery, forming a hard cap layer, and causing the roads to turn into muddy streams with each heavy rainfall.

As we drive to the organic farm, barely two miles away, he gestures through the truck window: 'Everything you can see now is organic. Organic spring wheat, organic oats, even organic sheep.' At the organic field, we see the same small rolling hills of green, knee-high winter wheat, but this crop is shorter, and the weeds noticeably taller – wild flowers (technically still weeds, from a farming point of view), including poppy, cornflower and dandelions. It's a pretty field, I can't help admiring it, but the crop itself is noticeably patchier. It's competing for nutrients. 'I can't put any weedkiller on it,' complains James. The organic field is ploughed to a depth of 15–20 cm every year, in part *because* you can't apply weedkiller – the weeds must be suppressed somehow. He may not be using chemicals, but the additional tilling means he is using four times as much diesel here than on the other fields. 'The carbon footprint is horrific,' he says. As we walk towards the centre of the field, our boots sink straight through blancmange-like soil. 'It's not bad soil,' he says, 'but you can't hold as much water; you haven't got the air pockets created by the root structure. The rain will run off easier.' With a changing water cycle of increasing dry periods followed by heavier showers, that is exactly what you don't want. He turns over the organic soil in his hands, showing me a clump of dark soil standing out against the red. 'That's compost that we applied last year. On the organic or conventional fields, we apply it, it gets ploughed in, and you'll still find it in this black form for about three or four years, because there isn't the biology to break it down. Whereas, if you apply that on the no-till field, the worms and everything will just eat that up.'

On the no-till field, where boots never get muddy, he has the opposite of compaction. You could, he says, drive a heavy tractor over it and the crop would spring back up as if nothing had happened. Its strength and structure has been created by many years of undisturbed root systems and microfungi. It's not just a trick of the eye, but, with exactly the same crop, the no-till field is noticeably greener and lusher than both the conventional and the organic fields we've just seen. 'This is *my* field,' he says proudly. 'Everyone comes to see this one! Whatever we plant in it grows well. It hasn't been ploughed for probably twenty to twenty-five years.' There's very little bare soil – down at the base of the crop are short grasses, mosses and clovers. I feel bad for making James dig into his twenty-year untouched field to bury some pants, but we press on. It's harder to dig too, compared to the brown blancmange of the tilled fields. He takes a lump of soil in his hands again, this one full of roots, plus a white, mould-like substance – 'The mycelium,' he says,

of the fungal root system that plays such an important part in natural soil structures. 'This is pure biology at work.'

We think of fungi as simply mushrooms and toadstools, but
those are only the temporary fruit; there's a whole tree of mycelium
threads beneath the ground. This mycelial network spreads throughout untilled soil in search of food and water. Merlin Sheldrake's
book *Entangled Life* explains how this mixes intimately with the
roots of plants to form mycorrhiza: a symbiotic association between
plant and fungus, exchanging nutrients, sugars and water. A plant
makes organic molecules and sugars via photosynthesis and feeds
them to the fungus; the fungus, in turn, passes water, along with
nutrients such as phosphorus, up the plant. Mycelial tubes are also
far thinner than plant roots and are able to access soil moisture that
plants can't tap into – an extremely handy partnership in water-
scarce times or places. Some plants remain entirely dependent on
mycorrhizae. The common bluebell (*Hyacinthoides non-scripta*),
for example, which paints Britain's woodlands with colour each
spring, relies on its mycorrhizal partners for nutrients and would
soon die without them.[3] Arable crops may not be as dependent, but
they can tap into the same benefits if we mimic natural systems and
allow the mycorrhizae time to form. Underground fungi have been
found to supply 80 per cent of a crop's nitrogen requirements and
up to 100 per cent of its phosphorus requirements, and to provide
water to crops in times of drought. This is why forest floors are
covered in life. And why food crops grown in no-till, uncultivated
farming systems keep soil moisture and nutrients for longer.

Steven Lear, a Buckinghamshire farmer with 1,200 ha of beef and
arable land, tells me they changed the arable 'overnight', from a full
ploughing regime to no-till, some seven years ago, using cover crops
to improve the soil. Given that around 25 per cent of England's total
land area is arable farmland, he believes that no-till could hold the
key to England's sustainable water future. He conducts tests to see
how fast rainwater infiltrates rather than just sitting on the surface,
'and our infiltration rates have gone up massively since we've been
doing no-till. All those wormholes are still in place, all the roots that

have put down over the years, the microbes that aggregate the soil – the fungi produce a substance called glomalin that'll bind soil particles together in little balls. When you get soil like that, you've got much higher infiltration.'

A cover crop must be killed before you plant into it, though, or the two crops will compete for nutrients. Ideally, this can be done using a 'crimper' roller to crush the crop, and extreme seasonal changes – heat or frost – will finish the job. But, in more temperate, soggy climates like England, 'things tend to survive,' says Steven, so he kills off his cover crops with glyphosate – also known as Roundup – before planting on top of them. He doesn't see the future being either using chemicals or not, but rather using fewer chemicals to get the same productivity. 'In organic systems, you have to plough, disturbing the soil health and putting carbon back into the atmosphere. Probably a mix of the two is where we need to be.'* Better still, however, is having animals do the work for you: a herd of sheep can graze off the cover crop, adding extra nitrogen via manure in the process. 'I mean, no natural system is without animals, is it?' It's such an obvious statement, but worth rereading, so divorced are modern farming methods from the natural system. 'Those plants that are growing are meant to be grazed. That's the end game, to graze off all the cover crops, and then they're recycling all that plant matter back in.'

Water companies are now cottoning on to the benefits of no-till and cover crops too. Research from Iowa State University shows that hourly water-infiltration rates of 66 mm in a conventional farm are more than doubled to 144 mm under no-till farming.[4] This, to

---

* In December 2017, French president Emmanuel Macron vowed to phase out the use of glyphosate within three years. However, recognizing the benefits of no-till farming, the new French Agency for Food, Environmental and Occupational Health and Safety (ANSES) rules in 2020 stated, 'glyphosate will be banned for use on arable crops . . . when the land has been ploughed between crops.' This not only created a loophole for no-till, but arguably incentivized the move towards no-till.

use the banking analogy again, is a huge down payment for water companies. Unsurprisingly, then, they are willing to pay (actual money) for it. On England's south coast, Southern Water are now paying farmers to leave crops on their fields over winter, rather than bare, tilled earth – effectively financially incentivizing no-till farming. This is not only for the groundwater-recharge benefits, but also to reduce nitrates from conventional farming leaching into the groundwater and rivers. Robin Kelly at Southern Water tells me that, 'nitrate concentrations in many of our groundwater sources are high and rising, and it is this trend we are focusing on and trying to reverse. The results clearly show a benefit of having continuous green cover over the winter.' In the first year of the scheme, farmers near Brighton were offered £35 per hectare of over-winter cover crops. The simple calculation is that it's more expensive for water companies to treat the water than it is to pay the farmers not to pollute it in the first place.

Since that first trial in 2018, Southern Water's subsidies have become more targeted, and more generous. The water company has identified fields that fall within 'priority catchments' for groundwater recharge, and a spring-sown cover crop here can earn up to £280 per hectare, while a new herbal grass ley (a rotation crop made up of legume, herb and grass) can earn £400 per hectare – similar to some staple food crops. 'Effectively, the priority rating reflects how quickly we are likely to see a reduction in nitrate at the point of abstraction,' Kelly tells me. 'Southern Water is starting to favour catchment management as a more sustainable option which can work alongside, or perhaps even replace, engineered solutions.' By which he means dams and pipelines. It's astonishing – and enlightening – that private water companies are now making the commercial decision to pay farmers. Kelly's conclusion, after three years of working with farmers, is that 'no-till or minimum-tillage systems can be beneficial in terms of soil conservation and water-quality outcomes – reducing soil erosion, run-off and leaching.' But there remains a question over the use of weedkiller. There is, says Kelly, 'ongoing research looking at the feasibility of organic no-till systems via the use of living mulches and mechanical weed control.'

Severn Trent Water in the Midlands and North Wales also offer grants for on-farm water-quality improvements, and in the five years to 2021 'distributed over 1,500 grants, worth more than £5 million, negating the need for £74 million of investment in our treatment processes.'[5] They've since launched a Farming for Water programme for priority recharge and river pollution areas that covers forty-four catchments and 432,000 hectares, working with two thirds of all farmers in their region. According to their sustainability plan for 2021, by doing this, 'we can reduce farming's contribution to phosphates in watercourses by up to 66%.'[6]

Collectively, farmers' fields span a far larger land area than any water company could ever hope to own or manage. Jake Rigg of Affinity Water, in the east of England, tells me they've also been working with an agricultural research institute to establish how much more water can go to the aquifer using no-till techniques. 'And they said, "You're talking about having an Abingdon Reservoir-sized amount of extra water in the aquifer."' I let out an involuntarily squeal. Such a volume would basically solve England's water-scarcity problem. You could remove all the problems mentioned in Chapter 6 through regenerative farming alone. The work of RAPID, the decades of grey infrastructure building that lie ahead . . . all could be solved overnight (or, at least, over winter) by farming techniques? 'By increasing the effectiveness of the rainfall, changing the way we manage land can be a really, really significant form of environmental management,' Rigg responds. 'We do a lot of work talking to Defra and Ofwat, saying, "Can you change the agricultural incentives you give to farmers [to incentivize no-till]?"'

Enlarging this out to a global scale, Sandra Postel, the 2021 Stockholm Water Prize Laureate, has said, 'Soils hold eight times as much water as all the world's rivers combined, but we rarely manage that soil reservoir as a water reservoir. So, if we can improve soil health, we're getting multiple benefits, better yields, less chemical fertilizer, more carbon in the soil and the ability to hold more water. One simple solution is to incentivize the planting of cover crops sooner during the off season . . . In the United States, maybe 6 per cent of all of our farmland gets cover crops. A state like Maryland, I believe, has more like 29 per cent, because it has incentives, working [with farmers to] encourage those kinds of solutions.'[7] The Stockholm Water Prize is as big as it gets in the water world – so Postel isn't espousing some niche, hippy idea, here. This is mainstream thinking, now: a water-secure future requires healthy soil and regenerative farming.

*

Ohio-based farmer John Kempf is a leading proponent of regenerative agriculture. A member of the Amish community, he was always destined for an agricultural life and is now a recognized expert in agronomy and soil science. Contrary to common misconception, he tells me, farmers in more liberal Amish groups do use pesticides and chemicals, and permit the use of technology for work. Kempf first became aware of regenerative farming when, back in 2002–4, no matter how much pesticide, fungicide and insecticide they applied on the family farm, crop failures were getting worse, to the point of losing 70 per cent of their main crop. His lightbulb moment came when they took over a neighbour's field that had only ever held livestock, with no previous chemical spraying. They planted both fields with cantaloupe; one crop came up with 80 per cent mildew infection on the leaves, the other with zero infection. 'There was a very pronounced, clear-cut line, right down the centre of the field, where the former field border had been,' he remembers. The irony being that it was the field expensively sprayed with the chemicals supposed to prevent such diseases that was now covered in mildew. After reading avidly and talking with agronomists, he concluded that the field fared better not only because it hadn't been sprayed, but also because it hadn't been ploughed – the soil had remained undisturbed for so long that it was full of bacteria, micro-fungi and retained moisture. 'I now think no-till is very necessary for the majority of agricultural acres,' he tells me. 'Soils have been mismanaged through our agricultural practices over the last seven decades. With the use of heavier and heavier machinery and larger tractors, soils now may have severe compaction down to a depth of ten, fifteen, twenty inches or further.' It's this compaction that is often so damaging for water tables, creating the hard 'cap' that prohibits water from percolating down into the groundwater and causing it instead to run off into streams or drainage canals, leaching nutrients along the way.

A guest on John's *Regenerative Agriculture* podcast, Dr Jerry Hatfield, from the USDA-ARS National Laboratory for Agriculture, commented, 'I always tell producers that it's not how much rain they get in a rain gauge – it's how much rain they get into the soil. If we're protecting that soil surface, we can infiltrate a great deal of water . . . That upper inch of soil is the gateway.' Hatfield gave the example of 'five-foot-tall corn in the middle of August in the Midwest.* That plant only has about eight days of unavailable water before it begins to be stressed. That's not very much.' Whereas, in no-till systems, the same corn plant 'can go for thirteen days without water. You've got five more days of available water for that plant to perform to its optimum, without stress.' That's potentially a huge difference in yield, and in the need for irrigation.

Kempf is convinced that the yield and lower-input benefits of regenerative agriculture will see it become mainstream. Farmers operate on tight margins, and this is a way to improve those margins. However, he also gives me a fascinating insight into why the decision to change methods is not always within the hands of farmers – it's also a parable for why *any* environmentally progressive change is so hard. 'Many farmers need to borrow money and have operating loans to plant a crop every year,' Kempf tells me. 'In order to qualify for their operating loan, the bank dictates that they must get crop insurance. And the insurance company dictates that, for them to get crop insurance, they have to follow the

---

* In the US, 'corn' refers to maize or sweetcorn; unlike in the UK, where 'corn' means wheat.

recommendations of the crop scouts and certified pest-control advisers, who tell them exactly which chemicals they have to spray. There's an unholy alliance, if you will – they are locked into this system.'

The California Sustainable Groundwater Management Act (SGMA, pronounced 'Sigma'), signed into law by former governor Jerry Brown in September 2014, was an attempt to holistically change the groundwater management system of the Golden State. Groundwater was protected under general California water law before, says Dr Helen Dahlke, at the University of California, Davis, 'but it wasn't very well specified. It only said that every overlying landowner has an equal share to the safe yield of the aquifer.' Without measurement or monitoring, what constitutes an 'equal share' is debatable.

SGMA was born thanks to 'a combination of political will and environmental stress,' says Newsha Ajami, who worked closely with the team helping it come to pass. 'The drought was real and everybody was suffering, but the governor was also willing to use his political capital to make this happen – and that was key.' SGMA requires that 127 high- and medium-priority groundwater basins in California (accounting for approximately 96 per cent of the state's groundwater use, and 88 per cent of the population) develop groundwater-sustainability plans (GSPs) that achieve sustainability within twenty years of implementation. The GSPs are then presided over by groundwater-sustainability agencies (GSAs). SGMA doesn't change landowners' water rights, nor does it define 'sustainable'; it only defines what *isn't* sustainable (via a list of six 'undesirable results'). In the first instance, says Ajami, it was meant to establish how many wells are out there, and which of them are over-drafted and in critical condition, so that plans can be made to make them sustainable. Prior to SGMA, there wasn't even any data on how many wells were in California, let alone how much water they were using. But there still isn't, as I write in 2022. Residences are required to have a water meter by 2025. SGMA also gave local groundwater agencies in critically over-drafted basins until 2040 to achieve sustainability. The CalMatters website reported in August 2021 (updated in January 2022), 'seven years later, little has changed for Californians relying on drinking-water wells: Depletion of their groundwater continues. Pumping is largely unrestricted, and there are few, if any, protections in place . . . Despite the law, about 2,700 wells across the state are projected to go dry this year, and if the drought continues, 1,000 more next year.'[8] The region-wide over-draft for 2003–17 was 2,969 GL per year – an increase, not a reduction, on the thirty-year average of 2,220 GL. Farmers with water rights may essentially still pump away, providing they demon-strate it's for 'reasonable (not wasteful) and beneficial uses'. Growing any food is likely seen as beneficial. While, adds Bloomberg Law, 'A pumping allocation or pumping assessment that does not equitably allocate the burden of groundwater management consistent with water rights is likely to draw a lawsuit . . . [and] invalidate the plan itself.'[9]

Jay Famiglietti was on the California State Water Board when SGMA was going through.[10] SGMA ensured that each of the 150-plus GSAs must have 'a sustainability plan approved by the State Water Board,' says Jay, so they do have to at least be credible. But, he says, 'There's only vague criteria, more or less "do no harm" – don't deplete the aquifers too much, don't impact the streams too much.' For him, it comes down to what's meant by 'sustainable', and who decides. 'That's the whole thing. The future of California depends on that.' The classic definition of sustainable, he says, is that 'you don't use more than is being replenished, right? But that's not happening in California.' I ask him what the end point of this is. 'Ultimately, the tennis ball reaches the bottom of the stairs. That's it.' So, no crops, no wine, no water in the taps? 'That's what it means. Maybe we'll figure out some master strategy. Maybe we'll just become super-efficient. Maybe we'll move some agriculture away. But the bottom line is it's basically just math. Right? There's a certain amount of water there, we're using way more than is being replenished every year, so you're gonna run out.'

In California's semi-arid climate, groundwater aquifers only rely on rainfall to recharge them during winter. A changing water cycle of more frequent droughts and floods means that nature needs a helping hand. The demand on groundwater resources *may* eventually stabilize, thanks to SGMA, but even if demand is reduced – and, as we've seen, it's a big if – there will be less reliable rain to recharge the aquifers. This may seem a bold statement, but it's built on the foundations of the crushed aquifers beneath Joe Poland's pole: California's water system will not survive without artificially recharging its aquifers. This, in the water business, is known as managed aquifer recharge (MAR). And there are ways of using farmland to recharge aquifers with more than just rainwater.

Dr Helen Dahlke's work is exciting because it combines a natural excess of water (winter floods) with agricultural demand (by far the highest water user in the state), along with perhaps the most finely detailed scientific mapping of underground aquifers in the world. As we've seen, California's water situation is in dire straits – but there remains an anomaly: it has an excess of water in the winter. As Dahlke writes in one of her papers, 'Demand for this water during the winter from the agricultural sector is relatively low', because – and sorry if this is obvious – that's not when the crops are growing. 'Reservoirs often make flood control releases of stored water in anticipation of large storm events . . . [which are] expected to increase in frequency and magnitude in the coming decades as a result of climate warming.' In short, she says, 'there seems to exist an abundance of surface water during the winter and wet years, resulting in flood risk for much of the Central Valley which could potentially be . . . captured.' When's the last time you read the words 'California' together with 'an abundance of surface water'? It certainly wasn't in this book.

'We've been flooding a vineyard as a form of groundwater recharge,' Dahlke explains. To accomplish SGMA's requirement for groundwater basins to be brought into balance within twenty years, she says, there are really only two options. 'Number one, grow less crops; or number two, find more water.' Helen's proposal finds more water *and* allows for growing crops. 'Finding more water includes these so-called "high-magnitude flows": heavy rain, in a short amount of time, when you see floods occurring, or river flows rising quickly.' Every other inch of surface water in the state is allocated (or over-allocated) due to the legal intricacies of the water-rights system – but not flood water. That is the only pool of H_2O that no one can lay a paper claim to.

Dahlke's recharge project therefore actively diverts water from the winter-swollen Cosumnes River onto farmland: 'we pump it out of the river in flood, and flood the vineyard,' she explains simply. When they first tried this, pumping at 7,274 L a minute, the ground just drank it up. It never pooled for very long, plummeting hungrily down to the aquifer below. Helen's research found that up to 90 per cent of the water recharges down into the aquifer, the other 10 per cent being 'lost' to 'evapotranspiration', otherwise known as 'feeding the crop'. That's 90 per cent being stored in the aquifer, as opposed to 100 per cent going to the San Francisco Bay (and polluting it with excess nutrients, causing algal blooms). Helen prefers flood irrigation over drip irrigation in agriculture because, for aquifer recharge, the infrastructure is already in place to flood the fields.

'The idea now is to do this all over California,' says Dahlke. The total overdraft in the Central Valley is over 197 km^3, which is, Helen tells me, 'about 1.3 times Lake Tahoe.' Helen puts it in terms that are hard to misunderstand: 'You could support several cities with that, for decades.' Decades. To start rebalancing the books, as SGMA requires, will take decades of paying back – but Dahlke's system provides a means of doing that. The region's water bureaucracy still stands in the way, requiring temporary surface-water rights from the water boards and irrigation districts, and staying in line with the California Environmental Quality Act and Department of Fish and Wildlife. But, after her pilots, Helen hopes this will now be a smoother process for others. The more farmers that do this, the less ground subsidence, the less crushed, lost aquifer capacity, and, crucially, the more water to draw upon in the dry season. It's a win-win. A 2020 paper estimated that flood water applied this way throughout California's Central Valley could recover around 22 per cent of the groundwater overdraft, while boosting the flow of 52–73 per cent of streams.[11] Similarly, when surplus water in the Ganges was diverted to canals in Uttar Pradesh, India, during the wet season, to flood crop fields for aquifer recharge (at 234 m^3 a second), it led to the reversal of declining water tables – average depth to groundwater went from 12 m to 6.5 m in ten years, reducing pumping costs, and increasing agricultural productivity.[12]

Currently, when reservoirs in California release water to make space for large storm events, this isn't done in communication with aquifer-recharge sites. Dahlke hopes that, in future, it will be: 'The law needs to change,' she says. Decisions are made all the time to release storm water during winter, and they are not always good ones. She remembers an occasion on Russian River, California, during one big storm in December: 'They drew down the reservoir, evacuated all this water out to the ocean, and then they had nothing left for the summer. It was the only storm they got that year . . . [But] they had to do that by law. They evacuated, like, 70,000 af [86.3M m^3] of water that they could have used.' With climate change, many farmers are calling for more reservoirs to be built, she says. But new reservoirs cannot compete on space or cost with the natural underground reservoirs beneath our feet. 'We have four times more room [in aquifers] than all of our surface-water reservoirs combined,' says Dahlke. On top of that, the other great reservoir – the snowpack – 'is going away,' she says. Scientists now predict that Colorado could lose somewhere around 50-60 per cent of its snowpack – the source of the Colorado River – by the year 2080.[13]

To work properly, on-farm managed aquifer recharge requires some state planning and – almost sacrilegious to say, in the US – some impinging of private water rights: 'You need to protect the locations where recharge is most feasible, where you have the fastest subsurface connection to the aquifer. You want to make sure you're not growing a fertilizer-hungry crop in that area.' To achieve that, there must be a subsidy programme for recharge.* To pave the way for this, Dahlke and her UC colleagues have created the STARR online index and interactive state map, which identifies 1.5 million hectares of California farmland with the best potential for aquifer recharge, based on soil type, land use, topography and other factors. There are nearly 3,000 locations where flooding-suitable agricultural land would recharge water for communities, covering a total of 372 km^2 – the majority of which are vineyards, which Dahlke has already shown to work well, with no stress to the crop.[14]

Groundwater recharge is essential to the long-term water security of California, and many other dry regions the world over. The barrier is not the science, but attitudes. An aquifer is not a clean, concrete-lined, engineered system. It's not even visible. That makes it hard for engineers to calculate and for politicians to champion. It doesn't keep the water 'in really well-defined boundaries,' admits Dahlke. 'Groundwater is constantly moving; it's not static, it's not sitting in the same place where you recharge it, it's going to move with gradient. If your neighbour starts pumping, the water will move in that direction. It also gives some back to rivers too.' That, she quickly adds, is a good thing. 'Some rivers have been disconnected from the aquifers for fifty years, now.' Any negativity towards aquifer recharge is 'really stubbornness,' she says. For some farmers, 'it's conservative thinking that this is how the system works: there is a reservoir upstream, they have a surface-water right, with a number on their licence saying they are entitled to X amount of water.' Groundwater isn't that precise – but its storage is potentially far greater.

Across the pond, Niels Hartog is Mr Managed Aquifer Recharge.†
He's principal scientist and geohydrologist at KWR Water Research

---

*   Surprisingly, this is a tick in the box for alfalfa, that much-maligned (by me, in Chapter 7) fodder crop. It doesn't require any nitrogen fertilizer to grow. Vineyards are good too, whereas almonds, tomato and corn are all on Helen's blacklist for requiring too much fertilizer.
†   My words, not his.

Institute, Nieuwegein, Utrect, whose glass offices sit, pleasingly, as if floating on an artificial wetland. Niels has seventy-one publications on MAR to his name and counting, and he sends me two of them as background reading before we talk. While studying geology at college, he read *Cadillac Desert*, Marc Reisner's classic history of water mismanagement in the American West, and 'it was a real eye-opener . . . it really made a big impact on my perspective on water scarcity and how to deal with it,' says Niels. I wonder if, being Dutch, he has a natural affinity with water management, given the country's history of hydrology? 'You know, the reputation of the Netherlands is mainly related to fighting against the water to keep it out!' he laughs. 'Now, we need to keep it back in!'

Managed aquifer recharge describes any intentional human intervention to recharge and store water in an aquifer. There's a whole bunch of methods to do this, from Dahlke's flooding of fields, to filtering through sand dunes, to reversing borehole wells to inject water back underground. Some countries grasped this technology with both hands when it first emerged at the turn of the twentieth century, while others still barely do it at all. An inventory of 224 MAR sites in Europe by Hartog and colleagues in 2013 found that most are in Germany (64), followed by the Netherlands (41) and France (21). These are not water-stressed countries, and their use of MAR in part explains why. Nearby water-stressed Belgium and England, for example, have just one major MAR site each.[15] Although, Niels tells me, the first reported MAR site in Europe was in Glasgow, UK, in 1810. The Netherlands copied this Scottish technique in the late nineteenth century, in response to cholera and typhoid epidemics. The canal-based city of Amsterdam 'was having a lot of trouble' with waterborne disease, until they discovered that the water beneath the nearby sand dunes was relatively clean. They began pumping water from the dunes, demand increased and soon there was no more water beneath the dunes. So, says Hartog, the next step was to artificially recharge the dunes with water diverted from the Rhine River – which became one of the earliest known examples of managed aquifer recharge to supply a city with drinking water. Today, Amsterdam receives about 60 per cent of its drinking water from such artificial recharge, including the sand dunes and over forty recharge ponds.[16]

'Rather than building a facility to store the water, the facility is already there underground – that's the simple story,' explains Niels. 'And you don't have the evaporation losses.' Evaporation losses from reservoirs shouldn't be underestimated – Lake Mead loses more than the city of Las Vegas uses each year. When water is stored underground, away from the sun, this isn't a problem. Because it now rains and snows less reliably and predictably, aquifers need a helping hand. Unlike reservoirs and rivers, which deplete during droughts, subsurface water can, if replenished, always be called upon. Nitrates can also be removed naturally, Niels says. 'The subsurface provides the space, time and reactivity.' He adds that a polluted subsurface can cause water-quality deterioration, though – so even more reason to manage the process and choose the right recharge area.

While the Netherlands may have the most MAR sites per capita, arguably no single city has embraced MAR as much as Berlin, Germany. There, at least eight active MAR sites contribute 67 per cent of the capital's total water supply.[17] Berlin largely uses infiltration ponds filled with water from the River Havel and Lake Tegel, topped up with treated wastewater – this filters into the groundwater below, and is then sucked back up through nearby supply boreholes. Niels doesn't believe there's anything unusual about Berlin's geology that allows it to do this; any city with aquifers, rivers and wastewater-treatment plants could do the same. Ironically, the largest MAR system in the world, in Abu Dhabi, is used to store desalinated water. This is more for national-security purposes, Niels tells me – it's harder to bomb, and provides a water source should there be an above-ground infrastructure breach. It sounds like an historical anomaly, an era when oil-rich countries could do crazy things like desalinate vast amounts of water and store it underground. Niels tells me he's been contacted by other countries in the region interested in doing the same.

The largest MAR site in the UK, Thames Water's North London Artificial Recharge Scheme (NLARS), also seems to be missing the environmental point of MAR. Rather than storm water, it pumps treated tap water back into the chalk and basal sand aquifers. It is technically MAR, but without any of the natural filtration benefits or storing anything that meets the definition of excess water. There are two sides to water scarcity, says Niels: quantity and quality. Infiltrating water back physically can help maintain water levels, but the strict water-quality legislation in some countries limits MAR deployment. Thames Water's groundwater manager Mike Jones agrees. His site was developed to take advantage of the storage capacity resulting from historical over-abstraction. Of the region's forty-eight boreholes that used to pump water out, around thirty are now equipped to inject water back in, he says. But it's almost alone in doing so because UK water-quality legislation requires that only fully treated tap water can be pumped back in. 'We store it there and pump it out during drought and dry-weather events – it's a strategic store, it's not used all the time.' They can't even reuse it directly – it gets mixed back in with the raw water reservoirs in the Lea Valley, to be treated all over again.

The irony – or hypocrisy – of this is that the UK sends wastewater effluent and even raw sewage out into its rivers every day. Why does regulation only care about groundwater? Obviously, polluted water should not be pumped into aquifers – but at least treated effluent, via sand or pond filtration, could be filtered and stored? 'There is a lot of unintended aquifer recharge,' says Niels. 'If you look at the chemical composition of the rivers, you see a major signature of treated wastewater. And, in some smaller streams, it can be up to 100 per cent wastewater.' As I saw in the chalk streams. 'Downstream riverbanks are likely doing [some aquifer recharge] without you knowing. There's also leaky pipes in the distribution network leaking into the subsurface. It's not something you should want. But it is happening.' Far better, then, to manage the process and top-up the stressed aquifers that need it the most.

To solve the world's water crisis, everywhere that can do MAR, should do MAR. According to the UN World Water Development Report 2022, MAR application has increased by a factor of ten over the last sixty years, 'but there is still ample scope for further expansion, from the current 10 km³/year to probably around 100 km³/year.' What's holding it back, Niels believes, is its public image, or lack thereof: 'For a long time, MAR has been the favourite option of hydrogeologists – the case from a hydrogeological point of view, from an environmental point of view, is clear.' But politicians like to have their photographs taken next to shiny new dams or desalination plants. The subsurface has 'a major communication challenge' by comparison, says Niels. There's also uncertainty, in terms of storage 'seeping out' into streams and rivers. 'But that is a good thing!' he laughs. 'I think that should be part of the positive communication around MAR: that we can "re-naturalize" – if that's even an English word?! – the water system.'

Taken to its most extreme, however, mega-MAR sites are the opposite of re-naturalized. In Phoenix, DeEtte Person, Central Arizona Project's communications strategist, takes me in a CAP-branded pickup truck to visit the Agua Fria Recharge facility. It looks like the dry bed of a reservoir: 15 ha in total, divided into eight basins, only one or two metres deep, connected to an inlet channel. Rather than connecting directly to the CAP canal, the inlet is a 6.4-km section of the Agua Fria River itself. If there were a river. The dry riverbed flows only artificially, when water is released from the CAP. This is known as 'water banking.'

The facility opened in 2002, and can recharge 37,000ML (30,000 af) of groundwater a year. The largest of the six recharge sites in the CAP system, Superstition Mountains Recharge, can store 69,700ML (56,500 af). CAP recharge engineer Justin Conley, who meets us on-site, tells me that a typical cycle will see the recharge site flooded with CAP water for five days, once a fortnight. One week on, one week off. Roughly 10 per cent of CAP's total 1.85 km³ (1.5 M af) of water is diverted into groundwater recharge sites such as this. Phil Pagels, CAP recharge manager, explains: 'Customers have rights to that water, and we'll schedule it to store at these facilities,' much like an actual bank. 'They then get groundwater credits for that. There are several active management areas within Arizona, and these allow for a customer to store water and either [spend] that credit in the same year or to store it to pump at a future date. So, year after year, you could be gathering credits, and then, at a future date, you can pump that groundwater out.' The single largest customer at this facility is the City of Peoria, which uses it for household water supply. CAP knows precisely how much water they release at the turn-off and inlets, and how much – minus a certain calculated amount for evaporation loss – is then stored underground.

In Arizona, the CAP is not just a canal, but an organization managing an annual budget of $321 million and nearly 500 staff, serving water to more than fifty cities, 141,639 ha of irrigated agriculture and eleven Native American nations.[18] The CAP's six underground recharge sites, collectively storing 370M m³ (300,000 af – the same, you'll remember, as Nevada's entire allocation) are central to its ability to service those needs. CAP-style groundwater recharge came about thanks to Arizona's 1980 Groundwater Management Act, which set a goal for the Phoenix, Tucson and Prescott regions to put as much water back into aquifers as was being withdrawn, by 2025, and required developers to prove they had a hundred-year assured water supply. When that requirement predictably proved difficult to achieve, the 1993 Central Arizona Groundwater Replenishment District allowed CAP to 'bank' it for them: developers could pump groundwater wherever they were, as long as the same amount was being 'replenished' at recharge sites like this one in Agua Fria. No matter if there wasn't any water beneath the development sites, or that replenishment came from the Colorado River, in a different state.[19]

The scale of CAP makes Arizona's recharge sites some of the largest in the world. I asked to visit the largest, Tonopah Desert Recharge, but CAP weren't keen. I later discovered Tonopah is mothballed and hasn't contained any water for years. It is a sign of what's to come.

At the time of my visit, CAP's recharge facilities are already below half capacity, only collectively storing 154M m³ (125,000 af) a year. When the seven Colorado River Basin states signed the Drought Contingency Plan, with cuts agreed according to 'tiers' triggered by elevations on Lake Mead, Tier Zero, triggered in January 2020, meant a cut in Arizona's annual CAP allowance of 236M m³ (192,000 af – see Table 1, p. 64). As aquifer recharge water is classified as 'surplus water', this was the first to be cut. On 6 August 2021, Tier 1 was reached, triggering a far deeper cut of 631.5M m³ (512,000 af) for Arizona, almost 20 per cent of its total Lake Mead allowance. Phil, Justin and their team now fear the worst for these sites. 'There's a large cut that's occurring here in Phoenix,' says Phil, staring out at the dry recharge beds. Now Arizona's recharge facilities will have to survive just by doing what aquifer recharge was originally designed for: storing excess storm water, the one or two times a year when it rains enough in Arizona to do so. Developers soon won't be getting their credits. And yet there was no sign of the housebuilding boom dying down when I was there – quite the opposite. Making hay before the sun sets for good.

What I see next, then, I may be one of the last visitors to see. We hit the road in two CAP SUVs to see the 'blow-off valve' – the site where the water is released from the CAP canal into the Agua Fria dry riverbed, which eventually runs into the recharge beds. As we bounce over rocks down a desert dirt track, tall saguaro cactuses litter the landscape like lost statues. Three apparently wild donkeys stare intently at us. We stop at a concrete structure cordoned off with barbed wire, with an empty rectangular pool made of thick concrete, containing a dozen chunky concrete blocks at one end, reminiscent of beach flood defences. Phil calls through and tells someone, 'Turn it on now.' An alarm blares for ten seconds, then a loud, underground rumble begins. 'So, we just turned it on for you,' says Phil, 'and we'll run it for about an hour at 150 cfs.' That's cubic feet per second. 'You're turning it on just for me?' I ask, surprised. 'Yeah,' shoots back Phil, with a grin.

The water begins to gush out violently, the concrete blocks now revealing their purpose: to take the impact. That 150 cfs is a lot of water. In an hour, it will amount to around 15.4M L (12.5 af). I'd got used to the term 'acre-feet' without really visualizing it; a dozen of them shooting out of a tunnel in an hour, battering, rocking and roiling into concrete, was one of the most awesome, in the truest sense of the word, water-engineering sights I'd seen. Each acre-foot, taking 4.8 minutes to surge out in front of me, would be equivalent to over 18,000 of my bathroom showers. I'm watching all the showers of my lifetime rush past my eyes in under five minutes. As the water bounces back off the concrete blocks (Phil calls them 'energy dissipaters'), it spills over a weir into the bed of the once-natural, and now lifeless, Agua Fria River. I then get to watch a dead river ghoulishly come back to life. It is, for the next few minutes, a full and flowing jerky automaton of a river. Soon, thanks to shortages and crippling cuts, the CAP will be for priority users only and this feat of engineering may no longer breathe life into the river and aquifer.

Many other water districts in the American West do groundwater recharge, too. 'We have several groundwater banks in the Central Valley – for example, Leakey Acres and Fresno,' Helen Dahlke tells me. 'Many districts have recharge basins or designated basins where they always put surplus water in the winter. But they are not enough to close the gap.' To build more, she says, 'you're looking at the trade-off of buying more land, which is expensive in the Central Valley, taking it out of production to put in an infiltration basin – you're looking at $30,000 per acre to buy the land, a good infiltration basin is 100–150 acres [40–60 ha] in size . . . That's why, in my opinion, on-farm recharge can better fill the gap.' I'm reminded again of Jake Rigg's 'ballpark' figure that, if all farming were regenerative in eastern England, it would have a recharge benefit equivalent to England's largest reservoir.

Not that groundwater recharge need happen on a grand engineering scale. Om Prakash Sharma, India director at the small Anglo-Indian water charity WaterHarvest, tells me of the *Chauka* system, in which a series of 23-cm-deep square ponds are cut in a row and bordered with 60-cm-high bunds. These slow down monsoon water near smallholder farms, collect surface water and recharge the groundwater beneath. As the first pond fills up, it over-flows into the next, and so on, in sequence, exactly like mini-versions of the Agua Fria recharge site. Om also tells me of groundwater-recharge dams in the Aravali Range – among the oldest mountains on Earth. There, small check dams or 'anicuts' of stone or concrete are laid across the ephemeral streams that run down the steep hill-sides. When the streams swell during monsoon, the water stored behind the small dams percolates down and recharges the ground-water. 'We have some villages where people are now growing three crops in a year thanks to this recharge . . . There were eighty dry wells in four villages; now, not a single well is dry.'

While some cities – such as Phoenix and Berlin – have large MAR facilities, there is a more decentralized approach that, like no-till farming and on-farm recharge, can have an even greater col-lective impact. It's called sustainable urban drainage systems, or SuDS. Imagine a field of grass in the rain: the field will absorb the rain. Imagine that same field covered in tarmac: the rain hits the hard surface and pools, or sheets off to the sides. That, essentially, is what our built environment does, funnelling and channelling rainwater to drains and rivers and out to sea, totally disconnected from our groundwater. SuDS, instead, are designed to make the urban environment more permeable, replacing tarmac with things like swales (grass ditches), tiled roofs with green roofs, and parking spaces with pervious pavements.* The idea is to mimic the way rain-fall drains in natural systems. SuDS also help with the combined sewage-overflow problem. For example, Enfield, North London, began in 2015 to replace concrete traffic-calming measures with green planting, aiming to stop highway run-off from entering the surface sewer system. Just five grassy additions to one road were

---

* There is, believe it or not, a 968-page *SuDS Manual*, produced in the UK by the Construction Industry Research and Information Associ-ation, which lists all the different available SuDS techniques. But, essentially, 'replace hard stuff with green stuff' covers most of them.

found to intercept a total of 1,000 m² of rain run-off surface.[20] Sewers are no longer inundated, streams are not polluted, and the aquifer below Enfield is recharged.

Cover a city with SuDS and it becomes a 'sponge city', an initiative begun in China in response to its increasingly devastating floods and droughts. President Xi Jinping declared that cities should 'act like sponges', and that the government would allocate CNY 20.7 billion ($3 billion) to sixteen pilot cities between 2015 and 2017. The goal was to change the pattern of lurching between natural disasters by instead soaking up heavy rain and releasing it slowly into aquifers, rivers and reservoirs. Using features such as rooftop gardens, wetlands, parks, permeable pavements and underground storage tanks, the ongoing plan is to absorb, reuse or recharge 70 per cent of the rainwater on four fifths of China's urban land.[21] Junyan Liu, of Greenpeace International, Beijing, tells me of some immediate successes. Taihu Lake, one of China's largest freshwater lakes, surrounded by the densely populated areas of Suzhou, Wuxi and Yixing, 'used to have a lot of concrete around the lake and, once there's floods or heavy rain, the ground cannot absorb any of the water,' says Junyan. 'But now they have taken away all of the concrete ground and they built a lot of man-made wetlands around the lake – it actually helps a lot.'

There are now thirty participating 'sponge cities' in China, each tasked to ensure that a whopping 80 per cent of their urban land includes SuDS features by 2030, able to retain 70–90 per cent of all storm water. In Wuhan, for example, 400 new SuDS now cover 38.5 km², including urban gardens, constructed wetlands, parks, rainwater pumps, repaired water channels, and artificial lakes, all designed to capture large volumes of water.[22] It's hoped that the sponge-city initiative will increase the country's water supply, restore depleted aquifers and reduce incidents of flooding, which have more than doubled since 2008.[23] According to the University of Leeds, Wuhan's sponge-city project was more than $600 million cheaper than upgrading the city's centralized drainage system would have been, and more effective. The significant co-benefits included 'improved local air quality, biodiversity and conservation benefits, health and lifestyle benefits, and increased land value'. The new Yangtze River Beach Park, for example, was found to have temperatures three degrees cooler than the city, causing land value in the surrounding area to double.[24]

India is following China's approach closely. In Delhi, Alpana Jain at the Nature Conservancy is calling for the integration of SuDS into city planning. Arguably, Delhi has few options left, with groundwater levels plummeting at a rate of up to 4 m a year.[25] The City of 1,000 Tanks programme, in Chennai, also hopes to shift policy from purely engineering-based solutions towards a sustainable groundwater-management approach, using SuDS and green spaces. The project aims to improve Chennai's water retention – and therefore water supply – by 200M–250M L per day, and thereby avert another Day Zero.[26] Since Chennai receives more water in the form of rain than it consumes each year, it should be possible to fulfil its water needs – if it can hold onto it. Historic stepped temple tanks that used to do just that, in the days of a much smaller population, are now being restored, while a network of bioswales is being added throughout the city. Constructed wetlands will treat captured grey water and recharge the groundwater too. The vision for Chennai is an interconnected green-blue network of water courses, rivers, lakes, ponds, parklands, managed wetlands and traditional temple tanks.[27] It's aiming to become the first wetland megacity, rebuilt on an understanding of how this former swampland functioned, and where its water must come from once again.

Alpana Jain argues that the big difference between India's initiatives and China's sponge-cities approach is that, in terms of development, India today is where China was twenty years ago. 'If China had thought at the time of integrating nature in their city planning, they wouldn't have reached the [perilous water] situation that they have today.' There's much to be learned from sponge cities, says Jain, but the cost would have been substantially lower had it been considered at the time of development. In India, she says, they're trying to build today for what's needed in the next twenty years.

If sponge cities sound unlikely to work in a North American context, then consider Seattle. In 2013, a Seattle City Council Resolution outlined plans to use 'green storm-water infrastructure' (this predates common use of the terms 'sponge city' and 'SuDS', but means much the same thing) to manage 1,500M L of storm-water run-off annually by the year 2020, and 2,600M L by 2025. The first progress report sums up the background:

> Before Seattle developed as a city, very little rain ran off the surface of the land as 'runoff'. Rather, the evergreen forest system that covered most of the Puget Sound Region managed rainfall in a way that slowed down the water, allowed it to evaporate back into the air or soak into the ground, filtered the water through plants and layers of soil, and replenished the groundwater that feeds our salmon-bearing creeks. As Seattle developed, this spongy forest system was replaced with a largely impervious built environment of roads, buildings, walkways, and parking areas; and an underground, piped system was built to carry stormwater runoff away.[28]

By 2021, Seattle's sponge-city-in-all-but-name was ahead of schedule, with 1.5B L (410M gallons) of run-off diverted into 'green storm-water infrastructure' that included 758 rain gardens, most of them recharging groundwater and around half of them filtering water into cistern tanks for reuse. Central reservations of many roads were replaced with planting, some even mimicking streams with water plants. In one of the nicest examples, a community-led effort turned an abandoned gas station into the Labateyah Youth Home, run by the United Indians of All Tribes, with an ornate rain garden and a 22,000-L rain tank.[29]

Other cities are looking at repurposing old infrastructure, turning former storm drains inwards. Los Angeles' iconic storm drains – famous as location shoots for numerous films, including

Terminator 2 – are designed to take storm water out to sea, but could one day be repurposed to capture storm water and take it to groundwater recharge sites – though, not easily, warns Newsha Ajami. Rainfall must be allowed to infiltrate immediately upon hitting the ground, so that it doesn't capture environmental pollutants that must then be removed. This means creating more spaces for the water to infiltrate, says Ajami. Basically, channelling it into canals full of trash is always going to be expensive to treat; SuDS and green spaces are much more desirable.

There is an urban storm-water system that combines the two approaches, however. In Singapore, the city-centre Marina Reservoir does capture storm water from the city's road drains. Being within the equatorial rain-belt, Singapore gets a lot of rain – over 2,300 mm a year, on average – and, unlike the Indian monsoon, it can rain at almost any time of year. Singapore wouldn't have a water problem at all if it wasn't for its dense population and small land mass. Professor Hu Jiangyong of the Centre for Water Research in Singapore tells me that they've devised a nature-based 'bioretention system' that can help to purify storm water upstream and then move it into the reservoir downstream. This uses a water garden adjacent to the reservoir, with drainage pipes beneath the plants and soil. When the roadside storm water drains into the system, 'the dirt as well as the nutrients like nitrogen and phosphorus will be kept by the soil, and the semi-purified water will leave through the drainpipes below the soil, flowing into the reservoir, while the nitrogen and phosphorus supports the plants to grow.' Which sounds great, except for all the microplastics, oil and tyre abrasion. 'They will be removed by the soil,' responds Hu. 'It's impossible for them to penetrate the soil. So, all the dust, the microplastic and those kinds of particle pollutants will accumulate in the soil, and then the clean water is going into the system.'

*

In the height of the English summer, I return to James Alexander's farm in Oxfordshire to dig up the underpants. It's results time. That shin-high green wheat is now waist-high and golden. Like all farmers at this time of year, James is on high alert – any day now, he will harvest, when the crops are ready and the weather gives a long enough dry spell. The barley has already been harvested, but it's been a poor year, James grumbles.* What I hear from elsewhere in the world – that drier spells last longer, but the rain comes harder – is James's reality now too. 'We get a spell of rain, and it will rain loads, and then it will be dry for a month. There is no rhyme or reason to the patterns.'

James expects that irrigation will play an increasing part in the future of British agriculture. 'We could build a big reservoir now,' he says. It's expensive to do, but can be offset by the increased earning potential of an irrigated field, and the technology 'is all done via apps, these days. So, yeah, maybe that time will come.' However, the problem remains of where that water comes from in the first place, and, with already over-abstracted rivers and aquifers in England, there's not a lot of slack from which the farmers can take. Better then, he feels, to change to no-till fields that require less drop per crop.

At the conventionally farmed field, we walk back to the spot marked by our red pole and dig up the Y-fronts. They come out crumpled and dirty, with a couple of holes, but if you washed them you could probably wear them (I've worn worse). They look merely preserved, soil-stored for safekeeping. James holds them up gingerly as I take a photo. We drive on to the organic field, where unwanted wild oats rise high above the wheat, interspersed with the shocking bright red of poppies. This low-yield crop is sparser, making our pole easy to spot. James digs and retrieves the pants. These are noticeably different to the ones in the conventional farm's wardrobe – there's no way you could wear this pair again – full of holes, chewed and munched by worms and bacteria. They are surprisingly clean, though; more dusty than dirty. I hold them up to the

---

* Though, admittedly – and I say this with farmers in my family – I have never heard a farmer say it's been a good year.

camera, rags blowing in the breeze. We finish on the no-till field. It's still the best-looking wheat field of the three. The ears of corn are fuller, the crop denser, the whole field glowing golden in the sun like an agricultural El Dorado. The pants James retrieves from here are truly a bedraggled mess. The striking difference is how much dirtier they are: dark soil clings to the holes, there's the white residue of microfungi, even purple patches perhaps left by nematodes or protozoa. A millipede jumps out of the formerly XXL underwear and scurries off. This soil is clearly teeming with life. And the ground, a firm sponge that hasn't been ploughed for two decades, has kept enough moisture and microfungi to grow uninterrupted through the recent dry spell. 'Yep,' says James, with satisfaction, surveying his golden field. 'I reckon this is ready to harvest first.'

CHAPTER 9

Making Water

Amid an array of dusty ocean-blue solar panels – over twenty rows, fifteen panels wide, stretching back through the otherwise silent desert scrub – I wonder what that sound is. There's a distinct hum. Tom, the site manager showing me around, points down to the base of one of them: 'You see there's a vent? That sound is three internal fans drawing air inside.' Because these solar panels aren't harvesting energy – they're harvesting water.

Looking more carefully, I see that the face of each 1.2 m x 2.4 m 'hydropanel', made by a company called Source, in Scottsdale, Arizona, is split into three: two solar-thermal panels, with a strip of solar PV sandwiched in the middle. The solar PV is there simply to run the 80 W fans, making the process entirely wireless and off-grid. They work, Tom explains, by 'making the space inside the panel as hot as we can, so that, no matter what, the air that's being drawn in is cooler than the air inside. Inside the machine, we're then able to affect passive condensation.' Even in the dry desert air of Arizona (and, if it can work here, at around 30 per cent relative humidity, it can work anywhere), each panel can pull a minimum of 2–4L of water out of thin air, every day. Across this field of 250 panels, that's around 1 m^3 a day, or 30 m^3 a month. In a humid tropical environment, it could achieve triple that, producing enough to fill a polytank daily. This pure water is then mineralized for drinking within the panel (pure H$_2$O being corrosive for our body, as well as for pipes), and runs into an internal 30-L storage tank, ready to drink or pipe or truck. The mineral cartridges need replacing every five years. Other than that, and a bit of dusting, they run themselves. But, just to make sure, each panel communicates digitally (via Wi-Fi, satellite, or even 'a $1 sim card') to the operations HQ in Scottsdale, no matter where they are in the world, giving live performance data.

I first spoke to Source's founder, Cody Friesen, for the BBC in 2018. Back then this was a promising technology, but, at $5,000–$6,500 per panel, the market was mostly wealthy off-grid houses and ranches. There were early signs of a potential new direction, however. Hurricane Maria had recently hit Puerto Rico, and Source (then called Zero Mass Water) played a small part in the relief efforts – providing hydropanels for a local fire station, to allow the firefighters to remain on site. One firefighter told Friesen, 'After the military goes away . . . the only potable water we'll have is gonna be this.' Other customers included schools in Mexico and an orphanage in Lebanon. 'My favourite project since is in Kenya,' Tom tells me, three years on, as we walk around the Arizona site. 'At a girls' school. The girls would be walking two hours or more every day to fill cans of water. Now, there's forty-plus hydropanels on their campus, meaning they can stay where they are, go to school, live their life.'

Back at HQ, Cody Friesen's ultimate goal – aside from, he freely admits, making a stack of money – is providing people with access to water. He believes the central-infrastructure model is on its last legs. And it's hard to argue, with a view from his sixth-floor office looking out over the CAP canal. 'It's hard to beat water in a lake,' he says. 'But, by the time you get it into the pipe, and you pump it over the hill and send it to your home, that cost structure is substantial.' My mind goes back to Lake Volta in Ghana – all that water, yet lacking the power and infrastructure needed to get it to the population. 'So, if you can [instead] access a resource that is abundant everywhere, and your feedstock is free – sunlight – plus, you can make that device ever more efficient . . . this could eventually be cheaper than municipal water.' He believes his Source hydropanels could follow the same trajectory as solar PV, which, according to the International Renewable Energy Agency, fell in cost by 85 per cent in the decade to 2020, becoming cheaper than coal.

All air, from the Arizona desert to humid cities, contains water vapour – globally, an estimated 12,900 km³ of water is suspended in the air around us. Pulling that water from the air isn't a novel concept – you might have a dehumidifier at home that does exactly that, and there are several companies adapting dehumidifier technology for drinking water. Water-from-air technology works by creating an artificial 'dew point'. Think of the droplets that form on a glass of iced drink on a summer's day – it's the temperature at which the water vapour in the air saturates (turns from a gas into a liquid) and condenses onto a surface. In nature, this often happens in the cool of the night; but, if you artificially heat the air and/or present a cold surface, then you can create an artificial dew point. A mechanical dehumidifier is essentially a refrigerator turned inside out. But the approach Friesen has pioneered is to warm the air first and then use a 'desiccant' to suck up the moisture instead of creating an (energy intensive) cold surface. Friesen explains: 'A desiccant absorbs water at ultra-low humidity, even 5 per cent humidity. For example, when you leave the lid off the sugar bowl, it gets kinda clumpy. Sugar is a natural desiccant, but it does that really slowly. Now imagine you have an engineered material that does that very fast.' The material developed in his Arizona State University lab is a trade secret, but he can say that it includes lithium chloride and 'organic ions'. The role of the solar panel is then to 'ratchet up the dew point inside. If you take a shower, for example, you push the dew point up so high that you get condensation on the mirror.'

Friesen believes that desiccant-based water-from-air technology has the edge over the refrigerator types, because 'for a refrigerant to work, you need relatively high humidity. You need an immense amount of energy to pull water from the air. And the water isn't clean, so you have to treat it.' His desiccant absorbs the water molecules, while heat is also used to drive the water vapour back out. 'It's like a double-distillation process.' Back in 2018, the panels generated water for about 16¢ per litre, and the idea of a hydropanel solar farm was just a dream. When I visit Scottsdale at the end of 2021, it's already down to 4¢ per litre and the solar farm is a reality that I get to visit. And now he's thinking bigger – much bigger. Climate change, in Friesen's way of thinking, is a 'major market disruptor'. One thing that increases with climate change, as we learned in Chapter 2, is water vapour in the air. In the milky late-December sun, Friesen tells me that they're beginning to partner with municipalities, reaching rural areas where pipes don't reach and Source panels are cheaper than transporting water by truck. I ask him if he can envisage a time when his hydropanels create water cheap enough to flush toilets with. 'Oh yeah,' he says, confidently. He compares it to desalination – which, despite its huge economic and energy costs, not only fills taps and toilets, but even ornate fountains in the Middle East.

The cost of water typically isn't its source, it's in getting it to the people. In Ghana, with the greatest reservoir in the world, I saw many people without a potable water supply. Plastic bottles and sachets of water are a lifeline, filling in the gaps where municipal water companies have failed. Such water is already expensive. 'The water in the Western United States is cheap because they've benefited from 150 years of infrastructure,' says Friesen. 'The most expensive water in the world is where women and girls walk long distances fetching water every day, going to non-potable sources. All that time lost from education [due to travel and]waterborne disease . . .' The average global price of bottled water is 55¢ a litre, Cody tells me, 'and 95 per cent of that market is in developing countries'. Seeing bottled water as 'bad' is a privileged position, Friesen argues: most people who drink bottled water do so because they have no choice. 'It's primary drinking water. It's like a package of sterility.' This reality in many parts of the water-scarce world is why Source now works with bottling companies (using glass and reusable bottles). One such partnership in Saudi Arabia will see Source water produced in a large hydropanel array and then bottled on site, at a rate of 300,000 bottles a year. No groundwater has been extracted, no sea desalinated, no lake depleted – you even produce the water without a mains power supply.

'Think how much it would cost to build the Hoover Dam today', says Friesen – it would be not only impossible, but also undesirable. He describes lakes Mead and Powell as 'stranded assets', stuck high and dry, up in the mountains, away from major cities. Source was founded to 'break that entire notion of water as an extracted resource', he says. 'In Warm Springs, Oregon, there's a community of some 5,000 people that had about six times the legal limit of arsenic in their underground water. So, we signed a contract with them and put in an array of 800 panels.' In Cape Town, the municipal water company installed 500 panels in a city slum. These examples offer a glimpse of an alternative future. Do we connect ever-expanding informal settlements to the grid? Or help them to live off grid, in a better way? These are open questions.

New water

Another way to make water is to recycle sewage into drinking water. That happens more than you might think. All sewage plants around the world are designed to turn our waste into effluent fit enough to return to rivers (or not, as we saw in Chapter 5), and are captured again in reservoirs downstream. But polishing a turd to the extent that you can drink it – now, that's a whole other story. 'Life is full of improbable stories', writes Kishore Mahbubani in his foreword to the academic book *The Singapore Water Story*. 'There are also very improbable stories. And then, of course, there are the truly miraculous, nearly impossible stories. Singapore's Water Story clearly belongs to this category.'

When Sir Stamford Raffles* first landed in Singapore in 1819, the water from inland streams and self-dug wells was enough to

---

*   That's not a made-up name. Fun Sir Raffles fact: as well as founding Singapore, he also founded ZSL, London Zoo.

maintain the island's few hundred inhabitants. By 1890, however, after rapidly developing as a strategically vital British port, over- crowding and pollution had contaminated the wells, and the Singapore River (which runs for only 2.95 km) was chronically pol- luted by pig and duck farms. When Singapore gained its independence in 1965, the water situation was dire. With just three reservoirs providing less than 20 per cent of the island's water needs, and the little available groundwater polluted, water rationing had been imposed in 1961. On 19 August 1961, there wasn't a drop of water available[1] – Singapore had reached Day Zero long before Cape Town coined the phrase, and now relied on its neighbour Malaysia for half of its water supply (and still does*).[2] But, on the day of Singapore's independence, Malaysia's prime minister told the British high commissioner that 'if Singapore's foreign policy was prejudicial to Malaysia's interests', Malaysia would cut off its water supply. Finding an independent source of water was therefore immediately top of the new nation's priorities: 'We were under ser- ious blackmail', inaugural prime minister Lee Kuan Yew has said.[3]

In the three decades between 1963 and 1993, the population rose from 1.8 million to 3.3 million, driving demand for water further upwards in this increasingly affluent nation; to make matters worse, per-capita consumption more than doubled, from 75 L to 173 L. The 1990s also saw a boom in the island's water-intensive IT semi- conductor industry. New reservoirs were built, but demand kept outstripping supply. Amazingly, the 1972 Singapore Water Master Plan earmarked 'unconventional water sources', including waste- water reuse, before the technology even existed at scale. A pilot wastewater-reclamation plant was dutifully attempted in Singapore in 1974, but the membrane filters quickly clogged up, and the plant shut down after just fourteen months.[4] The ambition remained alive,

---

* The country imports more than 50 per cent of its raw water under a deal with Malaysia that runs until 2061 – so the truth of every statement claiming Singapore as 'the first nation to achieve water self-sufficiency' can be divided by half.

however, and by the 2000s the technology had advanced sufficiently to try again. The project was to become known as 'NEWater'.

Professor Hu Jiangyong of the Centre for Water Research, Singapore, has been actively involved in the NEWater programme since its conception. Born in Beijing, she arrived in Singapore in 1996, when 'the water industry wasn't very active compared to today'. There was some water reclamation for aquaculture, but that was about it. In 1999, the go-ahead was given for the R & D stage of a wastewater-reclamation plant that could turn domestic sewage into drinking water. For three years, Hu and her colleagues in a fifty-strong research team worked to investigate the feasibility of achieving something that had never been done before on this scale. Water security and national security being synonymous in Singapore, there was political pressure too. In 2002, the R & D study concluded, and, in 2003, Singapore's first full-scale NEWater plant began operation. 'It's relatively small-scale compared to the newer plants,' says Hu, 'but that was the first one.'

NEWater was potable, but also needed to be palatable, so a huge education and charm offensive ran alongside the technology. Prime Minister Goh Chok Tong was filmed playing tennis and chugging bottles of NEWater between sets, happily pronouncing it 'good for your health'. At the 2002 National Day Parade in a packed-out Kallang stadium, 55,000 people were each handed a cold bottle of NEWater. Speaking in a 2018 documentary, Harry Seah, assistant chief of the national water agency, PUB, described that day as 'one chance to get it right – one chance to convince the population.' A PUB engineer had personally driven the tanker of NEWater to the bottling factory and overseen the entire process to ensure it all went without hiccup.[5] Not only did everyone happily drink it – and you would, if outdoors for hours in stifling humidity – but it became a source of national pride, as a truly independent new source of water.

The technological breakthrough that made NEWater possible where the 1974 attempt had failed was 'microfiltration, plus reverse osmosis [RO], plus ultraviolet disinfection,' explains Hu. The RO

filters simply weren't available in the 1970s. Other countries, including the US and Israel, also began mainstreaming this technology in
the 2000s. By 2011, five large-scale NEWater treatment plants were
in operation, meeting 30 per cent of Singapore's total water demand;
by 2060, it's expected to meet 50 per cent of demand. Achieving this
also required a fundamental redesign of the British combined sewage system. Separate pipes now lead black water (from toilets)
directly to the treatment plant, while all storm water and road-
surface run-off is collected separately in drains and channelled to
rivers, reservoirs or infiltration beds.

As NEWater has developed, so too has the technology, Hu tells
me, now using membrane bioreactors to combine filtration with
biological processes. Professor Hu and her colleague Professor Say-
Leong Ong also discovered the process was capable of removing
oestrogen found in toilet water, partly from the contraceptive pill –
a hot topic, given that fish and frogs had been changing sex in rivers
due to high oestrogen levels in the water from pharmaceuticals and
livestock. 'The good thing is that RO is able to reject more than 99
per cent of those compounds in the wastewater,' Hu tells me. 'We do
not see problems after going through this process.' The same goes
for microplastics. Strangely, then, raw sewage turned into drinking
water in this way results in 'purer' water, with fewer contaminants
than ordinary tap water.

Wastewater recycling has the potential to solve many countries'
water-scarcity problems. According to a report by CEEW, if India
were to treat 100 per cent of its sewage by 2030, it could meet all its
industrial water demand, or 27B m^3 a year, compared to only 8B m^3
treated today. One of the report's authors, Kangkanika Neog, tells
me, 'The Delhi government are trying to increase their wastewater
reuse by 40 per cent in the next few years. A study I did saw that
wastewater reuse could be a much cheaper option – especially for
industries.' Currently, they pay 65 rupees per kilolitre for fresh
water, which could be reduced to 18–20 rupees for wastewater
reuse. Though, she sees 'no streamlined work to get there' yet. Some
Indian cities treat up to 60 per cent of wastewater, cleaned and

returned to rivers or wetlands. But, where Neog lives, in Guwahati, Assam, 'of all the wastewater that is generated, only 0.1 per cent is treated in a centralized system. So, there's a lot of potential, but a long way to go to get there.'

Treated wastewater is emerging as an alternative to groundwater in Peru too, as a peace offering between agribusiness and local communities. Carla Toranzo of AWS, based in Lima, tells me that the agribusiness Sun Fruits actually restored a municipal wastewater-treatment plant in San José de Los Molinos because the local groundwater was polluted by around 1,000 m³ per day of untreated sewage (with a population of just 7,000). Restoring the broken-down plant would (according to the company's own figures) give Sun Fruits enough water to irrigate 30 ha of avocados *and* provide 365,000 m³ per year of water for the town's residents. The rebuild was completed in July 2019 and irrigating with wastewater for avo-cado production began in July 2020. Unusually, then, a private agribusiness invested half a million dollars towards public infra-structure, benefiting the town, the aquifer and the business.[6] 'That's the kind of approach we need,' says Toranzo, 'because it's evident that unfortunately the state – be it local government, regional gov-ernment or central government – can't do it alone.' This kind of solution, she says, 'is like a circular economy.' However, as she points out, at least Los Molinos had a wastewater-treatment plant in the first place – many towns do not. Government still needs to 'close those gaps', but Sun Fruits have 'built a way that could be a "copy paste" for others.'

Another circular water-reuse model that could be 'copy paste' can be found in the Las Vegas Valley. The valley forms a natural bowl that slopes down towards Lake Mead (formerly, of course, the unconstrained Colorado River). As Colby Pellegrino tells me in her office at the Southern Nevada Water Authority (SNWA), 'If you know anything about sewer planning, you want your shit to go downhill.' In the 1990s, by using a dried-up riverbed, a former tribu-tary to the Colorado, SNWA created the Las Vegas Wash: an urban river into which the wastewater-treatment plants can pump their treated effluent, using freshwater vegetation and sunlight to further treat it on its 17.7 km journey to Lake Mead, from where SNWA pumps its drinking water back out. It's almost a closed loop. As a result, SNWA cares less about how long people shower for (because that goes back out to the Wash), and far more about watering gardens (counted as 'consumptive use', which is lost and not returned). 'We take roughly 300,000 af [0.37 km^3] out of the reservoir, and 100,000 af [0.2 km^3] of that goes back in the Wash,' says Pellegrino. The inflow from the Wash now accounts for 2 per cent of Lake Mead's water. It also works as storm-water capture, if and when Vegas receives its 100 mm of rain a year.

I visit the Wash with Bronson Mack, SNWA's public-outreach officer. As we drive past a residential park in Henderson, Bronson spots a large ornamental pond he hasn't seen before. 'Good grief,' he sighs. 'That's Colorado River water.' Beyond the city, the only real vegetation are creosote bushes dotting the dry, rusty sand. Then we approach a lush line of green: the Las Vegas Wash itself. There's an 'ick factor' when you talk about 'toilet to tap', Mack admits.* That's why he likes taking people out here, to see the Wash for themselves. It used to be a natural wetland fed by the creek, until it dried up due to groundwater over-abstraction, becoming a managed wetland only twenty years ago, as 'nature-based wastewater treatment'. Invasive species such as African tamarisk were removed and native reeds replanted. Bronson pulls up the car and we walk through scrub† to emerge at a concrete weir. I'd expected a large stream or creek, but in fact the Wash is an impressive river, some 30 m wide. Just below the weir are concrete stepping-stones that span the width, put in as erosion control, but also allowing us to walk out to the middle. There's a tang to the water that hits your nostrils, taking me back to

---

* Although, the truth is that most places reuse water, just without advertising it. Effluent is typically treated, returned to rivers, captured in reservoirs, treated again and returned to taps.
† 'This is where the bodies are buried,' jokes Bronson, who adds, 'Have you seen the movie *Casino*?' And I realize he isn't entirely joking.

the effluent-outflow pipes at the Banbury sewage works – treated, but 'not yet water', as I was memorably told. 'We haven't had any rain,' confirms Bronson, as if it were necessary out here in the desert. 'So, 100 per cent of this [flow] is treated wastewater. This was all sewage about ten hours ago.' My morning toilet flush could well be flowing past me right now, along with the 830,000 m³ of water that flow through the Wash every day. But here it runs a clear, almost imperceptible green, and, at my head height, above the weir, I see waterfowl, ducks and coots happily swimming and hunting for fish amid reed beds. Extensive bird surveys have identified threatened and endangered species returning here. Native wetland plants are taking hold again. For a substantial number of flora and fauna, this 'not yet water' is water enough. And, rather than a green-washed drainage system (like the Cotswolds field where I saw sewage effectively dumped), this is the biggest, most impressive nature-based solution I see on all my travels. And in Las Vegas, the unlikeliest of all places. 'Shoot, if we didn't have the Wash, we'd either have to find another water source or Las Vegas would be a whole lot smaller,' admits Bronson over the roar of the weir.

Back at the office, Colby Pellegrino tells me that California is now looking to recreate the Wash's success, and partnering with SNWA to do so. Unlike Vegas, the much larger city of LA sends most of its billions of gallons of wastewater out into the Pacific Ocean. Now, the planned Regional Recycled Water Program will return 25,000– 30,000 af [3M–3.7M m³] back into its reservoirs. A Bill signed by governor Gerry Brown in 2017 also required the State Water Resources Control Board to come up with criteria for direct potable reuse of recycled water by the end of 2023. Los Angeles Mayor Eric Garcetti's Green New Deal, launched in 2019, aimed to recycle all the city's wastewater for reuse by 2035. Both political moves were signs of what's coming: water recycling must be part of a water-secure future. At the time of writing, water recycling adds just 2 per cent to LA's water supply.[7]

Meanwhile, the city of San Diego is looking more to Singapore's NEWater for inspiration. Its similarly named Pure Water will be the largest infrastructure project in the city's history, turning sewage water into potable water with the aim of supplying around 25 per cent of the city's water supply when completed in 2025, and 40 per cent by 2035. The various stages – which have been delayed for years by ongoing lawsuits – are a $356-million sewage-purification plant, a $123-million pipeline and a $110-million pumping station (costing over half a billion dollars in total).[8]

Recycling water for potable use doesn't come cheap. But water turned into effluent for agriculture is more readily achievable. On my visit to the Jordan Valley, I saw how wastewater could provide a new source of water there too. Next-door neighbour Israel is already the world leader, recycling 90 per cent of its wastewater, with 85 per cent of that used for agriculture.* Spain, believed to be second in the international league table, reuses about 20 per cent of its water, using 71 per cent of that for agriculture. Jordan, however, lags very far behind. But there is one shimmering, smelly oasis in the desert: the unassuming Tal Al-Mantah sewage works. One of only two small sewage-treatment works in the valley, Tal Al-Mantah serves a population of around 800,000, and is designed to treat 400 m³ of waste a day. EcoPeace's Eshak Al-Guza'a acts as my guide around the plant. 'Be careful!' he warns, as we walk past a huge cesspit. 'It's very deep! No one will pull you out of there.' Unlike my local sewage works, this site is entirely powered by a solar array, built with the help of German aid agency GIZ. It also uses (after the standard settlement tanks and oxygenation process) a final ultraviolet treatment, something most UK sewage works don't do either – much to Eshak's surprise, when I tell him. The end result is treated, reusable water. Much like

---

* Farmers like treated wastewater because it's full of the nitrogen, potassium and phosphorus that they normally pay for. According to the World Bank, farmers in Pakistan are willing to pay up to 200 per cent more for wastewater than for regular irrigation water. Wastewater also has the benefit of being available year-round, including during the dry season. (Richard Damania, Sébastien Desbureaux et al, *Quality Unknown: The Invisible Water Crisis*. World Bank Group, 2019)

the Karameh Dam, however, it is currently wasted. We watch the water flow out uselessly from the outflow pipe, down a sandy, rocky hill, where it benefits a few grateful weeds before evaporating in the sun. 'We *need* to confine this water and make a good use of it,' argues Eshak: 400 m³ of water per day, in an arid area like this, could be precious. Without this plant, thousands of septic tanks throughout the valley would have otherwise seeped into the soil, polluting groundwater and the local *side wadi* streams. But to see this water – treated with such care – simply flow out unused, while nearby farms are drying up through lack of irrigation, gives me a now-familiar sinking feeling.

'We need to find a source of water in Jordan, and this is a great alternative,' says the ever-positive Eshak. He has seen it in action, having previously worked at the much bigger As-Samra wastewater-treatment plant, near the city of Zarqa. There, he says, 400,000 m³ of wastewater are treated per day, and 202M m³ of it *does* go annually to agriculture: 'The amount is very impressive. That is showing what is possible.' Later, His Excellency Saad Saleh Abu Hammour tells me that 80 per cent of the crops in the Zarqa valley are now irrigated to some extent with As-Samra's treated wastewater. But As-Samra was built in 2008, for $97.9 million, with support from the US development agency USAID. For many water-poor countries like Jordan, affording one As-Samra is hard enough; two or three, near impossible. Attaching reuse pipelines to smaller plants such as Tal Al-Mantah, though, is surely a no-brainer.

Next, Eshak wants to show me a much smaller, off-grid solution, so we hit the central valley road again, heading north towards the Syrian border and EcoPeace's Jordan EcoPark. This covers 44.5 ha of previously dry and unused (unwanted) land, gifted by the Jordan Valley Authority to EcoPeace in 2002. It opened in 2004 to show-case EcoPeace's proposals for rehabilitating the valley via the replanting of trees and native plants to offer shade and water retention. Today, those trees are mature and a revitalized brook babbles through the site, passing a visitors' centre and numerous tourist eco-cabins. The temperature tops 40°C in the valley, but the tree canopy reduces the temperature by 7°C, so we can comfortably sit outside to eat.

After a traditional meal of rice, stuffed vine leaves, maqluba chicken and the ever-present shatta chilli sauce, Eshak is itching to show me the site's water-recycling system. Every guest lodge and staff building has separate colour-coded pipes for grey water and black water. Every grey pipe, primarily from showers and sinks, leads to two larger plastic tanks, each a cube of around 1.5m³.* The first is filled with small porous volcanic rock granules found locally (to demonstrate their abundance, Eshak scans the ground and quickly finds one to show me); an internal electric pump then mashes them up and down in the water, creating both a biological- and mechanical-breakdown process.† The water then passes slowly through a drip filter into another tank below and is distributed via irrigation tubing to younger plants and trees in the park. Eshak is clearly so pleased with the system that I take photos of him standing by his water-recycling baby. He shrugs; off-grid grey-water systems are pretty simple, he says. 'Last week, we installed 101 grey-water systems, smaller than this one, at local households in this area, using plastic barrels . . . We're spreading the word around the community and teaching them by practical example. It's not just words on paper.'

It's what happens to the black water, though, that is Eshak's true pride and joy. Black water from the toilets, as all my sewage-plant visits attest, is not so easy to treat. At the bottom of the valley, far enough to require jumping in the car again, the EcoPark's black-water pipes lead to an enclosed settlement tank, and then into a series of plastic-lined ponds, stocked with bulrushes. In Eshak's Instagram photos from the previous year, the rushes were just knee-high, but now they stand majestically at some 3.7 m, pumped with

*   Credit where it's due, this was installed with a grant from the Coca-Cola Foundation.
†   Using gravel to mechanically break down sewage via oxidation was first done in the 1870s by the English chemist Edward Frankland.

the phosphorus and nitrogen that comes from human waste. The water then passes into three further ponds, where more settlement and algal treatment clarifies it – in the last one, we see tadpoles. The final water passes down to the valley below, feeding a natural wetland that had all but dried up, with no natural rivers left, but is now recuperating thanks to this unexpected new water source. In the Jordan Valley, from the depleted Dead Sea to the pitiful River Jordan and the salty Karameh Dam, it's hard to see humans as anything but a pernicious influence on the landscape. But here, in the EcoPark, is a splash of hope. A sign that, when done right, even in this driest of landscapes, human activity can replenish nature.

Desalination nation

And yet: 'Desalination is the last hope for Jordan.' Or so I was told by Dr Raya Al-Masri, a Jordanian researcher at Surrey University's Centre for Environment and Sustainability, before my visit. 'If you ask any water official in the country, they will tell you the same thing,' she says. I did. And they did.

Jordan's plan for its first major desalination plant emerged from the ashes of the failed Red Sea–Dead Sea Conveyance (RSDSC). Jordan only has a very narrow strip of coastline in the far south of the country, where Aqaba meets the tip of the Red Sea. All its major cities, however, including Amman, are up in the north. The RSDSC was going to achieve the unthinkable: regional cooperation. Jordan would build a major desalination plant in Aqaba, but deliver the water a short distance across the border to Israel; Israel would then swap the equivalent amount of water from Lake Tiberias (Sea of Galilee) in the north, closer to Jordan's major cities. The ingenious part was that, together, they would build a pipeline for the highly salty brine desalination waste (a major problem for the desalination industry, as we'll see), direct to the Dead Sea, topping it back up and solving that particular ecological and tourism headache. The project, pencilled for 2018, even included a hydroelectric power plant, using

the 425-m elevation difference between the Red Sea and the Dead
Sea. Jordan, Israel and the Palestinian National Authority, as the
three beneficiaries, jointly commissioned a feasibility study managed by the World Bank. The environmental assessment included
the warning that it 'could change the unique chemical composition
of the Dead Sea.'[9] But that wasn't what sunk the project. Ultimately,
the unthinkable – regional cooperation – was unworkable, with talks
breaking down between the three parties. 'This project has been
cancelled,' confirms Abu Hammour to me brusquely, between figs
and phone calls, 'no more existing.' He served for a decade as the
RSDSC committee chair. 'They delay this project and they said that
the Israelis are not in agreement with us.' I'd heard that it was too
expensive to go ahead, but he quickly corrects me. No: it's a minis-
terial disagreement. He says the added benefit of the RSDSC was 'to
at least stop the shrinkage of the Dead Sea itself.' But it's no more.
Mark Zeitoun at the University of East Anglia (UEA) worked on the
Palestinian side of the deal and later tells me the RSDSC was driven
by the Jordanian side, but 'Israel not very much. They were just along
for the ride.'

Instead, Jordan decided to go it alone with the desalination
plant – keeping all the water, covering all the costs, and not rehabilitating the Dead Sea. Its proposed completion date is 2027–8.[10] The
now $1.8 billion Aqaba desalination project, aiming to generate
300M m^3 of water a year,[11] continues without any of the additional
benefits of the RSDSC. With no swap agreement, the desalinated
water will travel hundreds of kilometres uphill to reach Amman and
the major cities in the north. The brine will simply go back out into
the ocean. The overall water benefit to Jordan will be less and the
costs greater. But politically it's easier, because it doesn't involve col-
laborating with Israel, the hated neighbour that regularly bombs
Jordan's Palestinian brethren,* and so the project can therefore be
couched with undiluted nationalism (albeit, seemingly without

* My paraphrasing of the popular Jordanian viewpoint, not my own.
But 'brethren' is at least factually accurate: over half of the Jordanian

irony, using the name that Ben-Gurion coined – 'The National Water Carrier'). The tendering, due to complete in late 2022, came down to five international consortiums, two of them Saudi-led. Abu Hammour worked on the early cost analysis. 'The initial cost for each cubic metre received on the water meter of the customer is $3. No government can sustain this cost.'* Construction 'will take seven years from now . . . So, what are we going to do within the coming seasons? The expectation for the coming three years is drought.' I suggest the only option will be to purchase additional water from Israel's Lake Tiberias – the very deal that the RSDSC would have made easier and cheaper. Yet, even Israel, with its surplus of desalinated water, says Abu Hammour, can't keep on selling limitless water. 'Now, the only solution in front of us as a Jordanian people is to go for desalination,' he says, starkly, acknowledging the contradiction that it is unaffordable. 'We, as a country,' he says, have no more 'seas, we don't have rivers, we don't have anything.' The droughts are regional, and 'coming every year . . . climate change is affecting everywhere. Especially in the Middle East. So, no more solution, only desalination.'

Certainly desalination solved Israel's water woes, and surprisingly recently. On 3 August 2000, the Israeli government agreed to construct its first large reverse-osmosis [RO] seawater-desalination plant on its southern coast, capable of producing 50M m^3 of water a year. Just two decades later, by 2022, the state-owned National Water Company (Mekorot) operated thirty-one desalination plants, treating close to 1M m^3 of water every day, providing 85–90 per cent of municipal and industrial water supply.[12] 'The earlier technologies were producing water at $2 per m^3,' says Gidon Bromberg. 'The latest

---

population are from Palestinian origin, including the Queen, Rania al-Yassin.
* Gidon Bromberg also made this point to me separately, adding that 'the public pays 50¢. If the [Jordanian] government is subsidizing two and a half dollars for every cubic metre, the government is going to go bankrupt.'

desalination plant, Sorek 2, is just 44¢: There's also talk of moving to forward osmosis, which some believe will be even cheaper.

Desalination is a major, modern water success story. In 2017, the global water desalination market was valued at $15.5 billion; by 2025, it's expected to be double that. But let's get this out the way: it is not the silver bullet. In many ways, it's synonymous with nuclear energy: it's very expensive, produces a waste by-product that you can't get rid of, typically must be built on the coast, but it produces a steady, reliable baseline supply that utility companies and customers can count upon, come rain or shine.

For hundreds of years, desalinating seawater meant boiling it and capturing the vapour – known as thermal desalination.* In around AD 200, Alexander of Aphrodisias referred to a desalination process used by sailors, boiling seawater and capturing the steam with sponges.[13] But boiling large volumes of water is very energy intensive.† It wasn't until ultrafine RO membranes came along in the late 1990s – the very same that made high-grade water recycling and Singapore's NEWater possible – that desalination for municipal drinking water really became feasible. In natural osmosis, molecules move from a solution with few dissolved substances to a more mixed solution, equalizing or diluting. But, in RO, the dense salty water is pushed through a membrane towards a less salty solution – the opposite of the natural process. Because this is working against natural osmosis, RO requires high pressure to push the water through. The resulting fresh water is typically then further sterilized with ultraviolet and/or chlorine, and minerals are added to make it drinkable. All of which relies on a lot of mains electricity, and regular replacement of membrane filters.

The chief water economist at the World Bank told me for an

---

* Which is, of course, where all our water comes from, as water evaporates from the sea and condenses to form clouds.
† Albeit, as I type, I'm reminded of David Owen's mind-blowing stat that the energy pumping costs of Arizona's CAP canal are roughly equivalent to bringing the same volume of water to the boil – so, there's that.

article I wrote in 2018: 'On average, desalination is about five to seven times more expensive. The energy footprint is huge, and you've got to do something with the salt. If you look at aerial images around the coasts of Kuwait and Dubai [highly reliant on desalination], you'll see the havoc that is caused to marine ecosystems.' Given the current costs, both economic and ecologic, 'it really is only as a boutique solution in very rich places, basically cities,' he said. As such, most plants are located in high-income countries, the biggest being in Saudi Arabia, Australia, Israel and the UAE. But others are catching up – or getting desperate. In 2018, there were 16,000 desalination plants operating globally; just two years later, in 2020, there were already 22,000. The world is gambling all its chips on desalination. Some small island nations, such as the Bahamas, Maldives and Malta, already get all their municipal water from desalination. Saudi Arabia gets about 50 per cent, with one of the biggest plants in the world, the Ras Al Khair, producing 1.4M m³ a day. Leon Awerbuch, former president of the International Desalination Association, has predicted that 'mega-scale desalination plants in the future will get bigger than Ras Al Khair. The Japanese Mega-ton projects are already under consideration and a 1.5M m³ per day project in Saudi Arabia is already in planning stage.' Similar megaprojects are proposed for China, India and Abu Dhabi.[14]

In most desalination processes, however, for every litre of potable water made, a further 1.5 L of waste brine is produced. This is typically twice as salty as seawater, often with high concentrations of chlorine and copper, which is simply sent back out to the sea. This has consequences. Loïc Fauchon, president of the World Water Council, told Chatham House in 2019 that, when desalination first appeared in the Persian Gulf, 'the temperature of the sea was around 30°C, whereas today, it can be up to 40°C. This increasing sea temperature as a result of the desalination plants has contributed to changes in biodiversity. For example, we are seeing fish disappearing and even the growing population of giant jellyfish – which some desalination plants in Saudi Arabia are dealing with by using shredders.'[15] Saudi Arabia now produces 22 per cent of the world's desalination brine.[16] A multinational collaboration in 2021, including academics from Saudi Arabia, describes the Persian Gulf as having become 'more like a salty lake than a typical sea'.[17] With several of the world's largest desalination plants discharging into it, the Persian Gulf is now '25 per cent saltier than average seawater'. Estimates for population growth and groundwater limitations suggest that the Gulf countries will double their desalination capacity by 2030. Yet, the authors warn, 'unless the brine disposal is resolved, the seawater source will be too salty to accommodate the desalination cost'.

Ever saltier seawater is not only a catastrophe for ocean ecosystems, therefore, it also becomes more and more expensive to treat. The saltier the water, the harder it is to desalinate. Think back to the dead shores of Jordan's Karameh Dam – a desalination plant was added there to try to save face, but it was never functional because the water was too saline. That, in microcosm, is what might happen to the Persian Gulf and parts of the Mediterranean. Even off the west coast of the United States, 'Over fifty years, the ocean cannot sustain the damage,' Susan Jordan, executive director of the California Coastal Protection Network, told *The Washington Post*.[18]

Currently, the industry largely shrugs and says it's not a problem.* The following is actual advice from a desalination tech firm's website:

> In most cases, the easiest way to get rid of the important brine flow . . . is to discharge it in the sea via a brine outfall pipe. Brine concentration varies from 50 to 75 g/L and has a much higher density than seawater and therefore tends to fall on the sea floor near the brine outfall outlet (plume effect), creating a very salty layer which can have negative impacts on the flora and the marine life and any related human activities . . . to avoid the plume

---

* Not only a broad generalization, but also the literal response from one desalination site manager I talked to.

effect, the brine outfall should end within a strong sea current to aid mixing the brine with seawater.

Or, failing that, and somewhat missing the point of desalination, they suggest that you could dilute it by using a nearby 'natural freshwater stream'.[19]

Brine waste is something 'we don't talk about enough' at international policy level, Maggie White of SIWI tells me. She recalls a Maldivian minister at a COP meeting promoting desalination for the tourism industry. When she asked what they did with all the waste, 'He said, "Oh, well, we just dump it back into the sea." And I said, "Well, aren't you afraid that's going to imbalance the pH of the sea?" He said, "Oh, it's OK for the moment."'

By 2019, the world's desalination plants already discharged 51.8 km^3 of hypersaline brine a year – enough to cover the whole of Florida under 30 cm of brine.[20] Similar to peak oil, academics now warn of peak salt. The world could and should learn a lesson from the increasing salinity of the Persian Gulf. Instead, desalination is growing almost exponentially. In Europe, desalination was once only found in the south, but now countries like the Netherlands and Belgium are investing. Thames Water built a desalination plant in the Thames Estuary to serve London in 2010.* In 2022, there were twelve desalination plants in California, with ten more proposed. California is, however, at least attempting to address the peak-salt problem, passing the 2016 Desalination Amendment to the Water Quality Control Plan, tightening regulations for brine disposal and requiring 'mitigation measures feasible to minimize intake and mortality of all forms of marine life'.[21] Proponents gung-ho for desalination have complained that the Amendment is 'slowing the march toward a desal future'; ecologists and economists contend that that's a good thing. 'Desal should be the option of last resort,' as Newsha Ajami

---

*   However, it has rarely been turned on in that time; it is a £250-million white elephant of an investment, in a region which desperately needs wastewater treatment upgrades, new reservoirs and pipe upgrades.

has said. 'There are so many other inefficiencies in the system that can be fixed [first] to potentially harness more water.'

Israel, meanwhile, is now desalinating to such an extent that its National Water Carrier no longer needs to retrieve fresh water from the Sea of Galilee. Instead, the direction of the National Water Carrier is being reversed: it will take desalinated water and deliver it to the lake. 'Israel is investing a billion shekels [$300 million], the first country in the world to give desalinated water back to nature,' says Gidon Bromberg. 'Israel will start pumping 120M m^3, and then eventually 300M m^3 [to Lake Tiberias] annually. But not only for the purposes of the lake, but at least with the intention to sell more water to Jordan, [into] the King Abdullah Canal.' While Bromberg admits that the waste brine 'threatens to turn the eastern Mediterranean into a dead sea,' he argues the even bigger environmental impact 'is its high energy consumption.' Desalination in Israel is powered by fossil fuels. A 2018 paper by the Israeli academic and politician Alon Tal gives a simple calculation: 'desalinating 1,000 m^3 of water (1M L) per day consumes the rough equivalent of 10,000 tons of oil per year . . . This means that desalinating 1,000 m^3 of seawater could potentially release as much as 6.7 tons of CO_2 [per day].'[22] Israel plans to *triple* its already very large desalination capacity by 2050. But, 'Israel is failing to invest in renewables at the level that it should. Israel's commitment to Paris is 30 per cent,' says Bromberg. Current renewable energy capacity, when we talk, is stuck at 'under 9 per cent.' Tal's paper adds that the 'carbon footprint of seawater-desalination facilities can no longer be ignored.' In essence, desalination burns the very stuff that causes the need for it in the first place – fossil fuels – driving climate change, making the water crisis ever harder to solve. It's a global race to the bottom. The IPCC report calls desalination a 'maladaptation' to climate challenges.[23]

There is always, proponents say, a supposedly great-new-hope next-generation desalination just around the corner. The European Commission talk of 'microbial desalination cells' that use a sustainable low-energy process to desalinate seawater, with prototypes in

Tenerife, Spain.[24] Or the world's first wave-driven desalination system, the Wave2O, which could operate completely off-grid. Or the NEOM smart city being built in Saudi Arabia, which plans to use parabolic mirrors to superheat seawater by 2025, claiming to produce 30,000 m^3 of fresh water an hour at 34₵ per m^3.[25] Or the Floating WINDdesal in the Middle East, powered almost entirely by wind energy.[26] But, while there are good reasons to be excited about these, they are only a drop in the increasingly salty ocean of circa 22,000 desalination plants built or planned. Most plants are like the one I go to see in Teshie, Ghana.

Teshie: the 'silver bullet' that missed

The Teshie Desalination Project was built by Spanish contractors Befesa for the Ghana Water Company in February 2015 to address water-supply deficits in Greater Accra. Beset with problems from the start, it was shut down again in 2017, and remained closed for the next three years. But the water demands of the COVID-19 pandemic, with the increased need to wash hands, led to successful calls to get it running again in late 2020. To find out if it had solved the problems locally, my fixer Komi Vedomey suggests we begin by asking the 'local Assembly leader'. We meet him in a gloomy out-of-hours bar called Happy Days. An old Guinness sign hangs upside down above the door and the walls are made from loosely fitting wooden slats, perhaps from old pallets. Assembly leader Jonathan Adjei Adjetey is young and friendly, and starts to tell me of the myriad water problems his district has faced. When he was a kid, he says, 'we had to walk over a kilometre to the nearest well, carrying buckets, some carrying jerrycans on their heads. You didn't normally bath during the day.' This sounds like a tale from a rural village, but Teshie is close to central Accra, and Adjetey is only in his thirties. He says it was the year 2000 when municipal water supply arrived – and, even then, only for three days a week. When the desalination plant came in 2015, it promised to be a great leap forward, providing twenty-four-hour water supply. It turned out to be more of a stumble than a leap. 'We are encountering problems,' he says. 'The cost of production is high, so it flows for maybe three days, and then the other days from the [reservoir].' You can tell when they switch between the sources because the water from Kpong reservoir is overly chlorinated, he claims, while the water from the desalination plant can be salty. As for why it shut down for three years, 'When you go there, they will tell you that it was a technical problem,' says Adjetey, 'but I tell you, it was the production cost.'

Ghana Water Company (GWC) either couldn't afford the running costs or couldn't charge the real cost of water to the consumers. Since desalination returned in 2020, water bills have fluctuated a lot, Adjetey says. Each house that is supplied is metered, so should only pay for what they use, but he knows of a house that pays GH¢1,000 a month (£122 – far higher than my own £25–£40 water bill for a three-storey house in the UK), and he believes the average is around GH¢300 for a household (£36, and still more than my typical water bill). This isn't simply gossip – Adjetey's official role involves helping local people who are struggling with utility bills. But there's only so much he can do. 'You can't challenge GWC,' he says. 'If you do, their pipeline will be cut off.'

Desalination may technically be able to produce clean water in water-scarce regions, but the reality for poorer countries, says Adjetey, is that 'the cost is very high, which imposes a burden on the populace – you have to pay more.' Accra resident Richard Matey also tells me that the expense of desalination was covered by trickling the cost down to consumers – and that the water wasn't as good ('It's like, you are hungry, you've not eaten for some time now, and then you are served a half-baked, half-cooked meal, that comes with a big bill at the end of it!' Matey joked). Adjetey says that people unable to afford a connection often resort to stealing water from the pipes, or taking from dirty streams. 'We are recording more typhoid cases in the community because of this,' he says. 'We recorded high cases of cholera. This is Accra! If this is happening here, how much more in the rural areas?' He describes Teshie as 'an OK standard of living', although real poverty is not hard to find. 'But when you go further down to the traditional area, people there may go months with barely GH₵10 [£1.20]. So how do you expect them to buy water?' The government or NGOs will occasionally fill communal polytanks for free, but this is sporadic. As for the desalination plant itself, there is a low level of trust in the community. 'We don't normally drink it,' says Adjetey. 'We normally buy sachet water.'

Four days later, after much wrangling with GWC, I get to visit the Teshie desalination plant myself. It is, I find, highly political in more ways than one. Razor wires loop atop the high walls and heavy, impenetrable metal gates. The site is still run by the Spanish company Befesa, and the site manager, Francisco Galindo Medrano, is stressed out and sweating. He clearly doesn't like press nosing around and wouldn't have agreed to this visit if it were up to him. But he is kind enough to give me a full site tour, and I am, to my surprise, allowed to take pictures. While state-of-the-art is an overused term, and this is by no means the largest desalination plant in the world, for a modest-sized plant built in 2015 it *is* impressive, and Medrano runs a tight ship. Among Accra's infrastructure, it's the most gleamingly spotless site I've seen, rivalled only by the eerily omnipresent Shell gas-station forecourts.

Medrano begins with a brief presentation of facts and figures that he reassures me he'll email to me later (but, despite repeated follow-up emails, never does). I learn that this is a £130-million ($150-million) site, built and run by Befesa on a twenty-five-year lease, after which it will be fully handed over to GWC.* It has the capacity to produce 60,000 m³ of water daily (though I'm told it more typically runs at 40,000–50,000 m³), serving half a million people in east Accra. The site requires 13 MW of electricity to run ('Far too big for solar,' says Medrano dismissively, when I meekly ask about the possibility of using renewables in this reliably sunny, equatorial country).

* Known, wordily, as a 'design, build, operate, maintain, own and transfer (DBOT)' contract. The World Bank's risk insurance arm, MIGA, gave an investment guarantee of $179 million to the project.

Outside, a concrete jetty runs into the sea, housing the inlet pipe some 500 m out into the surf, sucking in 7,000–8,000 m³ of seawater per hour. A little further away and not visible is the outlet pipe, which pumps the waste brine back into the sea, around 300 m out – 'calculated precisely', Medrano says, to be dispersed by the natural currents. He claims that local fishermen even like to fish near it. At the inflow-pipe entrance on shore, I watch the water rushing in, with a solid-waste filter catching mostly plastic bags. Two white carbon-dioxide towers, as high as three-storey buildings, stand proudly nearby. The CO_2 is added to form carbonic acid during the water treatment.

Inside the plant, the noise of the pumps and RO tubes make it too loud to hear what's being said, which Medrano doesn't seem to mind. Huge colour-coded pipes of bright green, orange, blue and purple give it something of a Willy Wonka factory feel. Back outside again, as I struggle to keep up, I'm shown the re-carbonization process, which involves adding crushed calcite ('We have to get this imported from Spain,' says Medrano, it's hard to tell whether ruefully or not). Next are three lab-based tests of the water to check for water quality – one test on site, one by GWC, and a third by an independent lab. I ask why the local Teshans complain about the water tasting salty, and Medrano looks bemused. 'There is no salt, zero. You try it.' There's a testing tap at the end of the factory and he invites me to turn it on and drink. I do, and agree that there's no discernible salt – it's reassuringly flat. Adjetey had believed that chlorine denoted water from the reservoir, but in fact it is most likely from here; overly pure water from RO can be corrosive to the pipe network, so chlorine is added at 0.23 mg/L to pass through the pipes. The very final pipe, close to a metre in thickness and painted a deep blue ultramarine, has a comically small sticker on it reading *potable water*, and an arrow pointing outwards. After passing this modest exit sign, Ghana's only desalination plant distributes highly treated water to a corner of Greater Accra.

Something that Befesa and GWC *really* don't like talking about is money, and all my questions on that topic are given a straight bat.

However, reporting by Ghana's Citi FM radio station in October 2017 found that GWC was buying the water from the desalination plant at about GH₵6.5 (57p) per m³, but could only sell it to the Accra residents at the then-regulated rate of GH₵1.5 (13p) per m³.[27] Monthly costs to the GWC also included an estimated GH₵1.5 million (£130,616) electricity bill. 'The GWC is currently unable to pay the Electricity Company of Ghana, which is also struggling financially,' found Citi FM. The report found that the desalination plant was run for GWC at a loss of GH₵6 million (around £500,000) every single month. The plant was shut down for 'technical issues' just two months after the Citi FM exposé, in January 2018, and £500,000 a month in losses is technical enough.[28] It didn't open again until late 2020, when the COVID-19 pandemic made water shortages untenable, with President Akufo-Addo additionally announcing free water bills for everyone for three months – generous, but likely rubbing salt into the GWC accountants' wounds.

When I visit in September 2021, the pandemic is still very much ongoing and the government is continuing to bear the financial burden. But desalinated water, as in the three years prior to 2020, may simply prove too great a cost in the long run, and Medrano and his Spanish colleagues may be sent home once again. Even then, the GWC may have to foot a considerable bill. According to the Ghanaian news website, Graphic, under the contract agreement, GWC is obligated to pay Befesa GH₵1.4 million (£121,800) a month for twenty-five years, whether the desalination plant is operational or not. It is, says Graphic, 'a financial albatross . . . on the neck of the Ghana Water Company Limited.'[29] And, therefore, around the neck of the people.

On a visit to GWC's Kpong reservoir and water-treatment plant later that week, site manager Yaw Adjei sharply inhales when I ask about the Teshie Desalination Project. He then asks, pre-empting my own question: 'Do we need it?' I suggest that the Teshie residents previously without water might say yes. 'Look, is it about the desalination plant, or is it about needing water?' asks Adjei. Needing water, I say. 'Good. Then we don't need it. We have the water here.' It's hard to argue, as we talk just miles away from the world's largest reservoir. Build the conventional treatment plants first, suggests Yaw – he informs me that a second plant at Kpong has been on the cards for years, but never materializes – then, if that doesn't work, perhaps look at desalinating seawater. He doesn't say it, but desalination is – and always should be, given its huge capital, environmental and energy costs – a last resort. I suggest that maybe the Spanish-built plant, run by outside contractors for twenty-five years, was a good deal. 'For them!' laughs Yaw. 'If the amount of money invested there had been invested here [at the reservoir], that would have served [Accra] all right.' The RO membranes 'come with very high costs,' he says. 'The membranes have to be replaced . . . But the sand filters here . . .' He gestures to the treatment pools outside. 'This one has been there since 1995. And it is working all right.'

The Kpong water-treatment plant and reservoir that Yaw Adjei presides over produces 138,000 m³ of treated water per day, over three times the output of the Teshie Desalination Project. Combined, the five plants on the Akosombo system produce 345,000 m³ of water per day, or eight times the Teshie output. They do this at a production cost, Adjei tells me, of GH₵1.86 (16p) per m³. When I follow up with the GWC to ask for the comparable figure from the desalination plant, I am told the Teshie plant produces water at GH₵8.78 (76p) per m³ – almost five times the cost. They do, however, add that the pumping costs from the Kpong reservoir to Accra take its cost up to GH₵4.14 (36p) – but that's still under half the cost of desal, with a treatment process that is far less energy intensive or environmentally damaging.

Reap what you sow

Some sense another silver bullet, too: that we can simply start to mine deep-sea aquifers for the vast reserves of fresh groundwater beneath the seabed. Like all overly engineered solutions, you can see why it's initially tempting. Described as 'fresh water's final frontier' and long theorized, they were discovered in the mid-1970s off the north-east coast of the USA. An article in *Nature* in 2013 – 'Off-shore fresh groundwater reserves as a global phenomenon' – kicked off a resurgence of interest, especially given California's by then acute drought. The *Nature* authors cited 'overwhelming evidence that vast meteoric groundwater reserves below the sea floor are a common global phenomenon'. The paper included a world map showing vast offshore groundwater reserves near every continent, including off the coast of Florida; Suriname; South Africa; the Niger Delta; Perth, Australia; and the East China Sea.[30] While they do state that 'offshore groundwater is not the answer to global water crises,' they also say that it holds 'a strategic value that should be acknowledged [and] weighed against other options'. I ask the University of Arizona's groundwater guru Dr Jennifer McIntosh to 'weigh' it for us: 'There *are* extensive freshwater aquifers beneath the sea floor, off the east coast of the United States, and off of South Africa,' she tells me. The aquifer under the continental shelf of north-eastern USA, for example, is believed to contain a volume of fresh water equivalent to Lake Ontario. 'But, in order to extract that fresh water there's a lot of environmental pushback. It's the same as oil and gas; you'd need to have offshore drilling to extract that fresh-water resource.' The costs of extraction and delivery would be prohibitive, and the fossil water would almost certainly need treating for heavy metals before using. Not to mention, exactly like oil and gas, the fields are finite: once mined, they're gone. It's another short-term fix, not a sustainable strategy.

There are offshore aquifers near the UK too, laid down during the previous ice age. But, despite Thames Water's desperate need for new water sources, as outlined in Chapter 6, their groundwater expert Mike Jones tells me that offshore aquifers are a non-starter. 'They are more of an academic interest,' not a practical one, he says. I ask if he could foresee a time, as we get ever more water scarce, that they become an active interest for water companies. Even then, 'My gut response would still be no, not really,' he says. 'You would probably just have more desalination, more effluent reuse, something that's, from an infrastructure perspective, easier to construct, rather than offshore platforms with pipelines back to shore.' A world once again hooked on mining, using up a non-renewable resource, learning none of its past mistakes, is a grim prospect indeed.

A grim prospect that has already been realized is cloud seeding. Cloud seeding is a form of weather modification whereby you fly over clouds (or fire from ground-based cannon), dispersing chemical particles into them – typically silver iodide or potassium chloride – to form nucleation points, causing water droplets to cluster and rain to fall prematurely. Cloud-seeding experiments were first conducted in 1946 by American scientists Vincent Schaefer and Bernard Vonnegut (brother of novelist, Kurt). A 2016 article in *Chemical & Engineering News* stated that more than fifty countries worldwide now participate in cloud-seeding. Abu Dhabi Sustainability Week in 2018 included a three-day forum on 'rain-enhancement science', a euphemism for cloud seeding, culminating with a $5-million grant awarded for drone-based cloud seeding – 201 entries were received, from sixty-eight countries.[31] Cloud seeding forms a big part of the UAE's climate-adaptation plans. The UAE's cloud-seeding operation began in the 1990s; by 2017, 242 cloud-seeding missions a year were carried out by the Emirates. 'Our operations have focused on towering cumuliform clouds, which are the most common rain-bearing clouds in the UAE,' its national weather centre told *The National*. 'Based on our previous seeding operations, we estimate that cloud-seeding operations can enhance rainfall by as much as 30 to 35 per cent in a clean atmosphere, and by up to 10 to 15 per cent in a turbid [dusty] atmosphere.'[32] (The difference being that, in an already dusty atmosphere, there are plenty of existing particles to condense around.)

The Chinese government has long been an enthusiastic proponent of cloud seeding, using it to serve agriculture, protect from hailstorms, and even 'improve' cultural events such as the 2008 Olympics.[33]* Between 2012 and 2017, China spent over $1.34 billion

---

* Albeit a worldwide event like the Olympics is far more justifiable than the reported use of cloud seeding ahead of the Royal Wedding of Prince

on weather-modification programmes.[34] In 2020, it announced its intention to drastically expand cloud seeding across 5.5M km^2 – more than 1.5 times the size of India. This isn't just to increase rainfall, but also to reduce extreme weather events, causing rain to form earlier, spreading out over a wider area to ease its impact. The State Council issued a statement saying that 'China will have a developed weather-modification system by 2025', and that, 'by 2035, China's weather modification should arrive at a worldwide advanced level in terms of operation, technologies, and services.'[35]

Often the target for cloud seeding is snowfall, adding to the frozen snowcap reservoirs that so many rivers rely on. A University of Colorado Boulder study in 2020 found that cloud seeding 'can boost snowfall across a wide area'. Using aircraft over the mountains of Idaho, the study measured increases between 0.05 and 0.3 mm* of snow precipitation generated by cloud seeding.[36] In a 2008–13 study, the Wyoming Weather Modification Pilot Project concluded that cloud seeding had an impact of just 1.5 per cent on annual precipitation – which, compared to natural variability, failed 'to reject the null hypothesis that there is no seeding effect'.[37] It also found that most clouds don't meet the criteria for seeding. It requires Goldilocks conditions of everything being just right. Cloud seeding is, in summary, a lot of effort for very little reward.

But even if cloud seeding does work – is it ethical? The very obvious downside is that forcing rain in one region means a lack of rain that should otherwise have fallen downwind. Within large countries such as the United States, China and India, this could be benefiting one state and prolonging drought in another. It also has a transboundary impact on neighbouring countries. Aparna Roy, climate change and energy lead at the Centre for New Economic

---

William and Kate Middleton in 2011, with one UK company subsequently offering the service for any other happy couple not wanting rain to spoil their wedding pics for the princely sum of £100,000 ($123,000).

* Which brings to mind Andrew Tucker's phrase, 'Well, 0.25 mm is going to do fuck all.'

Diplomacy, writes, 'Speculations are rife that China may use its weather-modification technology as a strategic weapon against India to distort weather and unleash floods and droughts.'[38] There is precedent, here: America did exactly that with Operation Popeye during the Vietnam War. A 2017 paper by the Department of Geography, National Taiwan University, also argues that 'weather-modification activity can lead to charges of "rain stealing" between neighbouring regions', while also raising 'moral issues regarding state intervention in natural processes, the equitable distribution of resources, and best practices for weather modification to minimize harm'. It can also go very wrong. The Taiwanese authors cite a blizzard induced by cloud seeding in China, in November 2009, that caused over 50 million yuan (£7.5 million) worth of damage and forty deaths. Farmers in North Dakota have also campaigned against cloud seeding, claiming it made their situation worse, not better.[39] One farmer announced in a Town Hall meeting that a long-awaited drought-ending storm was actually dispersed by an attempt to seed it, complaining, 'They killed that storm before it did anything [and] they used three planes to do it!'[40]

Either way, it looks likely that cloud seeding will increase, as water-scarce states become more desperate. Rick Ledbetter of the Roosevelt Soil and Water District in New Mexico has run a cloud-seeding pilot scheme, telling The Guardian in 2021, 'I believe that there will be no choice in the future but to look at weather modification,' and that he was 'very hopeful for significant funding'.[41] Cloud seeding forms part of Wyoming's Drought Plan, with other states expected to follow suit. Many cite the University of Colorado Boulder study as evidence that it works. Yet the author of that study, Katja Friedrich, has repeatedly urged caution. She told Discover Magazine, in one of many such interviews, 'We can't mitigate drought just with cloud seeding . . . in order to really mitigate drought you need to address a lot of other issues.'[42] As is so often the case, the cautious reason of scientists is likely to be ignored.

CHAPTER 10

Tough Choices, Smart Solutions

By 2020, the Murray–Darling Basin in Australia had suffered its driest three-year period on record, worse than the Millennium drought it had only just recovered from. Bush fires in March horrified the world, burning unprecedented areas of forest and spewing smoke that circumnavigated the globe. A much smaller, lesser-known fire burned too. Angry farmers piled up and set alight printed copies of the Murray–Darling Basin Plan – a plan they felt left insufficient water both to farm with or to prevent such tinder-box conditions. Phillip Glyde, CEO of the Murray–Darling Basin Authority (MDBA), speaks with me in early 2021 from his office in Canberra – but it may as well be a bunker, such is the political crossfire he's caught in. He understands the anger, because 'it's this generation who'll pay for the reform. That's the thing that keeps me awake at night, actually. Water's a natural resource owned by the government. It's governments that have created [this problem] over the last hundred years by allocating out much more water than we ever had. And climate change is going to reduce it further . . . now, it's this generation having to sell the family farm. The economic pain is now.'

The Murray–Darling Basin is the size of Germany and France combined, spanning three of Australia's five mainland states. Its 9,500 irrigated farms generating more than AU\$8.6 billion a year of food and fibre, accounts for around half of all the country's irrigated agriculture. On average, 90 per cent of the rain that falls there evaporates; some soaks into soil and into the ground, leaving just a small percentage to flow into streams and rivers. The case for allowing 'environmental flows'* for conservation was only made clear via catastrophe: in 2010, the mighty Murray River no longer reached the sea, and its mouth began drying up. *The Sydney Morning Herald* described a 'tragedy', with 'forests of red gums, black box eucalypts and coolibah trees dead', and was clear on where the blame lay: 'the waters of the Murray–Darling rivers were notoriously over-allocated by overzealous government water agencies. So many dams were built by taxpayers that they could capture 130 per cent of the average annual flow of all the system's rivers.'[1] Needless to say, the Murray didn't have 130 per cent to give. Simply building more dams, continued the *Herald*, 'doesn't get you more water in the same way that opening a bank account doesn't get you more money'. The link to over-irrigation put the concept of 'drought' in a new light too – as not entirely climate-driven, but also man-made. The dead forests of the farmland-choked Murray River also helped to fuel the subsequent catastrophic bush fires.

In 2007, Prime Minister John Howard had declared that 'the old way of managing the Basin has reached its use-by date', as he passed Australia's National Water Act (2007). The Murray–Darling Basin Authority (MDBA) was set up as part of this and required to develop a plan. The subsequent Basin Plan (2012) aimed to manage the Basin as a connected, sustainable system. Crucially, it set the amount of water that can be taken from the Basin each year, while leaving enough for the rivers, lakes and 30,000 wetlands (including sixteen deemed of international significance), to be independently reviewed by scientists. In 2012, the Australian states signed up to the Basin Plan to stave off environmental disaster. Managed by the commonwealth government and the MDBA, headed by Phillip Glyde, its aim was to recover and then protect at least 2,750 GL (a huge 2.75 km^3) of water annually for the environment. Broad

* Also known as 'e-flows', this simply means water being allowed to stay in the river ecosystem and not be extracted.

reforms ushered in by the Basin Plan saw water heavily priced, water trading introduced and 20 per cent of water previously used for agricultural irrigation reclaimed and returned to the river as 'e-flows'.

The only way to reclaim that 20 per cent for the environment was for government to buy back water entitlements through the water-trading system. This quickly proved unpopular with communities, however, as farms were closed and jobs lost. According to MDBA figures, since the start of the Basin Plan, overall water abstraction from the river came down from 14,011 GL to 10,621 GL – an annual saving of more than 3,000 GL (3 km³) over six years. In a speech in 2019, Glyde said the figures proved that the Basin Plan is working: 'In my view, the Basin Plan means we are in a better position to deal with the current drought than we were during the Millennium drought or any previous major drought in this nation's history.'[2]

Due to a strange quirk of the Plan, this is partly thanks to one person. In terms of water, the Commonwealth Environmental Water Holder is possibly the single most powerful individual in the world. The position established by the Water Act, explains Glyde, means that the Environmental Water Holder controls over 2,000 GL of water entitlements that the government has purchased off farmers, to be used for environmental benefit. The MDBA sets the science and environmental targets, and the Environmental Water Holder's office houses about a hundred staff, but ultimately the decisions about delivery (which wetlands get flooded, and which get bypassed downstream) are made by one person alone – vested under the legislation 'deliberately, to keep her out of the reach of governments. It's a curious arrangement,' admits Glyde, with typical Aussie understatement, and the Environmental Water Holder has 'a fearsome responsibility'. But it works.

With consumption from the rivers reduced by a fifth, the Murray reliably reaches the sea once again. That was 'a really, really significant achievement' that has benefited the environment all the way down the river, says Glyde. 'It's flowing out into wetlands, it's bringing nutrients into the water, the vegetation and bugs, the fish, the birds.' The Plan's water-trading market is also used to purchase water rights between irrigators. The theory being that water trading helps to balance out agriculture in a way that favours wetter regions, and away from drier regions. 'In a lot of countries around the world, water and land are tied to one another; if you want to buy more water for your farm, you need to buy the next-door neighbour's farm,' explains Glyde. 'But the [Australian] Water Act agreed that it would be better to maximize the use of water to its highest value. So, the farmer that can get the biggest return from a megalitre of water will buy off the farm that's getting the least return.'

However, the political pain caused by the Basin Plan may yet take it down. As Glyde suggests, if you've over-abstracted for so long, people's presumption of what's 'normal' doesn't necessarily tally with reality. Saving water and fending off drought will always be highly political; a tug-of-war between short-term relief and long-term sustainability. Market signals that incentivize farmers to use water more efficiently, because every drop wasted costs money, favour higher-value crops, such as nuts and grapes. Glyde says this causes tensions between industries, with rice and cotton farmers being squeezed out of the water market. 'We do a lot of processing of dairy milk in this country, most of which goes overseas in the form of cheese or skimmed-milk powder. And the same in the rice industry . . . those secondary industries and processing are the big employers. So, there's a real fear among the rice and dairy process- ing industries, for example, that the water market is taking away their ability to get more product.' With AU\$2 billion to AU\$3 bil- lion of water being traded, it's also attracted foreign investors and speculators. A water-markets inquiry in February 2021 uncovered 'trading behaviours that can undermine the integrity of markets, such as market manipulation,'* and found that 'opportunities are best understood and leveraged by professional traders and large

---

* The inquiry also found that some farmers wanted a return to water rights tied to land rights – the very thing that caused the problem in the first place.

agribusinesses with the time and knowledge to analyse and navigate them.'[3] Anecdotally, says Glyde, a lot of pension funds and multi-nationals are moving in and buying up farmland: 'When I was in the agriculture department, Ferrero Rocher chose Australia to grow hazelnuts. It took them ten years to get through a quarantine with their trees to be able to do it. So, you're thinking, why are they doing that? That seems such a big investment. But I think . . . I'd *like* to think that the security of Australia's water markets and governance around water in the Basin is seen by countries, certainly outside of Australia –' he laughs – 'as a positive.'

Ironic, then, that the success of the MDBP, water trading and environmental returns has been such that it has attracted foreign investment to the detriment of local family farms. And that's why it's so difficult politically, causing what Glyde calls 'social disloca-tion': 'I've been working on natural-resource management issues for the last forty years, and water is probably the most controversial one I've worked on,' he deadpans. The Basin Plan is next reviewed in 2025, and one thing everyone can agree on, he says, is that no one wants to go back to the way it was before: everyone accepts the need for change. 'We're trying to change a hundred years of over-allocation,' he says. The difficulty is that everyone would rather that change came for others. 'Why am I, my community, my industry, my wetland, the one to be suffering?'

Less than a year after we spoke, Phillip Glyde took long-term leave, then announced his retirement.[4] The National Irrigators' Council, with which he was sometimes at loggerheads, issued a statement thanking him 'for his service to the very challenging water-reform process', adding that 'water policy and river manage-ment are complex, and everyone has an opinion.'[5] The latest row was over the final 450 GL of e-flow returns needed to meet the Basin Plan's legal requirement of 3,200 GL returned to the rivers by 2024. The Conservatives and Nationals were pushing for an extension – that is, to rewrite the law. The Greens and Labour were keen to honour the original Plan, but admitted the only way to do so was more compulsory 'buy-backs' from farmers, the last of which happened in 2018.[6] On 1 April 2022, in what some may have mistaken for an April Fool's joke, Phillip's successor was announced: oil and gas lobbyist, and former chief executive of the Australian Petroleum Production and Exploration Association, Andrew McConville.[7] The *Herald Sun* ran with the headline, 'Nationals put "upstream mates" in charge of River Murray'.[8]

The Jal Jeevan Mission

On the steps of Delhi's Red Fort, Indian Independence Day 2019, Prime Minister Narendra Modi announced an extraordinary ambition: to provide a tap-water connection to every household in the county, supplying at least 55 L per person, per day, in just five years, by Independence Day 2024. The scale of the task seemed impossible. As he spoke, according to the Jal Shakti Water Ministry, 83 per cent of rural households (161 million households, in a country of 1.3 billion people) had no mains water supply. Connecting them all in five years seemed crazy. But Modi had already achieved something similarly ambitious with the Swachh Bharat Mission, which, from 2014 to 2019, provided toilets to 60 million Indians to end open defecation.* Close to 100 million toilets were eventually distributed, making 567,489 villages 'open defecation free'.[9] The country went from 39 per cent toilet coverage to 99.3 per cent, within timescale and on budget. Now, India had an even bigger moon-shot goal to rally around: the Jal Jeevan Mission.[†]

On the frontline of achieving the Jal Jeevan Mission (JJM) are people like Shivani Sharma, project manager at the Women + Water

---

* Modi said, 'There were two things extremely close to Mahatma Gandhi's heart – the independence of India, and sanitation. But, given a choice between these two, he said that, for him, sanitation was even more important.'
† *Jal Jeevan* means 'Water is life'. Incidentally, 'climate change' in Hindi is *Jal Vayu Parivartan*: *Jal* (water), *Vayu* (air), *Parivartan* (change), recognizing water at the very heart of climate change.

Alliance in Bhopal. Sharma works across 600 villages in two of Madhya Pradesh's fifty-two districts. The furthest village she visits is an eight-hour drive from her office in Bhopal. The state relies on the monsoon rains more than groundwater. 'Whenever there is a good rainfall, we do not have much water stress,' Shivani tells me. 'For instance, in 2019, we had a very good rainfall, close to 1,400 mm. And, whenever such a thing happens, we usually do not face any water shortage, because we have a lot of reservoirs.' But the rain, thanks to the changing water cycle, is increasingly unreliable. As of 2019, her work with the Women + Water Alliance (part funded by USAID and Gap Inc.) has largely been incorporated into the JJM. Rather than ignore or replace hundreds of NGOs working on water in rural India, the Modi government chose to partner with the more effective ones and envelop their work. Shivani describes the JJM goal as 'very ambitious' – a word I hear from literally every Indian I talk to about it. 'So, we are assisting them.'

Typically, in rural Madhya Pradesh, the current water supply comes from hand-dug wells and ponds, or above-ground storage tanks connected to groundwater pumps. The pressure for quick results has sometimes meant that the villages with existing infrastructure are tackled first. That said, the success of JJM is already undeniable, achieving what no previous Indian government has been able to do, while being entirely transparent. Anyone can log on to the online dashboard and interactive map at ejalshakti.gov.in to see the progress to date, broken down by state, by district, even by individual villages. When I spoke with Shivani, in March 2021, the interactive figures for Madhya Pradesh showed 28 per cent of households connected to tap-water supply – a 17 per cent increase, or over 2 million households, since JJM launched in 2019. By mid-2022, the latest figures for Madhya Pradesh show 40.2 per cent of households connected to tap-water supply, precisely 3,562,561 more than when the JJM started in 2019. One district, Burhanpur, has already reached 100 per cent tap connection, and another, Indore, is not far behind (84.2 per cent, despite having started from just 16 per cent). Shivani's team is also involved in monitoring where and how much work has been done: 'the government sends out the proposal, then there's bidding, then a private company comes into the picture, gets the project and starts building it. We do observe lags here and there . . . but there is a lot of construction being done.'

India has been here before, with previous ambitious, wide-reaching schemes. There was the 1951 First Five Year Plan, the 1972–3 Accelerated Rural Water Supply Programme, the 2009 National Rural Drinking Water Programme. But there is a feeling in India that this time is different. 'It all comes under one umbrella, now,' Shivani says. 'All the different departments are engaged in the same project.' There's a feeling that 2024 is too early to be achievable, but Shivani makes the point that it should try to be as early as possible. 'Because not having water is a lot easier said than felt. I have visited villages during one o'clock in the afternoon, and there'll be a lot of people standing, waiting in line, in scorching heat, for water. It is pathetic.' And she means 'pathetic' in the true, compassionate sense of the word. Perhaps JJM's goal is too big, it's target too soon. But it is the exact opposite of dragging heels, kicking cans down the road, shrugging that nothing can be done. At the time of writing, halfway through the JJM, the percentage of households with tap-water connections in India has leaped from 16.75 to almost half (49 per cent) – it is looking more achievable by the day.

However, a major concern is where all that extra water supply is going to come from. It's one thing installing taps and pipes, quite another ensuring there is sufficient water to come out of them. Om Prakash Sharma of WaterHarvest India (another of the hundreds of small water NGOs on the ground), in Uidapur, Rajasthan, believes the JJM is achievable. 'Our prime minister is very good at setting these dreams which are pushing our boundaries,' he says, tactfully. But, 'we know that there is not enough water resource. The resources are drying up.' The JJM can only work, he says, in conjunction with groundwater recharge, grey-water reuse, demand reduction and community engagement. 'Even just Rajasthan is impossible,' suggests Om. 'The hilly regions, you cannot connect; the Thar desert, you cannot go into remote households with piped water.' Instead, he says, decentralized solutions are the way forward for remote communities. But he's clearly torn. 'I'm positive, you know – let's try it,' he says of the JJM. He describes working with the government on its previous Swachh Bharat Mission as like 'riding with tigers'. The same problem occurred – many toilets were built, but there wasn't the water to flush them or wash hands. But what both missions have undeniably built, says Om, is a momentum for change: an understanding that cleanliness, water and sanitation are human rights and national rights. I wonder if the JJM wording of 'piped water' may get dropped and replaced with 'water supply', to include off-grid water supply and rainwater harvesting. 'I totally agree,' says Om. 'Or "piped" could be connected to local groundwater-recharge solutions.'

Shivani, too, sees a bigger role for both water conservation and grey-water recycling. She says there are education efforts in place around water waste and water scarcity. For many, these are not unfamiliar concepts: 'They've experienced village-level migrations in which the water would dry up and they have to go to different villages to settle.' The communication challenge is to convince people who've never had something to treat it as scarce once it flows abundantly into their home. In the 600 villages Shivani personally works with, 'village-level grey-water management does happen.' Usually, household wastewater from laundry and dishwashing goes into drainage in front of every house, which usually leads to a big pond where it's all collected. Shivani and her team are now working on planting reed beds into these grey-water ponds to provide nature-based filtration, so the water can be sufficiently treated, pumped and used for farming. Funds for such grey-water projects are currently harder to come by, but, she believes, 'the government will have to do it' due to the scarcity of supply options. In its first year, the JJM 'majorly focused on pipe-water supplies and water quality', but, as it progresses – at least in the villages in Madhya Pradesh where Shivani works – they plan to 'focus more on grey water.' It seems an inevitable, natural progression – but it wouldn't happen without committed professionals like Shivani and Om on the ground, trialling and promoting these projects.

In Delhi, Raina Singh at the Climate Centre for Cities, within the National Institute of Urban Affairs, is helping urban areas to translate the JJM into action. 'This is quite an ambitious plan,' she says, echoing all others. 'I think one of the messages we are trying to advocate is that you cannot look at water supply in isolation; it has to be an integrated approach. Cities cannot just talk about 24/7 water supply and not look at conservation and rejuvenation of water bodies, not talk about groundwater recharge, about green areas . . . I think that is largely missing.'

What price, fresh water?

When Gary White co-founded Water.org with Matt Damon (yes, *that* Matt Damon), his big idea was to use microloans, not charity, to supply water in the poorest countries. Headquartered in Kansas, Missouri, Water.org's work is almost entirely focused on the Global South. The website blurb says it 'enables access to critical financing for household water and sanitation solutions for those living in poverty.' Its case studies include Grace in Kenya, who, with 'a small, affordable loan from a Water.org partner, installed a water storage tank on her farm . . . Grace can [now] feed her chickens, grow food in her garden, and use the water to cook, bathe, and do laundry.' Gary explains, when we speak: 'I would meet people, particularly women, who were paying 125 per cent interest to loan sharks, so that they could build a toilet at their home . . . What we did was create, basically, the on-ramp for microfinance institutions to start focusing [on loans] for water and sanitation.'

When we speak over video call in 2020, we both have what became known as 'lockdown hair'. Typically clean-cut and clean-shaven, Gary's chin now bristles, and his hair flows in numerous, unfamiliar directions; while mine, unkempt at the best of times, has made a resolute and sustained bid for freedom. I brush my fringe out of my eyes and ask him where it all began (the water career, not the hair). Gary laughs, telling me that, 'after studying engineering, I decided that the solutions actually all came down to finance!' This must be the first time I've heard an engineer admit that. If you really break down the solutions to water and sanitation, he says, 'from the NGO perspective, it's predominantly been to raise money to help people drill more wells . . . But, if you look at the scale of the problem, you're burying your head in the sand if you think that's going to solve the crisis.' As co-founder Matt Damon told *TIME100*, 'Gary and I were in Ethiopia and we looked at this beautiful, state-of-the-art well that an Indian NGO had raised money for and built . . . and it was gathering dust because it was so high-tech it had failed and nobody in the village was taught how to run the thing . . . Jim Kim told us when he was running the World Bank: "I can [reach] 85 per cent of the people – it's the last 15 per cent that are the hardest to reach."'[10]

Gary believes that, to build the water infrastructure to serve the 2–3 billion people in the world without adequate water and sanitation, it would cost between $1 trillion and $2 trillion, plus far more with annual maintenance costs. The reality, as I saw in Ghana, is that water is most expensive for the poorest people. People already pay a lot for – and highly value – clean water. Rather than work against that, White and Water.org work with it to provide affordable (rather than loan-shark) loans, that can be more easily repaid and the money reused. In turn, this can also attract the interest of the capital markets and social-impact investors seeking to put their money towards social good, ideally with a bit of interest on top (also known as ESG investment). White's 'Aha!' moment was when he realized he could matchmake these Wall Street financiers with people earning $5 a day. 'We saw the potential for hundreds of millions of people without money to tap into finance as a solution, so they still get the services that they want, but they're paying far less for it, in a much more regulated environment.' Their fund is expected to return around 8–11 per cent for investors, which moves it into the category of mainstream investments such as pension funds and insurance companies. Suddenly, like the JJM, that $1 trillion or $2 trillion starts to look achievable.

This 'bottom of the pyramid' investing is the type of thing money lenders have been getting excited about for years. Articles perennially pop up in the *Harvard Business Review* along the lines of 'Profits at the Bottom of the Pyramid', targeting 'the more than 4 billion people who individually earn less than $1,500 per year', and describing how 'value chains . . . need to be reconfigured to reach low-income buyers'. Such talk of turning poverty into profit makes a liberal environmentalist like me twitchy. Poor people's lives often get reduced to patronizing, even predatory terms, such as 'untapped markets' or 'the last frontier'. And yet millions of people, from Prampram to Rajasthan, *are* trapped in a vicious cycle of paying too much for water. If 8–11 per cent interest earned from the world's poorest seems exploitative, it's a godsend to those previously paying 125 per cent, or spending half their earnings on one jerrycan of water. Cheap loans make water supply more affordable for more people, which in turn creates the markets to sell, fit and maintain the pipes. 'The informal economy already exists', says White. 'The money is in the system, it's just moving around in an inefficient way.' The water market was often overlooked by microfinance institutions (MFIs), he says. 'They didn't recognize that, if somebody gets a water connection, literally overnight the savings that they capture are dramatic. If they were paying a water-vendor truck $50 a month for water before and now their loan repayment [towards their own water connection] is $5 a month . . . all of a sudden, they've got much more money in their pocket.'

Unlike loan sharks, MFIs only give out loans that can be repaid, with fail-safes for those that can't. Across Water.org's loans, 99 per cent are repaid, reaching (at the time Gary and I spoke) 2 million people every quarter – 8 million people per year. Gary's approach doesn't replace or compete with charity. 'There's literally 1,000-plus, probably far more, NGOs focusing on water, from the behemoths like WaterAid, all the way down to the local church', he argues. It's not really treading on anyone's toes to try a different approach. Even so, he fundamentally believes this is a better solution. In economic terms, it's a 'demand-side approach'. Supply-side means giving anyone a latrine; demand-side means helping people who want a latrine to buy a latrine. He explains the difference by recalling a visit to a rural community in Honduras where many people had three pit latrines at their house: 'one they'd store grain in, another they'd store tools in and a third they'd actually use as a latrine. I asked, "Why do you have three latrines?" And they said, "Because three different NGOs came along handing out latrines and we said yes."'

Initially, Water.org simply smoothed the path for the MFIs to lend money, but increasingly they help to provide funds through philanthropic capital too, because 'access to affordable capital is a consistent barrier to expansion, particularly in India'. Water.org is also directly investing in water and utility companies. 'Utilities are notoriously capital constrained,' says Gary. 'They don't have the access to capital that they need to expand the infrastructure into some of the poorest neighbourhoods.' There are efficiencies of scale when working with utilities. Providing a water connection to a whole street is cheaper and easier for a centralized utility company than it is to provide loans to every householder individually. 'That's exactly what we hear from the utility in Lima, Peru – one of our most successful water-credit programmes,' says Gary. The utility only responds when there's a 'critical mass' of people willing to sign up for connection in one neighbourhood. Water.org gives such utilities access to slightly cheaper capital on the proviso they'll connect all the households, including those that might otherwise go unserved – creating what Gary calls a 'virtuous circle of capital'. It could be a substantial circle: when we speak, Water.org's partner network has already loaned more than $2.6 billion in microloans.

Ultimately, Water.org may need to move into the biggest water ticket of all: reservoirs and canals. Ask a charity what they can do about national water supply, and they will – rightly – deflect it as a government issue. Ask Gary White, and it's open for discussion. 'The short answer is . . . this has got to be solved, because it's only going to get worse with climate change; water stress is going to increase. We can't ignore that infrastructure area.' That said, he also makes the point that water stress is often caused by unequal water use, with some using thousands of litres per day, while others get none; improving agriculture's water efficiency by just 2 per cent could supply everyone with their tens of litres of water per day.

Water.org's microloans are also playing a part in the JJM. India requires its banks to lend 40 per cent of their assets to 'priority sectors'; in 2019, Water.org worked in partnership with the Reserve Bank of India to reclassify water and sanitation as one such priority sector. This caused a huge uptick in funding. 'That really was a breakthrough,' says Gary. 'If the banks had already maxed out their priority-sector stuff, then they were going to stop lending for water.' This 'enhanced liquidity' could ultimately help the JJM to succeed. The Indian government recognizes how costly piping water to everyone will be, he says. 'So, being able to aggregate more capital is a big plus for them.' If the JJM succeeds and 100 per cent of India gets water supply by 2024, is that Water.org's mission over in India? 'No,' is his obvious and short answer. But his reasoning is that water access isn't a simple end goal; it's more of a hierarchy of needs. After you get water supply, you get a latrine, then a concrete floor so your feet don't get muddy, then a bathing area. 'What we're trying to do is to create an enduring capital market around water and sanitation access, so that people have the ability not just to get one loan and they're done, but to climb the water-and-sanitation ladder.' (It's a ladder we're all on, by the way, whether it's wanting a rooftop water-harvesting tank, a nicer toilet or an Ikea water-recycling shower.*)

If water is overpriced for the global poor, then it is underpriced for the rest, argues Torgny Holmgren at the Stockholm International Water Institute (SIWI). 'You realize the value of water the day that you don't have water – then the value is skyrocketing,' he says. 'But they don't put that value into the pricing mechanism, and that boils down to the possibility for extracting funding from banks and the

---

*   We'll get to that.

financial sector.' In other words, if water is cheap, then the market signals there's no problem, and nothing to invest in. In Pakistan, for example, says Maheen Malik of AWS, based in Lahore, water pricing is desperately needed, but culturally very difficult to achieve. 'Some people say that water is a right. Whenever any minister, for instance, comes in and says that they want to raise the price of water, he comes under fire. Water pricing is a huge problem, because we don't value water the way it's supposed to be valued.' This has included, she says, drinks companies heavily abstracting groundwater for free and then selling it on for a profit in plastic bottles. Pakistan is by no means alone, in that regard.

In England, too, water is cheap. A review by the UK water-efficiency body WaterWise found that 'the low price of water is a barrier [to water saving] . . . it creates little or no financial incentive for homebuyers to demand, and developers to install, water-efficiency and water-reuse measures. Financial savings on water bills are very low compared to savings made by energy efficiency and carbon offsetting.'[11] As for things like rainwater harvesting and grey-water reuse, you can forget it if the price of water remains low. Feargal Sharkey has concluded that the only way to sort out the country's mess is via big expenditure on water infrastructure, which will necessitate a higher price for water. He questions why water isn't subject to a tiered pricing model: 'If you've got an issue with universal access to water, fine,' he says, frame it this way: 'We think* that you should be able to live on 100 L of water a day, so we'll charge you a cheap flat rate for your household for 100 L of water per person. After that, if you want to fill up your swimming pool, go ahead . . . but we're going to charge you seventy-five fucking grand.'

To solve water scarcity and water stress, 'pricing is something we can't do without,' agrees Rick Hogeboom of the Water Footprint Network, in the Netherlands. 'I think there will always be a percentage of the population that will just not care or will not be persuaded to change their behaviour, or their diets, without that.'

* And so does the WHO.

In countries renowned for solving water supply despite relative water scarcity, such as Israel and Singapore, users do indeed pay accordingly. 'The price of water in Israel is very high,' Gidon Bromberg of EcoPeace tells me. 'Water use is quite low and that's because of price.' As many, from Feargal to the World Bank, propose, this works via a 'stepped' or tiered pricing structure: 'The first quantity of use is very affordable, but, beyond that, it becomes very expensive,' summarizes Bromberg. In his 2015 book *Let There Be Water: Israel's Solution for a Water-Starved World*, Seth Siegel writes that the Israeli Water Authority made this decision in 2008.[12] To charge the 'real' price for water, water bills increased by 40 per cent. 'The public howled, and logically so,' writes Siegel. 'If a mayor wanted the parks in his city to be watered every night, he could do so, but would have to pay for it out of his municipal budget. There would be no more "free" water.' Farmers were given a similar quid pro quo: they could have as much water as they liked, but they'd have to pay the cost price.* During past droughts, farmers would see their water allocations cut; now, they could be assured of delivery, as long as they paid. This instantly drove on-farm water efficiency and drip-irrigation technology: 'The effect of introducing real pricing for farms and homes almost immediately changed usage levels . . . Farmers didn't need a phased-in, multiyear step-up pricing schedule to give them time to transition to new crops. They began changing their water-use patterns in the first growing season after the announcement was made.' Similarly, household consumers 'found ways to save nearly double the amount of water.' It turned out, writes Siegel, 'that price was the most effective incentive of all.'†

---

* 'Cost price' more typically refers to a saving, stripped of any tax or profit mark-up. But water is heavily subsidized by states the world over, meaning the 'cost price' far exceeds the bill that lands on your doorstep.
† Meanwhile, Newsha Ajami at Stanford University told me, 'I live in a four-person family house, and we pay about $80–$100 a month for water *and* sewerage. I pay about $100 a month for my cell phone, just one cell phone. That shows you how undervalued water is [in California].'

Singapore has been using water pricing to drive efficiency for fifty years. In their book *The Singapore Water Story* (2013), authors Tortajada, Joshi and Biswas write that, in 1973, 'water tariffs were modified and an increasing block tariff system was introduced to raise the [water utility company's] total income and prevent the main wastages'. It is based on 'an escalating cost for every block of 25 m³'. Water consumption dropped 6.1 per cent the year it was introduced. A further water conservation tax was introduced in 1991, applied to domestic consumption above 20 m³. Professor Hu Jiangyong at the Centre for Water Research, Singapore, tells me that, as recently as 2020, 'we had another water-price increase by 30 per cent . . . This does drive people's behaviour, especially industrial consumers, to recycle their water. They see the benefit of doing that; it's going to save them a lot of budget. For domestic users too, people install water-saving devices.' High water users can still unrepentantly use water, but by doing so they contribute extra money to the utility company to pay for much-needed infrastructure. Tiered pricing means both more water and more money to go around.*

Big business and water stewards

Not that water's value can be captured by price alone. The international Alliance for Water Stewardship is something of a Fairtrade for water. AWS helps businesses to become responsible water users, step by step, and be accredited for it. Its two-pronged approach appeals to both the basest nature of business (saving water

* However, The UN-Water report, *Valuing Water*, in 2021, does warn that areas of extreme scarcity must be careful not to be too generous with lower tiers. As Day Zero approached Cape Town in 2018, the 'water utility was saddled with a complex Increasing Block Tariff (IBT), which failed to send customers a clear price signal that Day Zero was approaching and that everyone needed to conserve water . . . most customers in Cape Town were still receiving price signals that water was cheap and plentiful'. The lesson learned, they write, is that in times of severe demand, 'the cost of water provision' must rise accordingly.

equals saving money) and the higher-minded (look virtuous while doing it).

'We're a group of organizations who are committed to advancing sustainable use of fresh water,' explains CEO Adrian Sym. 'And we're a voluntary standard system that provides a market-based mechanism to enable major water users to improve their practice.' The idea being that any organization, no matter its size, can follow the same five steps: 1. Research and understand their water use; 2. Commit to a plan; 3. Implement it; 4. Evaluate it; and 5. Communicate it openly. Each step comprises detailed criteria with minimum requirements and 'advanced' indicators. Either way, continuous improvement is part of the deal. It's a water audit, essentially, including supply chains and public stakeholders, paying 'particular attention to traditionally disadvantaged and potentially less-vocal groups, such as indigenous communities, women, children and the elderly.' Like Fairtrade (where Adrian did in fact work prior to running AWS), it uses 'a voluntary mechanism to try and bridge the gap between sustainability and policy or regulation. Regulation often only defines the minimum requirements and is often poorly enforced. Sustainability, generally speaking, needs a lot more than that. Systems like ours try to fill that gap.'

The AWS standard works with asparagus farmers in Ica, Peru – the very arid region we visited in Chapter 7, with just 8 mm of annual rainfall and thirsty export crops that have dire local consequences. In 2020, six large producers in the Ica Valley committed to the AWS standard. Shortly after, AWS signed an MoU with SUNASS, Peru's National Authority for water and sanitation, to apply the AWS standard across the municipality of Pueblo Nuevo, becoming the world's first local-government authority to do so. 'Pueblo Neuvo is a very important place,' explains Carla Toranzo of AWS in Peru. 'SUNASS did a hydrological study that found that [it's] important for the recharge of the aquifer. The idea is that Pueblo Nuevo could close the gaps between water and sanitation.' This could include reforestation, wastewater treatment and aquifer recharge. The hope is, she says, that 'for the first time, you will have the public *and* private sector aligned in a region.'

Pakistan's AWS coordinator Maheen Malik is riding the high of helping her country's first textile factory gain AWS certification just months before we talk. 'They started putting in water meters and implemented a computerized SCADA system* so that they have the data sets. They started changing their washing systems for their jeans; they did stakeholder engagement with communities; they also committed to develop a rainwater harvesting system for their facility.' Ongoing AWS events then helped to promote that best practice throughout the textiles industry in Pakistan and abroad. 'Interestingly, when they got their certification, other industries really felt the competition and wanted to step up their game,' says Malik. 'It wasn't that they wanted to just get themselves certified; they want to do it at a better level. So, competition was also a driving factor in all of this, which is a good thing.' International buyers and supply-chain pressure is 'a huge influence'. And that begins with consumer pressure, which Malik says should not be underestimated. She recently attended a small-suppliers sustainability summit organized by Levi's, where targets for 40 per cent water use reduction, among other things, were announced. 'Levi's essentially said that, if these goals were achieved, they'd do business; but suppliers who couldn't deliver were out. So, there is a lot of pressure from the international markets that translates into what these industries are doing,' says Malik. Doing anything to request better water practices of your retailers – whether via AWS, the Partnership for Cleaner Textile (PaCT), the Better Cotton Initiative or another standard – has a trickle-down effect to the practices in supplier countries.

The Water Funds Network, founded by the Nature Conservancy in 2014, is another attempt to bring both private and public sectors together to work on river catchment restoration. Downstream users

* Supervisory Control and Data Acquisition (SCADA): an industrial control system used across manufacturing.

have a direct incentive to invest, but so do public utilities, private foundations and donors. Kenya's Upper Tana-Nairobi Water Fund, for example, invested in planting trees, riparian buffer zones and teaching farm-terracing techniques to improve water use. The leader of Greater Cape Town Water Fund, Louise Stafford, further explains: 'We need to have a public–private partnership because the problem is so big that government just doesn't have the resources – there's a lot of competing priorities. The private sector is dependent on water, if you think of the big bottling companies, agriculture – if there's no water, there's no business. So, it's in the interest of big water users who depend on the water for their survival to help turn the tide. But it's not the private sector's responsibility to maintain that over the long term . . . it's [the] government's responsibility to maintain it.' The Cape Town Water Fund is based on private-sector funding for the first six years, then working with the city authorities and national government thereafter. In 2021, there were forty-three such Water Funds globally, primarily in the Global South, with thirty-five more in development.[13]

The reason for working with business, rather than campaigning against them, is made clear by Conor Linstead at WWF-UK: 'Business are big taxpayers, and often have a very strong voice with government,' he says. 'If we can make a joint case with business around fair water allocations that reflect environmental needs, then our case is much stronger.' In the Doñana National Park wetlands in Spain, for example, WWF works with (and puts pressure on) major retailers and growers to collectively influence how water is managed. Doñana is an important wetland on the migratory route of many wildfowl species coming from Africa to Europe. But it's also been used for water-demanding berry crops that are irrigated from groundwater, and that's affected the wetland. Most UK supermarkets import berries grown in Doñana and 'there's been a whole number of initiatives to try and influence how farmers are using water in the area,' says Linstead. But it continues to suffer from over-abstraction. Illegal wells keep popping up and are hard to trace directly to the grower responsible. Sometimes illegal boreholes appear in the protected area itself: 'They go in at the dead of night with a tractor, bury a pipe and connect up to their farm, kilometres away. So, it's very difficult to trace.'

In 2022, a year after we talk, Doñana hit the headlines again when the Andalusian government decided to legalize those illegal wells, effectively granting an amnesty across 1,900 ha of land near the National Park. Money talks, and berry farms provide jobs and boost GDP. Spain's national minister for ecological transition, Teresa Ribera, urged the Andalusian government to abandon the plan, warning of large fines from the EU if this protected ecosystem was damaged further. Andalusian President Juan Moreno Bonilla dismissively retorted, 'The park cannot be a fishbowl which people go to look at. It must also clearly be an element in the area's economy.'[14] A land-use plan for sustainable water use had previously been agreed by all parties in 2014, but remains unimplemented. On World Water Day, 22 March 2022, companies – including some of Europe's biggest retailers (Aldi, Asda, Coop, EDEKA, Lidl, Migros, Sainsbury's and Tesco) – co-signed a letter with WWF, addressed to Moreno Bonilla, stating that 'we would like to see the area prosper from both a social and an environmental point of view' and noting concern that the proposed amnesty for illegal abstraction endangered 'the sustainability of water . . . in the long term [and] the reputation and the long-term development of the region.' It called again for the implementation of the 2014 plan. Similar co-signed letters were sent in 2014 and 2019. Linstead says this is an example of good water stewardship. But it also raises the question: when are strongly worded letters no longer enough? Retailers have continued to buy from Doñana in full knowledge that the agreed land-use plan has yet to be implemented. By the end of summer 2022, the Viñuela reservoir near Malaga was down to almost 10 per cent of its capacity and the municipality were siphoning off half the water from desalination plants to farms to keep crops alive[15] – an expensive sticking plaster to cover a gaping wound.

When the first assessments that formed the Water Footprint Network were commissioned by the likes of Coca-Cola, Heineken

and Unilever, Rick Hogeboom tells me, 'One of the major surprises
to most of them, especially in beverage, was to see that most of the
water, over 90 per cent, was in supply chain.' That was intended to
be a first step, like the AWS's Step 1. Research and understand. But,
over a decade on, Hogeboom, despite working closely with many
multinationals, is still waiting for the second step: 'So far, I feel that
there is a lack of taking ownership of the water in supply chains by
these companies . . . I would have liked to see much more action
after those first 2007 reports.' His feeling is that progress was stalled
when the true scale of the challenge became clear.

Applying labels

There are other ways to help encourage, or shove, companies
towards best practice. Consumer-facing water labels – stuck on
products to tell you how water intensive they are – could be one of
them. In 2009, Finnish food company Raisio became the first to vol-
untarily put a water footprint on a product packet. Its box of oat
flakes showed that 101 L of water were used to produce 100 g of oat
flakes, of which 99.3 per cent was in growing the oats, 0.57 per cent
in manufacturing and 0.16 per cent for packaging. You'd have
thought that Arjen Hoekstra, the inventor of the water-footprint
concept, would have liked that, but in fact he later wrote of the Raisio packet, 'The label itself gives no insight into whether this
footprint is good or not, so it does not offer a basis for conscious
product choice.' It couldn't help you to compare apples with oranges
(or oats with oats from a different water catchment). Even so, Hoekstra's successor, Rick Hogeboom, believes that water labels are 'a
step in the right direction. Because, right now, as consumers, we're
just groping in the dark.' I ask, if he was asked to design one tomorrow by the Dutch government, what his preferred choice would be.
He's clearly already given it a lot of thought, with such a phone call
being a distinct possibility: 'A label that says, "This is the water footprint and this is the unsustainable share of the footprint." So, this
chocolate bar cost 1,000 L, 600 L of which are unsustainable (that

part's in red) and 400 L are sustainable (that part's in green). I think that's the level at which both policymakers and consumers operate – they don't want all the nuance; they just want to see "good" or "bad". Standards like AWS could step in to help define and decide what is 'sustainable'.

Where water labels make most sense are on domestic appliances: your tap (faucet), shower or washing machine. In the US, the Environmental Protection Agency's WaterSense label is available on toilets, showerheads, garden sprinklers and even entire new-build homes. WaterSense-labelled products are typically 20–30 per cent more water efficient than the market standard. The label appeals not only to water-conscious consumers, but to dollar-conscious ones too, with rebates offered in twenty-two states. The city of Brighton, Colorado, has offered rebates to residents since 2016, including $125 for a high-efficiency washing machine and $100 for a 4.2-L-per-flush toilet.*

When my own Bosch dishwasher, inherited from the previous house owner, gave up the ghost in 2020, I looked for the most water-saving replacement I could find. Without a similar water label in the UK, this wasn't straightforward. Many retail websites don't list water consumption, and, if they do, the metrics vary (some state 'per year' water use, which, unless you keep a daily spreadsheet of your dish-water use, is pretty pointless). After a lot of searching, to my surprise I found that a new Bosch SD6P1B only uses 6.5 L per cycle in eco-mode. Compared to 12 L a minute from my kitchen tap, a three-hour wash cycle for a whole load of dishes would use less water than forty seconds of running the tap.† I contacted Bosch to ask what the water

---

* A similar flush rate – 3.5–4.5 L – has been mandatory in Singapore for over twenty years.
† I've talked to a number of water-saving experts and they all confirm that, counterintuitively, most modern dishwashers use a lot less water than washing dishes by hand. Even some of the more wasteful modern models I found used 10–14 L per cycle, which is still barely more than a minute of running the tap.

consumption of my old model was, for comparison. The answer: 'Due to the age of your appliance –' which, by the way, can't have been more than ten years – 'we do not have the data of the annual water consumption.' In short, they never knew. Making appliances water efficient, even in the 2000s, just wasn't a thing. Andrew Tucker, water-efficiency manager at Thames Water, confirmed that's, 'Absolutely the norm. The thing that we are missing the most in the UK compared to other places around the world is a mandatory water label. We have the A–G European energy label, but there's no equivalent for water. There's nothing driving the market for development and there's nothing driving consumer-purchasing behaviours. We're working very, very hard to get the government to introduce that.' Tucker, who spends his professional life reviewing the effectiveness of water-efficiency measures, believes that 'a water label is the only thing that's going to influence every device in every home and every business . . . Nothing else will do that.'

A report by the Energy Saving Trust into long-term reductions in water demand in the UK, commissioned by Waterwise in 2019, found that mandatory water labelling of 'fixtures, fittings and appliances given a rating based on their water use in a similar manner to the energy use' would reduce PCC (per capita consumption) by 6.3 L per day within ten years, and 31.4 L per day in twenty-five years.[16] The label scheme in Australia, WELS, running since 2005, gives a star rating out of six, plus the amount of water used in litres per wash for dishwashers and washing machines, and per minute for showers and taps. An independent review of WELS found that it had saved around 20 L per person, per day, saving Australia as much as 204,000M L by 2030 (equivalent to achieving half of that elusive remaining e-flow return required by the Murray–Darling Basin Plan).

Smart choices

But why stop with minimum requirements? Just how low could water use go if we really pushed the technology further? The UK

Energy Saving Trust report found that that a per-person daily demand of just 49 L was possible (lower than Cape Town's strictest rationing during Day Zero) with 'high-tech solutions such as water-less toilets, integration of "smart" devices, innovative tariffs and "pay-per-use" services'. In 2020, a group of businesses led by Procter & Gamble set about trying to achieve this. Calling themselves the '50 L Home Coalition', the stated ambition is to create urban homes (or, more specifically, the domestic appliances within the homes) that can run on 50 L a day. With pilots underway in Los Angeles, Beijing, Mexico City and Delhi, Frantz Beznik, associate director at Procter & Gamble, told me that 'the inspiration came from Cape Town'. Rather than view low water use as restrictive, he instead wants to (slightly nauseatingly, but stick with it) make 'the sustainable irresistible'. The challenge for Beznick and the 50 L Home Coalition is to make 50 L 'feel luxurious', going from using 50 L 'because you have no choice' to making 50 L 'feel like 500'. Achieving this largely includes reusing and recycling water – domestic water systems that take a small amount from the central water supply each day, but then reuse and recirculate it again and again without the householder even noticing.

Which brings us back – as ever, in water circles – to the Dutch. A Netherlands start-up called Hydraloop has created a water-recycling system that collects household wastewater, cleans it, and sends it back into the household pipes. The company claims that the system could reduce a household's water use by 75,000 L a year. Similar in size to a fridge-freezer, fitting well in garages or utility rooms, Hydraloop works without a filter, instead using sedimentation, UV disinfection and an aerobic bioreactor. Once re-cleaned, the water can be reused in toilets, washing machines or for any other non-potable purposes. One customer, David Mertens, who lives with his two children in Grimbergen, Belgium, claims that his household use (combined with Flanders' mandatory rainwater-harvesting tank) is now only 13 L of mains water per person, per day.[17]

There are water-recycling showers on the market, too. Whereas a typical shower could use upwards of 12 L a minute, and an EPA WaterSense-certified showerhead rains down 7.5 L per minute, the Flow Loop – built by a Danish company, partnering with IKEA – uses just 1–2 L a minute. The system creates a circulating loop of warm water, essentially sending the same 16 L of water round and round for each eight minutes of shower time. Instead of letting all the warm water go down the drain, Flow Loop recovers it from the shower floor and passes it through a filter and a UV system before recirculating it back to your head, cleaned and ready to be reused. This meets Beznik's vision of making less water feel like more. Whereas that WaterSense-certified 7.5 L shower works much like the low-flow regulators I've fitted to my own shower, effectively making each shower half as satisfying, the Flow Loop pelts it back out at you at what feels like 12 L a minute.

The smartest appliance of all, though, may just be the smart water meter. My own water meter is a 'dumb' meter, with a similarly dumb online account that just shows me my quarterly usage – I have no idea when that usage goes up or down by hour, day, week, even month. Smart meters instead track consumption on a minute-by-minute basis, showing exactly when your water use is highest and lowest, encouraging lower usage and immediately notifying of leaks. Initial trials of smart meters in the UK have shown they reduce consumption by 7 per cent. A study by Water UK suggests that smart meters could reduce water use overall in the UK by between 368M and 482M L per day – more than RAPID's Grand Union Canal Transfer, Southampton's proposed desalination plant and the London Effluent Reuse scheme combined.

Andrew Tucker oversaw the smart meter rollout in London for Thames Water, with an eight-year plan to get them to as many customers as possible. In four years, Tucker says, they've installed over 400,000 smart meters across the capital. Whereas dumb meters rely on manual readings taken every six to twelve months, 'We're now getting over 10 million reads a day.' All that data 'is game changing. For example, with dumb meters, you might not realize there's a leak inside your property until you get your bill, but with a smart meter the water company will be able to detect something's wrong within twenty-four hours and give you the options to fix it. It's giving us incredible insight and power to act faster and more proactively to stop water loss.' This has already uncovered surprising levels of constant leaks in households: 'Approximately one in ten, or 10 per cent, of all the homes never stop flowing. That's not water being used, it's precious water being lost,' from invisibly leaking loos to dripping taps. Similarly, the Indian Smarter Homes WaterOn device – a combined smart-metering and automated leakage-prevention system – has been fitted to thousands of apartment buildings in India, and is claimed to have reduced water consumption by an average of 35 per cent.

The smartest solution, meanwhile, may be to solve the water crisis and the energy crisis at the same time. India began shading canals with solar panels in Gujarat, simultaneously reducing evaporation and producing renewable energy, back in 2012. A 1-MW pilot project was launched by a then regional minister called Narendra Modi, covering a 750-metre stretch of the Narmada Branch Canal with solar panels, preventing evaporation losses of 9M L of water annually. If just 10 per cent of the state of Gujarat's canal network were similarly shaded, it could produce an estimated 2,200 MW of solar power and save 20B L of water. Land is also expensive in India, meaning that these linear solar farms could be put up with no extra land-purchasing cost. In 2019, the Narmada Main Canal was given the green light to fit a full 40-km section, producing 100 MW of electricity, the same as the gas-based power station in Sepahijala District, Tripura.[18] Algal growth is also drastically reduced by the shading, while the water helps to cool the solar panels above, increasing their efficiency by around 2.5–5 per cent.[19] California is now following suit. Project Nexus, a public–private–academic partnership involving the University of California, is fitting the first-ever solar-panelled canal in the United States. The $20-million pilot, due to be completed in 2024, will cover just a couple of kilometres. But the UC believe the pilot will show that covering all of California's 6,440 km of canals could save 227B L of water annually, comparable to the residential needs of more than 2 million people, plus generate 13 GWh of power (more than six Hoover Dams).

Reservoirs can be fitted with floating solar panels, too. China, as ever, can lay claim to the world's largest floating solar array: on the surface of a reservoir near the city of Huainan, installed in 2017, are some 166,000 floating panels, producing a combined 40 MW. A 2018 World Bank report estimated the global potential for floating solar arrays on man-made reservoirs could exceed 400 GWh. That means that floating PV on dammed reservoirs could exceed the energy output of the hydro-dams themselves. If just 6 per cent of Lake Mead were covered in solar panels, for example, it could theoretically produce 3,400 MW – considerably more than the Hoover Dam's 2,074 MW. While, if just 1 per cent of Africa's megadams were fitted with floating solar, it would double Africa's hydropower capacity. Forget the arguments about solar farms replacing agricultural land; putting them on our dams and canals, reducing water loss and improving the solar efficiency should be the priority.

CHAPTER 11

Simpler Solutions

Chloe Whipple was hard to find. She performed her one-woman show *Life on Water* in Exeter, England, in 2016, but when I contact her five years later, I find her running a vegan pizza restaurant. 'I guess my career's taken a bit of a U-turn,' she admits, having trained in London in contemporary theatre, 'but it's been a good adventure.' *Life on Water* was an immersive piece describing her year spent living with just 15 L of water per day. The 50 L Cape Town Day Zero rations would have been a luxury for Chloe. A project to match local artists with scientists saw her pair with Exeter University's water expert Dr Peter Melville-Shreeve, who told her that the average Brit uses 150 L per day. Finding it hard to believe that people used that much water, she explored the idea with her theatre director of perhaps living with half that? Maybe even less? 'Fifteen litres seemed extreme enough for that to be a potential struggle or have an awareness,' she tells me. A year later, she stood on the Bike Shed Theatre stage with her three 5 L water bottles, a measuring jug and a plastic bucket, and recounted her experience.

I wonder how many times over the year she regretted landing on that arbitrary quantity, or wished it was higher. 'Never, really,' she says. Most days, she had to use less than 15 L, maybe 8 or 9 L, in order to save up enough to do her laundry at the weekend. She'd started out handwashing her clothes, but decided after a couple of months that her washing machine, using 43.5 L on an eco-wash, could wash more clothes than she could manually. Initially, a lot went on toilet flushing too, but a small grant helped her to get a 200-L rainwater butt plumbed into her toilet. Collected rainwater was counted as free to use. She reused water from her dishwashing bowl, and 'I would always bring a bucket into the shower, so I could reuse that too . . . It's really interesting, all of the opportunities that water has to have an extra life, instead of just coming through the tap, hitting something and then going away again.' Even so, her showers were three minutes max, and only every three days, helped by the poor water pressure in her one-bedroom flat. 'If I had one of those power showers, they would have probably only been a minute long!' she says.

Her 15-L limit didn't stop when she left the house, either. 'It was wherever I was – at work, at home, visiting friends, going out, on a holiday or whatever. That did lead to some interesting conversations, especially at work, which included a theatre tour that year.' Specifically, she couldn't flush a toilet when she was out because she didn't know the water-use capacity of each toilet; so she'd use the toilet before or after friends to share 'their' flush. 'I had to break a kind of taboo conversation,' she acknowledges. Her monthly period was another. She used a Mooncup and needed enough water to wash that hygienically. 'It made me think about what it's like for girls going to school – in this country, but also in other countries, where maybe the toilet facilities aren't up to scratch, and the impact that has on their self-esteem or on them being able to have access to education fully,' she says. 'Water flows through our life in such an intersectional way, it affects so many aspects of our life.'

As the year progressed, and she researched water issues more, the nature of her project changed from one of restriction to one of abundance: 15 L a day was hard, but surprisingly achievable. She felt she had proved that 50 L could, in fact, be a lot. And that 150 L, the UK average, is needlessly wasteful. She says, rather than having to give everything up, that 'changing our behaviour a little bit' can have a big impact. We maybe don't need as much as we think we need. It also changed her relationship with the rain. In the UK, despite living on the relatively dry south coast, 'There is so much rainfall,' she stresses. 'I mean, obviously, there's more and more dry periods now. But there's so much that we could be harvesting and storing.' Talking to Peter Melville-Shreeve made her frustrated with the lack of engagement from developers about potential solutions. 'There's just literally no one making decisions for the long term, it's so short-sighted.' In the UK, she says, 'We get annoyed when it's raining,' when really, 'We should be like, "Oh my God, it's raining! This is amazing!" It definitely changed my relationship with rain. Rain is good.'

You might think, then, that Chloe's message would be that everyone should do more to reduce their personal water use. It is partly that. But her main message is quite different. 'I feel like, a lot of the time, the onus is on us as individuals to completely change our behaviour, and buy all the stuff we need to be self-sufficient – buy this, do that – that it's our sole responsibility . . . When really it's a much bigger picture.' The waste that happens in our water systems, the pollution of our rivers, the leakages in the underground pipes, the building codes that allow developers to build water-inefficient houses: all that happens before we can even turn a tap on. The wastage, the 150 L a day, is built into the system. Chloe *also* found there is a lot we can do as individuals, but those things almost all boil down to rainwater capture and grey-water recycling, both of which can and should be integral to water-company practices and building codes. In Belgium, for example, you don't *consider* installing a rainwater-capture tank in a new-build home – you *have* to. 'Yes, individual action is really, really important,' she says. 'But bigger systemic action is the missing part of the puzzle.' Remember the guy who installed Hydraloop and now only uses 13 L a day? That's 2 L less than Chloe, with none of the hardship.

When I speak to Chloe's academic collaborator, Dr Peter Melville-Shreeve, he has just sold his shares in the rainwater-harvesting company he founded and has returned to full-time lecturing at Exeter University. Therein lies a story in itself. He's a firm believer in the need for rainwater harvesting to achieve water security in the south of England. Yet, when he tried to commercialize retrofit rainwater systems for domestic houses, the market just wasn't there. My own downstairs loo, for example, would be perfect for a rainwater system: it unusually has an outdoor cistern from an old/odd extension that pre-dates us. I connected a 110-L wall-mounted water butt to the outside wall, hoping that it could later be linked to that toilet somehow, to save on mains water. With a 7-L flush used roughly five times a day, that could save 35 L a day (albeit less during summer months, when the water butt might run dry) and perhaps up to 10,000 L a year. But at least three 'rainwater-harvesting' companies I spoke to told me it couldn't be done – that such a system doesn't exist. If I wanted to rip up my garden to bury a several-thousand-litre tank-and-pump system, fine; or they could post me a small garden butt for watering the plants. But there was nothing in between.

At one of the companies I contact, RainWater Harvesting Limited, Ian is happy to answer my questions via email. He says the issue with storing a small amount of water above ground is twofold: first, there's the potential that it could freeze in winter; second, 'and more importantly', there's the potential that it will heat up in the summer months, 'causing bacterial growth . . . where you get legionella.' When I push him on the likelihood of that, he replies, 'Unfortunately if toilets are flushed and there is bacteria present it then becomes airborne, so it is an issue.'* I ask Ian if he's getting more queries like mine from homeowners worried about drought. Is demand increasing? No, he says. 'The vast majority of systems supplied are for new builds and unfortunately the authorities do not seem to be on board with harvesting – developers add a 200 L water butt to a new house and call it harvesting and authorities accept it, which really is a nonsense and doesn't save anything.' His views are that 'the average householder is oblivious to the water scarcity that is facing us', while retrofitting is stuck at the starting line because of the 'high cost associated'. And this from someone who runs a

---

* Three years on, and my wall-mounted water butt has never frozen in winter, while the water remains cool in summer.

rainwater-harvesting company. No wonder Peter went back to academia.

Peter confirms that this lack of motivation to retrofit is typical in the UK. When I put the legionella risk to him, he says, 'How many people have you seen putting their head down the toilet when they flush?' If that seems glib, he adds, 'If it's a concern, you just install a UV filter.' The barriers in the market, he says, remain 'inertia' on the part of water companies and local authorities, 'and the low price of water.'

Again, it requires systemic change. Charging a fixed rate for all water customers currently means that Person A, who lives next to the reservoir, pays exactly the same as Person B, who lives 80 km away from it, with high energy costs to pump the water and a lot of water lost through leaky pipes on the way. While it is fair and just that they pay the same, from the water companies' perspective, why not ensure that Person B has a supplementary rainwater system installed? The flip side of this is wastewater. In most parts of the UK, when there's a rainstorm, the storm water fills drains and enters CSO systems, which at best fills storm tanks at the sewage works, but at worst (as we saw in Chapter 6) combines with sewage and directly pollutes rivers. If all houses had large rainwater tanks – as they do in Belgium – more water would be collected at properties instead of entering CSOs, and would slowly be released over the coming days and weeks as households used it in toilets and washing machines. According to Peter, in some locations, where there's high pressure on the combined sewage network, 'a water company might spend £10 million, £30 million' to upgrade a sewage works. Instead, there could be a decentralized scheme to retrofit rainwater systems. 'You will benefit by saving water, but there's a larger holistic benefit to having a rainwater system for storm-water management.' That benefit is more for the local authority or water company than the individual householder; and at the moment it seems neither is willing to foot the bill.

Having hit his head against this brick wall for years, Peter is very clear on why the water companies *don't* do this. 'To break it down simply, in any housing development, half of your paved surface is roadways and parking and the other half of your paved surface is roofs. If you installed rainwater-harvesting systems, you can halve your storm-water run-off because you've captured it off the roof . . . the cost benefit is: is it cheaper to disconnect all streets from storm-water run-off, or all the roofs? To disconnect the streets, the wastewater company can install one expensive project in one location that it has control over. It's easy to build an investment plan to say that's a good idea. If I disconnect all of the roofs into water butts or below-ground tanks, those assets are decentralized, dispersed and left in the community, and no longer under their control.' From an engineering and business point of view, it's messy – and engineers and accountants alike don't like messy.

But there is a third way. Between 2018 and 2020, South West Water gave Peter funding to pilot his own design on a hundred properties on St Mary's in the Isles of Scilly. The year-long pilot evaluated the benefits of rainwater collected from the house roofs and stored in 600–1,000-L internet-connected tanks, using Peter's patented system which communicated real-time data. When the tanks were full, any additional rainwater spilled back to the drainage system; and, if the tanks ran empty during dry periods, they would automatically top-up with mains water, meaning that householders never noticed a change in service level (what Peter calls 'fit and forget'). The system could even drain the tanks in anticipation of a major storm event, to avoid overwhelming the sewers. The water savings alone – not even taking into account the wastewater and storm-water benefits – came to a staggering average saving of 54 L per day, per house (20,000 L per year), and there was minute-by-minute data to prove it. As small islands with limited water supply, the Isles of Scilly are often linked with potential investment in desalination. Peter proved that the water coming freely from the sky could fill that supply–demand gap and more besides.[1] All that talk among water companies, with whole conferences devoted to how to reduce that 150 L PCC figure, and the solution was there on people's rooftops all along. The pilot was regarded as a success, but the project was discontinued, and is probably gathering dust some-where on a South West Water desktop. Meanwhile, water companies still put forward desalination-plant proposals – such as Southern Water's ill-fated £600-million Fawley scheme to provide up to 75M L of water per day.[2] You'd get similar volumes, at far less cost, if you fitted all households with Peter's system.

Rainwater harvesting and storm-water tanks should clearly form part of the solution to the water crisis. Even greater savings can be made when these solutions are combined with smart appliances. Another project Peter was involved with saw over a hundred Pro-pelair low-flush toilets installed on the Exeter University campus. Using an air tank as well as a water tank and an air-tight toilet lid, working in a similar way to an aeroplane toilet, the Propelair uses just 1.5 L of water per flush, compared to the 7–10 L of a conven-tional toilet, or even the 4.2 L of a US WaterSense toilet. I ask Peter the 'If you ruled the country . . .' question: what would he do to make such changes nationwide, rather than university trials? He says, if he was designing new housing developments for 100-L PCC, he would install an appropriate rainwater system alongside Propelair air-flush toilets and ensure there was a covenant in the deeds stating that householders must always have 1.5-L flush toilets. And that's it, he says. 'Boom. Job done. They can have a nice shower; they can even have a washing machine. But you can get down to 100 L easily with a bit of technology.' Peter believes the barrier for both air-flush toi-lets and rainwater harvesting is the lack of scale. 'If we introduce scale and go beyond early adopters to sell 10,000, 50,000 a year of a product, then the price point comes crashing down. If the policy shifted, then investors would pull the money in to make it happen.' In other words, if building standards and specifications call for it, it will happen.

At the moment, it's only the early eco-adopters. In Manchester, England, where it rains for an average of twelve days every month, Manchester Metropolitan University's Birley Campus includes a 20,000-L underground grey-water tank. Rainwater is used for all the building's toilets – with zero cases of legionella – lowering their water bill by 60 per cent. In Buckinghamshire, Steven Lear uses rainwater harvesting to provide drinking water for his 400-plus beef herd. 'I don't think we should be feeding cows chlorinated water,' he says, with a firmness that makes me wonder whether we should be feeding ourselves with it, either. Rainwater run-off is collected from the shed roofs, sent to underground tanks, then put through UV and particulate filters. 'It works really, really well. And that's probably saved me £9,000 a year in water bills.'

Away from the UK, Germany has a much more established market for rainwater harvesting. Most family houses and apartments are fitted with rainwater tanks, and approximately 75,000 new household rainwater systems are installed every year. Regulations introduced from 1989 to 2011 made it possible, with a combination of tax breaks and charges per m^3 of rainwater sent down the drain. A 2013 German study found 'significant savings' from using rainwater for non-drinking purposes including toilets and washing machines, saving households hundreds of euros a year and around 54 L of drinking water per person, per day (exactly the same as Peter's Isles of Scilly trial).[3] Mandatory rainwater-harvesting tanks make Belgium's national PCC just 100 L, too, a figure that UK water companies would kill for. Marijke Huysmans explains, 'The toilets are flushed with rainwater, not with tap water. We even use it for the washing machine, and it works very well. These things are very common here in Belgium.' It is, she says, 'a very easy and cheap solution; it has no disadvantages. We pay much less for our water because of this.'

In Australia, 26 per cent of households have a rainwater tank. Almost all rural properties have it as their main source of water, storing rainwater for months on end, typically combined with sand filters to treat it before use. Some cities, such as Melbourne, have installed rainwater-harvesting systems in over 30 per cent of urban residences too. According to the Rainwater Harvesting Association of Australia (yes, there is one – an indicator in itself), rainwater harvesting provides Australians with an estimated 274B L annually: that's over 10 per cent of all residential water in Australia, saving some AU$540 million in municipal water costs. By contrast, the largest desalination plant in Australia, Wonthaggi Desalination Plant in Victoria, cost AU$3.5 billion just to build, before ongoing running costs, and can deliver 'only' 150B L of water a year.

In 2019, the mayor of Mexico City, Claudia Sheinbaum, started the *Cosecha de Lluvia* (Rain Harvest) programme to install domestic rainwater-harvesting systems. A lot of public resources were invested to make it happen, says Anaid Velasco, an environmental lawyer in Mexico City. 'It has worked well, especially in those neighbourhoods like Iztapalapa, where they do not have access to water.' The Mexico City environment secretariat stated in January 2022 that, 'From 2019 to 2021 we have installed 31,239 harvesters, benefiting almost 150,000 people directly, which has allowed a decrease in spending on the purchase of pipes [and] better use of time by providing water autonomy to families.' The programme operated with 50–100 per cent subsidy, recommending that the water be used for washing patios, bathrooms, clothes, and even, perhaps surprisingly, 'for personal cleanliness such as bathing, washing hands and dishes.'[4]

Concern over the water quality of harvested rainwater is something of a Western privilege. In many countries, harvested rainwater is the *most* trusted source, because it hasn't yet had the opportunity to become polluted. Nirere Sadrach of End Plastic Pollution Uganda tells me, 'Literally every house in Uganda will have a tank on the roof or a basin to collect it, then you can use it for most domestic purposes.' This is premium water, rather than just something to flush the toilet or water plants with, he says. 'For us, we grew up fetching water from the roof; you feel more comfortable . . . if you get your water from the roof. Ah, you feel better!' he sighs.

Storage, however, is a problem – people in poverty cannot afford large tanks, let alone sterile ones. Working with rural communities in the volcanic hills of Nicaragua, the local charity Nuevas Esperanzas helps build rainwater-harvesting tanks using ferro-cement structures. A basic yurt-style shape is formed from chicken-wire mesh, and cement mortar is then hand-trowelled over; building one needs only basic construction skills and materials that can be carried to the site. With closed roofs, these ferro-cement tanks are designed to let rainwater in but keep mosquitoes out. According to the charity, the tanks can last for thirty years or more, holding up to 40,000 L of rainwater. For one community in northern Nicaragua, it used to be a 1.5-hour return journey on horseback to their nearest well; now, they can just turn the tap on.[5]

In Rajasthan, India, Om Prakash Sharma has been a director of the small Anglo-Indian water charity WaterHarvest since 2000. He tells me that, in the last thirty years, he's come to see successful water-infrastructure planning as small-scale and locally managed. When WaterHarvest was founded, it was called Wells for India, and focused on digging village wells in rural Rajasthan. But, with ground-water levels falling as water extraction increased, they had to keep deepening the wells. Eventually came the realization that they were not just fighting against the tide, but contributing to its decline. Harvesting water and restoring groundwater levels would be much more effective. WaterHarvest now build rainwater *taankas** to catch and store the rain, providing domestic water all year round. These underground cement tanks are either connected to the guttering from rooftops or are fed into from sloped on-ground channels. They look much like the Nicaraguan yurt-style structures, but are partially buried, with just the roof poking above ground. Each of the 1,700 *taankas* WaterHarvest has constructed typically sits within the centre of a rural village and fills with around 20,000 L of water during monsoon season, designed to provide 8–10 L of water per person, per day. The *taankas* are made as watertight as they can be, with no exposure to sunlight, to avoid biological growth. There is some unavoidable bacterial contamination, says Om. But, compared to all other local water sources, it's the best quality available. The groundwater wells in the Thar region are typically 'highly saline, and fluoride levels are often very high,' says Om. 'So, usually people [prefer] catching water on the surface. Such *taankas* have been used for 500, 600 years – we just add the modern technology, like

---

*   The English word 'tank' actually comes from the Hindi, *taanka*.

concrete. And it's not just in the desert; we can do this in the hilly regions too.'

Rajasthan is also known for *beri*, a larger tank prevalent in areas with a layer of clay or gypsum that can form concrete-like channels; these can hold up to 500,000 L.[6] In total, Om says he has documented forty-five different traditional rainwater-harvesting techniques in Rajasthan alone. Many of these have been revived in modern times by Rajendra Singh, known as 'the Water Man of India' – also a WaterHarvest partner and advocate for localized solutions. Alongside the Jal Jeevan Mission is Jal Shakti Abhiyan ('catch the rain'), which Narendra Modi relaunched on World Water Day 2021, making the link clear that 'the better India manages rainwater, the lesser the country's dependence on groundwater . . . To save the country from a water crisis, it is now necessary to work rapidly in this direction.'[7] Shivani Sharma, in Madhya Pradesh, tells me this has initiated a government drive to build rainwater-harvesting systems on every municipal building roof in local villages.

Not that rainwater harvesting always requires rooftops. Some cities have built giant underground storm-water reservoirs. Pakistan's cities are flooded every monsoon season, only to become water-stressed again for the other ten months of the year. To address both issues, Lahore built its first underground storm-water storage system in 2020 to collect rainwater for reuse. The Lahore Monsoon Underground Water Reservoir was built at a cost of about 150 million rupees (£1.6 million) and stores 6.37M L of rainwater, using nearby Bagh-e-Jinnah park to provide some additional filtration. The water agency, reports Pakistan's Gulf News website, 'is keen to replicate the system throughout the city . . . Most of the collected water will be used to irrigate the Lahore city's gardens and parks. The project will also help recharge the irreplaceable groundwater.'[8] They even built a tennis court over the tank to provide residents with additional sports facilities.[9]

The Municipal Corporation of Greater Mumbai, India, has also earmarked 3 billion rupees (£33 million) in annual spending for climate resilience, including the construction of three floodwater tanks with a collective capacity of over 26M L. Underneath the public parks of Melbourne, Australia, storm-water tanks – filtered by the grass and planting above – can capture 230,000 L of rainwater at a time, returning it to the Yarra River and nearby storage ponds. In total, the system captures from a catchment of 34 ha, taking in 47M L of rainwater each year, at a cost of just AU$1.6 million (£900,000).[10] In 1998, Berlin built a storm-water tank underneath Potsdamer Platz that holds 3.5M L. The largest in the world, though, is in Tokyo, Japan. The G-Cans Project took seventeen years to build, at a cost of $2 billion (£1.8 billion), and is designed to withstand a once-in-200-years flood. It's popularly known as the Temple, thanks to its fifty-nine 20-m-tall internal columns spanning the 177-m tank. If it were ever to fill up, which seems unlikely, it would hold over 250M L of water.[11]

Ancient wisdom

The sixth-century Roman Byzantine Empire, however, had already built something arguably more impressive – and more ornate – than Tokyo's G-Cans. The Basilica Cistern was the largest of the many water cisterns that served Constantinople, now Istanbul. It was used to store water to supply the city and was built large enough to withstand long sieges.[12] There was even a filtration system. Supporting the ceiling are 336 Corinthian columns, each topped with ornate Roman capitals. The cistern is 138 m long, 65 m wide and 9 m high, with an astonishing storage capacity of 800M L – bigger than the present-day underground tanks in Melbourne, Mumbai and Lahore put together. The structure still stands, and is very high on my 'must visit' list.

In fact, storing rainwater in underground tanks is an idea as old as civilization. The famous red-stone facades of Petra, Jordan, were built by the Nabataeans – an ancient Arab culture between 400 BC and AD 106 who knew a thing or two about water engineering. It's not what every visitor to Petra looks for, but as you pass through the narrow rock canyons of the Siq, anticipation building for the famous Treasury facade that will appear around the corner any moment, you notice a shelf-like groove cut into the rock walls at around waist height, snaking along the route. Astonishingly, these rock troughs once carried sealed, pressurized water pipes, for both hot and cold water: clay pipes for cold, lead pipes for hot. I saw remnants of these pipes on display in the Jordan Museum in Amman. Water filters also removed lime particles.

The very name Nabataean is believed by some historians to derive from the Arabic *nabat*, meaning to extract water. In fact, given the water scarcity in the region then and now, you could argue they knew more than us about how to be water sufficient. Petra, known to the Nabataeans as Raqmu, had eight natural springs, plus over twenty small dams. The dams and channels controlled the flow of rainwater and protected Petra from flash floods – and still do, with one as recently as 2019 largely tamed by the Nabataean system. The channels lead to underground cisterns that stored enough water to supply the city for months. Some of the larger ones had vaulted stone roofs held by stone columns, much like the Byzantines' Basilica Cistern, built some six centuries later. Many cisterns in this area are still used by local Bedouins. One open Nabataean cistern, which looks like a small rectangular swimming pool, is still used in the local village of al-Humayma.

The water in Petra not only served the local population of circa 20,000, but also provided agricultural irrigation and water for the passing trade caravans. Archaeologists have so far recorded more than 200 water-storage cisterns at Raqmu-Petra alone, with a total capacity of about 40M L, enough for some 100,000 people. Aqueducts were also built to transport water from distant springs and carry run-off water from hills and mountains. Nabataean farmers practised flood irrigation – just like Bart Fisher and the Palo Verde Irrigation District. The canals followed natural land contours, while the pipes within the city descend according to precisely engineered gradients – the Siq wall pipeline I walked alongside is at a 2-degree slope, calculated to have released water at a flow rate of 23 L a second (twice that of my kitchen tap).[13] According to the Museum of Jordan, the complexity and extent of the water systems imply that they were under central-government control. A piece of papyrus restored in 1995 records a legal dispute between two Nabatean neighbours over the water spout on their connected roofs: Stephanos had built an extension which took more of the water, and Theodoros did not approve. Rainwater was well worth suing your neighbour over.

The tradition of surface-water engineering didn't come from nowhere. The famous Mesha Stele, a polished black block of basalt about a metre tall, carved with thirty-four lines of writing, is dated to around 840 BC. Written in the name of King Mesha of Moab (now in modern Jordan), it contains an interesting glimpse into ancient water management:

(21) . . . I built Qarboh: the walls of the parks and the walls of
(22) the acropolis. I rebuilt its gates, and I rebuilt its towers. I
(23) built the palace and made the retaining walls of the
reservoir for the spring inside
(24) the city. Now there was no cistern inside the city in
Qarhoh – so I said to all the people, 'Make your
(25) selves each a cistern in his house.'

Amazingly, though, both the Nabataean and Moab water systems were preceded by some 3,000 years in the north of Jordan. Bronze Age Jawa, close to today's Syrian border, used a sophisticated water system designed to store winter rainfall in reservoirs, with a series of deflection dams that directed flash floods through stone-lined canals into storage reservoirs. Constructed 5,000 years ago, it is one of the earliest known water-harvesting systems in the world.

Why am I telling you about ancient water engineering? Because it makes a simple point: we have a lot to learn from old techniques. Many pre-modern humans lived and thrived in regions we now consider water scarce, from the desert cities of Arabia to the Hohokam of the Sonora. Modern nineteenth- to twenty-first-century water engineering offered miraculous (and profitable, for construction companies and politicians) short-term solutions, but left a time bomb of long-term problems for future generations – most notably, as we've seen, dams silting up and groundwater pumped to extinction.

King Mesha ruled a hydrologically advanced kingdom almost 3,000 years ago. Jawa's technology enabled the desert town to thrive for around 1,500 years. Arguably, the Nabataean system is still going strong some 2,000 years on, with its cisterns and flood defences in use to this day. By comparison, the King Abdullah Canal, the backbone of present-day Jordan's already struggling water system, is barely sixty years old. The Glen Canyon Dam may not reach a hundred, with strong calls already to pull it down. It's hard not to look at modern water systems – and our relative lack of rainwater harvesting – and wonder what went wrong.

A display in the Museum of Jordan tells me that the average Jordanian today uses 90 L of water per person, per day. This, says the information sign, compares very favourably to their more wasteful neighbours: 652 L in UAE, 570 L in Bahrain and 438 L in Kuwait. There is clearly pride in being the region's most efficient water user (more efficient even than neighbouring 'water superpower' Israel). There's a hint of victimization too: in Jordan, the numbers seem to say, the water crisis is not their fault. And yet the typical rooftop water tank – visible on the flat roofs of every building in Amman – holding around 3,100 L, is refilled either by mains pipes or by water-truck delivery. None of them, or at least very few, are connected to rainwater harvesting, like that of Stephanos or Theodoros (whichever it was that owned it – but let's not start that fight again). In Amman today, the precious few times it rains, it floods and runs down the drain.

We also talk of organic and regenerative farming as if they are modern inventions. Yet prior to the Green Revolution of the 1950s and 1960s, which introduced modern fertilizers and agrochemicals, *all* farming was organic and regenerative. James Alexander, in Oxfordshire, told me that his grandfather, soon to turn ninety, remembers ploughing with horses. 'They only ploughed down to three inches,' said James. 'Now, we plough to six or eight inches. They always used to do less damage. *And* there was more crop rotation. *And* there was livestock on the fields [to naturally fertilize].'

Many regions around the world are now looking to rediscover old techniques, as we move away from expensive, polluting, centralized projects towards more resilient, localized, off-grid solutions. For hundreds of years, in south-central Bangladesh, floating vegetable gardens called *dhap* or *baira* would rise and fall with the flood waters. Given Bangladesh's ever-present flood problem, where 80 per cent of the country is floodplain and 48 per cent of people are landless, modern farmers are returning to this old practice, which uses rafts made from dried water hyacinth or paddy stalks, typically about 6 m long, and up to 55 m.[14] Journalist Archana Yadav writes that the *dhap* are found to be ten times more productive than on-field farming, with the organic beds constantly watered and rich in nutrients. Ingeniously, when the floods recede, the platforms turn into nutritional compost on the ground for the plants to keep growing.[15]

The fields of La Huerta around Spain's third-largest city, Valencia, are a maze of channels, ditches, weirs and floodgates invented by the region's Muslim rulers 1,200 years ago. Eight main irrigation channels, or *acequías*, funnel water from the River Turia, which then gravity-feeds a series of smaller branches, distributing the water to thousands of small growing plots. The amount of water each plot receives isn't measured in terms of volume, but rather on how well the river is flowing. The unit, known as a *fila* (from the Arabic, meaning 'thread'), represents an individual's right to a proportion of water over a period of time; the irrigation cycle usually lasts a week, but when the river's low, the cycle is extended. Journalist Keith Drew writes, 'It's an incredibly efficient system. Each plot receives the same access to water for the same amount of time, no matter where they are in the mosaic, and there are no water shortages, even in periods of drought . . . the result is an incredibly diverse crop yield.'[16] The Water Court of the Plains of Valencia, which adjudicates it all, established around the year 960, is believed to be the world's oldest judicial body.

Peru's Ica Valley, of asparagus fame, used to be known for a similar system of sustainable irrigation called *pozas*. A nineteenth-century traveller to the Ica Valley observed that 'all fields are inside boxes made of the same earth . . . one vara [83 cm] in depth. The seasonal rains from the mountains fill these boxes that contain up to 2,000 plants. They [the farmers] leave water bounded. Until harvest, they irrigate once . . . without fear, for the yield will undoubtedly be abundant.'[17] By 'irrigate once', he literally meant once a year – and the region was no less of a desert back then. Modern anthropologists have found smallholders continuing to use *pozas*. Visitors in 2008 described farmers' complaints about the declining water table caused by modern agribusinesses, causing the water bounded by their *pozas* to now infiltrate too quickly. Here was a system, hundreds of years old, at risk of being destroyed in a couple of decades by industrial over-abstraction; yet it showed how sustainable agriculture could work there all along.

Peru could yet return to older ways and be a world leader in regenerative, aquifer-recharging agriculture. In 2018, the country enacted a Framework Law on Climate Change. Article 3 states that it, 'recovers, values and uses the traditional knowledge of indigenous or original peoples and their vision of development in harmony with nature' and 'protects, restores and sustainably manages the hydrological cycle . . . through management and land use planning that anticipates their vulnerability to the effects of climate change and guarantees the right to water.' In late 2019, journalist Alberto Ñiquen Guerra sought out the *qocha* system in the high plains, 4,300 m above sea level. There, check dams and reservoirs believed to date back to the Incas store rainwater, reducing run-off and soil erosion, increase infiltration and conserve wetlands and springs. 'Now everyone is enthusiastic [about *qocha*] because we all want water!' a local told Guerra.[18] The Climate Change Adaptation Program now run training schemes on the construction and management of *qochas*.

In Iran, Rastynn had shown me pictures of the *ab anbar*, the white-domed village water tank, similar in design – and age – to the *tankaas* of India. But even older and more substantial are the Iranian *qanats*.* These comprise a series of vertical shafts interconnected at the bottom by a clay-lined tunnel of running water, typically cut into hillsides. *Qanats* serve a dual purpose, both tapping into the groundwater as a source and directly feeding rain into the aquifer to provide an early form of managed recharge (sorry, Glasgow, consider your MAR record beaten). The Iranian Qanats and Historic Hydraulic Structures website explains: 'The first shaft (mother well) is sunk to a level below the groundwater table. Shafts are sunk at intervals of 20 m to 200 m in a line between the groundwater recharge zone and the irrigated land.' This gives an idea of just how large these systems are, with an average length of about 6 km (3.7 miles). Much like La Huerta in Spain, subsequent water use from the mouth of the *qanat* is tightly managed according to custom, with shareholders taking turns to irrigate their fields every twelve days. According to the Iranian Ministry of Energy, Iran has over 36,000 *qanats* – if you lined them up end to end, they would go round the equator nine times (how's *that* for mega-infrastructure). There are now attempts, given Iran's extreme water crisis – potentially the world's worst – to fully revive the *qanat* system, most of which has fallen into disrepair. A paper in 2018 by Amin Mahan at Islamic Azad University, Iran, found that modern deep-well drilling had meant 'many *qanats* have been totally destructed and are no more operable'; yet, due to their centuries-old, zero-emissions, gravity-fed systems, 'it seems wise to regenerate *qanats* and reuse [them] on a big scale'.[19]

---

* In fact, the *qanat* is not unique to Iran – with different local names, they can be found in Oman, where there are more than 3,000; North Africa; the north-western desert of China, Afghanistan, Pakistan and Central Asia – and many still supply water. According to the Middle East Institute, *qanat* technology can be found in more than thirty-four countries.

In 2016, *qanats* were collectively given UNESCO World Heritage status.[20] In his speech presenting the award, Gary Lewis, UN resident coordinator, Iran, addressed the significant decline in the use of *qanats* for water supply. 'In order to meet the growing demand for water, large-scale pumping of groundwater by modern machinery has occurred. This has led to the near-complete eclipse of Iran's older and *more sustainable* system of harvesting underground water.' Instead, he urged, 'Reactivation of *qanats* can solve some of these problems. They can offer part of the solution to Iran's looming water crisis.'[21] In 2021, the *Tehran Times* reported that the centuries-old Behjat-Abad *qanat*, feeding into central Tehran, was indeed to be restored, and quoted an official saying, 'as we are facing the shortage of water, reviving these *qanats* could be a good solution.'[22]

Small scale, big ambition

The lesson from all these ancient water systems (and those mentioned above barely scratch the surface) is that multiple, decentralized systems are often more sustainable than the megaprojects of today.* The problem, says Om Prakash Sharma at WaterHarvest India, is that, despite the work of people like him and Rajendra Singh, traditional techniques are still being spurned in the name of progress. 'Look at Uidapur. It was the city of rainwater harvesting. Now, the whole city is dependent on lakes that they keep filled up to keep tourists happy.' The Indira Gandhi Canal, which runs through the Thar Desert from the Indian Punjab some 470 km away, was a megaproject built following the Indus Water Treaty (1960) between India and Pakistan. It made the desert bloom. With it,

---

*   Albeit some ancient systems can rightly be considered megaprojects – the 80 km brick-vaulted Zarch *qanat* in central Iran, built around 1300, is believed to be the world's longest underground aqueduct. Even so, with 36,000 *qanats* in the country, the point still stands that they are many and localized, not few and centralized.

'popular interest in [traditional] small-scale systems went,' laments Om. 'Now, 23 million people who live here are dependent on these large-scale systems. [But] you cannot survive forever like that.' Northern India's water woes – like Pakistan's, like Libya's Gaddafi-built follies, like Arizona's CAP – are proof of that. 'We are literally destroying 5,000 years of water wisdom in the whole region [within] thirty to forty years,' says Om. Water used to be 'treated as a god' in the desert regions of Rajasthan. 'No one was allowed to contamin-ate catchments – these things were community managed . . . Now, water is a commodity, you can buy it, you can throw it away, you can contaminate it, and you think that technology will solve it.' If that happens, 'We are heading for disaster.' It pains Om to talk nega-tively. A naturally optimistic, practical person, he balances it with a positive. The famous step wells of Jaipur – the thousand-year-old ornate temples with tiled rainwater-capture wells – are being revived. This, he says, could perhaps turn the tide towards a wider revival movement.

While there's an awful lot to be learned from clay and clod approaches, modern techniques can still be high-tech while learn-ing the decentralized lesson. In Ghana, I got in touch with Christian Siawor primarily to arrange my visit to the Teshie Desalination Pro-ject, at the total opposite end of the water-solutions scale. Siawor's a senior manager at GWC, the government department respon-sible for all Ghana's water supply and infrastructure – the epitome of the centralized, municipal approach. And yet, like many entre-preneurial Ghanaian's, he has a side hustle – and I enthusiastically take him up on the offer to go see it.

From Teshie, it takes us an hour to get to the site, once a semi-rural plot of land in Ablekuma, near to the Densu River, but, like so many parts of Greater Accra in the last two to three years, now swallowed up by the urban sprawl, with the clog of traffic that comes with it. Chris and his driver listen to an evangelical sermon on the car radio on the way. 'Without *faith*,' says the preacher, 'it is impossible to please God.' When we eventually pass into a narrow side lane, a heavy, solid metal gate slowly opens to reveal something quite surprising. Despite being a senior manager at GWC, with his own time, money and engineering skills, Chris has designed and built a mini off-grid water-treatment plant here. A 3-m-tall water-intake and mixing tank stands in gleaming stainless steel next to two equally tall, impeccably welded chemical tanks. An air-conditioned control room is currently being wired inside a reused shipping container, the roof of which also houses the water-intake tank. Water, pumped up from the river, is then gravity fed, step by step, through the treatment process. There's even space on the plot for lawns, a beautifully planted border of plants and flowers, and a fishpond – all fed by the wastewater. The solid-waste sludge is dried in the sun and sold for agricultural manure, much as I saw at Banbury sewage works.

'About 400 m from this place, we have the Densu River,' says Siawor. That's where the Weija municipal reservoir takes its raw water from. 'My intake system draws the water to this facility, then it will actually go to the top.' He points to a blue tank on top of the shipping container, where the chemicals will be added. 'The entire facility is designed by myself, and the alum dosing system is actually my patent,' he tells me proudly. 'You will not get this design anywhere else in the world.' It then goes through aeration, chlorine, and a dash of soda ash to adjust for pH. 'I trained as a biochemist,' Chris explains. 'I wanted to identify a chemical process that would optimize the water-treatment process. Then I moved into engineering and construction.' This included doing a master's at Delft, in the Netherlands, that spiritual home of water engineering.

Siawor has been nurturing this side project for some time, mostly under the radar. His gravity-fed dosing system won a global award at the International Water Association Awards in Kenya, but he didn't tell his colleagues about it. As soon as the plant is up and running – which he hopes will be soon – he plans to run his company, Blue-Grey Aqua Technology, full time. The water he treats will initially be sold to private water tankers, but his medium-term plan is to provide low-cost water to the community. He's keen to stress that it will cost 'less than the water sold by the GWC . . . I'm not doing this for money. I'm just doing this to serve the community.' Most people can't afford GWC's approved tariff – or, if they can, it's a significant part of their monthly expenditure.[23] The huge water-treatment plants I visited in Kpong, near the mighty Lake Akosombo, produce excellent, high-quality water, but expensively pumped over many kilometres of hills. 'You do not need that volume of money,' argues Chris, as we look around his small plant, which could be reproduced several times over, close to point of use. 'If I can build this with my own salary, it means that, if government supported private industry, we could really move to develop the country.'

This site in Ablekuma is both a pilot to show what can be done, and a working, ready-to-go water-treatment facility. Chris believes it would take under a month to build replicas of the same unit – which has a ground footprint of barely 8 m x 15 m, and could be compacted further, if you lost the flower beds. His initial test run produced 30 m^3 of treated water per day, but he believes it has the capacity to handle 130 m^3. This could see many such small treatment units built near rivers and lakes throughout Ghana, easing reliance on the 1960s megaplants now straining under the weight of ever-growing demand. Chris's system would enable water-treatment expansion at the same speed and growth as the urban sprawl of Accra itself. It seems far better suited to the reality of Ghana and many other developing countries.

Rather than pick a relatively clean water intake, such as the Volta system on the east of the city, Chris chose the Densu specifically because it's so polluted. 'It's one of the most difficult waters to treat in Ghana,' he says. If he could prove the concept worked here, he could prove it works anywhere. 'I am developing this facility with a mindset that I will not engage a lot of people to operate it. It is semi-automated, so operational costs are low. I'm an industry person who appreciates such issues.' He's also used premium-grade materials, so it will be low maintenance. Crucially, close to source, the energy costs are low too, with his patented gravity-fed dispensing system requiring no power.* I think of Teshie and wonder whether having many small-scale treatment plants such as this is a better solution for countries like Ghana than desalination. 'I think so. I honestly think so,' says Chris.

Chris's ideal solution would be to develop his mini treatment sites *for* GWC, rather than competing with them. He doesn't want this to be a nice little addition to the system – he believes this should *be* the system. 'I'm so surprised they are not developing [infrastructure] like this . . . you could actually dot facilities like this around the city many times.' He hopes one day the site could be entirely run with solar panels, too, improving not only the energy footprint, but also reliability in a country where power outages are common and disrupt even the mighty Kpong plants.

There's space for many such scalable, off-grid solutions to address water challenges. In Dakar, on the west coast of Africa, a Bill and Melinda Gates Foundation project aims to be a mini NEWater factory. The Janicki Omni Processor takes up the space of a basketball court, and turns latrine and cesspit waste into electricity, water and compost. Faecal sludge is heated, the steam both turning an electricity turbine and being captured and condensed (triple filtered) to produce drinking water. The first commercial-sized unit in Dakar produced 7,000 L of clean water per day, while processing an impressive 30 metric tons of faecal sludge. It has also since been installed at dairy farms. Bill Gates wrote on his blog, 'Unfortunately, rich-world solutions aren't feasible in poor countries – they require too much expensive infrastructure. The idea behind every Omni Processor design is to solve this problem.'[24]

Back in Accra, Chris says with calm, understated pride, as we look around his immaculate site, 'No one knows about this.' It's quite some secret to keep. I'm excited for him, but I know he'll face

---

* Chris tells me its patented name is Constant Head Gravity Fluid Dispenser, saying, 'If you google it, my name will come up.' And it does. (https://patentscope.wipo.int/search/en/detail.jsf?docId=WO2014 006429)

a lot of barriers and resistance in taking it to market in a country where central government control of infrastructure is deeply ingrained. 'It is something that will actually surprise a lot of people here,' he admits. In the car back to central Accra, the radio news announces overcrowding in the Greater Accra region. An expert is quoted, saying, 'The Greater Accra population density is now over 1,658 [people per square kilometre] – or two people per square metre . . . think about this from the perspective of how diseases can easily spread.' Especially, I think, when people are not connected to water and sanitation.

CHAPTER 12

Restoring and Rewilding

It's not unusual on a long countryside walk to end up carrying a stick. But this is no ordinary stick. The bark has been stripped from it with fresh, broad-toothed bite marks, the likes of which haven't been seen in this part of England for 800 years. The base of this willow branch, bitten clean off the tree, clearly shows the grooves of three convex teeth. This was bitten by a beaver. And, until just six months ago, no beavers had been spotted in Derbyshire since they were hunted to extinction in the twelfth century. So, why have two breeding pairs now been returned to the Willington Wetlands, near Burton-on-Trent, with the help of water company Severn Trent?

When I grew up in the Midlands, not a million miles away from Willington Wetlands,* wild beavers in England were unthinkable. They were exotic creatures, the stuff of North American storybooks. Or from a time long ago, akin to rhino or crocodile bones turning up in archaeological digs. However, unbeknownst to me and most British schoolchildren, the Eurasian beaver hadn't gone extinct – though, it very nearly did. At the end of the nineteenth century, just 200 remained in Germany and a smattering across Eastern Europe and Russia. Slightly bigger than the North American beaver (yes, American readers, our beaver is bigger than your

---

* Sixteen miles, to be precise – though I'd never heard of it, because it was still a working gravel quarry back then. The wetlands were flooded and given to the Derbyshire Wildlife Trust in 2005.

beaver), it was always easy to hunt, with large, easy-to-spot river-side lodges. Its pelt was fashionable, its glands used for castor oil, its meat popular – even the tail was a traditional meal to have on Friday nights.* But, by the early twentieth century, Friday-beaver-tail-night was no more (unless you lived on the Elbe river, in Saxony). Reintroduction programmes to save them began in Sweden as early as the 1920s, and in France and Switzerland from the 1950s. But never in Britain.

The last beaver in Scotland is mentioned in the 1526 *Chronicles of Scotland*. However, in 2007, a licence application submitted by the Scottish Wildlife Trust and Royal Zoological Society of Scotland to undertake a trial reintroduction of beavers at Knapdale, Argyll, was finally approved. In 2009, the splash of beavers was heard on the lochs for the first time in 500 years. Around the same time, there were sightings of beavers in the River Tay – over 160 km from the approved beaver site – likely released by vigilante conservationists. Attempts to capture and rehouse the Tay beavers failed when it became clear that there were too many of them (they may have been there for several years), so, in 2012, the Scottish government decided to 'tolerate' and monitor them too. That's when the water compan-ies started taking an interest. According to the first progress report to the Scottish government, the beavers were providing 'ecosystem services', including 'increased groundwater storage, flow stabiliza-tion and flood prevention', along with a marked 'improvement to water quality.' In particular, 'during dry summers beaver dams and canals have been shown to hold more than 60 per cent more water (including in the water table) than comparable environments with-out beaver activity, and the delivery of standing water during times of drought might help alleviate the negative socio-economic impacts of droughts.'[1] South of the border, having barely survived the

* It's thought that Friday-night fish-and-chip suppers replaced beaver tail, because a flat, battered beaver tail looked much like a flat, battered fish.

drought of 2012 without resorting to standpipes, these words mois-
tened the parched lips of the English water companies.

Early progress was tentative but determined: a reintroduction at
the River Otter in Devon; another in Cornwall; then one in Wales. In
2019, an organization called the Beaver Trust was founded. By
2021–2, there was a beaver boom. Reintroduction sites appeared all
across England. London saw the return of beavers in an enclosed site
in Enfield, 'roughly the same area as Windsor Castle'*.[2] And one
among the boom, near my old childhood stomping grounds, and the
next train stop along from the brewery town of Burton-on-Trent,
was Willington Wetlands, run by the Derbyshire Wildlife Trust.

In April 2022, I meet Henry Richards from the Derbyshire Wildlife Trust at the end of a long gravel track, flanked by the white
spring blossom of blackthorn and serenaded by birdsong. I find
Henry waiting in his bright red, Trust-branded four-by-four ('*Transforming the Trent Valley*'), its back doors almost bursting with
wellies, coats and conservation equipment. Young and knowledgeable, his dad used to take him out on nature walks, passing on
expertise such as how to distinguish mink poo from otter poo (the
latter, apparently, smells like jasmine tea). In conversation, Henry
regularly stops mid-sentence to identify nearby birdsong. But beavers, more than any other animal, have occupied his every waking
hour in recent months. And I got in touch with him just in time,
because today is the very first public tour since the beavers' arrival
in September. 'Our licence from Natural England is for two family
groups of beavers,' he tells our little group of a dozen, largely retired
birdwatchers, plus a family on an Easter-holidays day trip. 'These
beavers came down from Scotland. Our site is about 46 ha, whereas
other sites have been much smaller . . . It's possible that we could get
more beavers here if the Beaver Trust needed somewhere to place
some more.' But the current plan is to see how these two pairs get
on over a five-year trial. One pair, keeping a low profile, may have
already had their first litter of kits, Henry tells us, excitedly.

*   The English, of course, measure things according to castle-size.

We walk on to goggle at the release site, a pleasant grass bank in a sheltered corner of the gravel-pit lake, bordered all around by reed beds. A video of the beavers' release shows them taking to the water, happily gliding around for the camera, before heading off to explore the wider reserve. They always stay close to water, Henry tells us, for fear of predators – although their main predator, the wolf, no longer exists in England. Even so, the conditions of the reintroduction licence require the entire site to be enclosed with robust metal fencing through which the beavers can't escape. This cost around £200,000, a major expenditure for a wildlife charity, paid for with assistance from the water company Severn Trent. The nearby River Trent – one of England's major rivers – is only a few hundred metres away, and they can't risk a wild population getting in there. Not yet, anyway. The long-term aspiration – and the longer-term benefit – would be realized when truly wild beavers can enter the river too.

This whole site, before the gravel quarry began making big holes in it, was once a floodplain for the Trent – and even now it still gets a couple of deluges a year when the river spills over, in spite of its raised man-made banks. But, for thousands of years, this would have been a real wetland, managed by beavers. In a natural system, Henry explains, the River Trent would be allowed to move. 'And, as it moves, it would erode and leave behind oxbow lakes,* which would, over time, turn into reed beds, then willow, then wet woodland and eventually dry woodland. That would be the natural succession.' There's been a drive to reintroduce reed-bed habitat in the UK over the last twenty years or so, with active management from the likes of the Wildlife Trust and the RSPB, but reed beds should regularly flood to wash away the dry brush and sediment. When that no longer happens naturally, reed sites must be actively managed by volunteer clearance teams, including removing fast-growing willow that quickly outgrows the reeds. Beavers, however,

---

* A term I hadn't heard since school geography lessons, meaning when a river meander closes at the narrowest point, creating a freestanding, U-shaped pool.

take care of both problems, coppicing the willow and clearing channels through the reeds. Henry and his team have already seen these
benefits in action, in just six months, with just four beavers. A small
brook passes through Willington Wetlands, before going on to the
local village and feeding into the Trent. Eurasian beavers, unlike
their brasher American cousins, prefer to build multiple small dams
rather than one large show-off one. The hope is that the beavers will
start to dam the brook, which will reduce flooding in the village. If
they were one day released more widely, they would likely dam
many of the smaller tributaries to the Trent, such as the Dove and
the Derwent, holding flood water back, recharging aquifers and
slowing the rate at which the Trent rises when in flood.

Graham Osborn, principal ecologist for Severn Trent, agrees:
'This project isn't just about the reintroduction of a lost species. It's
about the regeneration of healthy wetland habitat . . . We're always
looking for more nature-based solutions to reduce the risk of flooding, and beavers can help by capturing and cleaning nearby
waterways.' As the wetlands are surrounded by small, young woodland, there is a chance the beavers might help to re-wet these
floodplain woodlands too. Beavers, Henry tells me, like to swim
whenever they can, and will dig small canal channels to reach new
sources of willow, rather than risk walking over land. Severn Trent's
sustainability report therefore describes beavers as 'wetland engineers' and a 'keystone species, meaning their activity can actually
create wildlife habitats in wetlands, bringing a wide range of ecological and economic benefits.'[3]

At the other end of the lake, following a muddy path and keeping
our eyes out for beaver poo (it looks like a mini version of horse poo),
Henry stops at the bankside by a patch of reeds and mature willow.
'This is one of their favourite spots,' he says. He points out willow
that has been chewed through and the bark stripped, but there's also
new shoots – they don't kill the tree, but rather manage it as a food
source. On the ground are several twigs with the bark meticulously
stripped, replete with teeth marks. He suggests we all take one as a
souvenir. 'When we first saw one of these from Scotland, we all

passed it around in awe, he says. 'Now, we just give them away!' I put two small sticks in my backpack. One of the young girls from the family group takes a large one and uses it like a walking stick – the first child to do so in the Midlands for 800 years.

At the brook, which is mostly concreted and straightened by the former quarry, you can tell Henry is impatient for nature to take over. At night, the beavers have been stalking along the side of the fence, desperate to get into the brook. It's interesting how drawn they are to moving water, given that they dam water specifically to make it still: their natural instinct appears to be to get damming.* 'From a human point of view, they're a really useful tool in managing flooding,' Henry tells us. 'And they can alleviate droughts. By holding water back and creating these wetlands higher up in the catchment, the water is held here for longer, rather than having streams straightened with concrete edges, where the water shoots straight down to the main river.' If there's no rain for a while, the beavers' habitat holds water in place – so, when it's dry, there's water, and when it's raining the water's held back. For water companies with beavers in their catchments, 'this might mean that their reservoirs stay more consistently full through summer, with cleaner, reed-filtered water,' says Henry. He admits he's getting ahead of himself, imagining 'a catchment full of beavers', but even small-scale releases such as this are showing immediate impacts.

When I speak to Jake Rigg at Affinity Water in the south-east, he says, only half-jokingly, 'We need to hire a load of beavers . . . we're talking about funding a study to look at the impact of beavers on slowing flow and making rivers more resilient in low flow conditions.' One of the things they'll investigate is whether building more wetlands can at least partly return rivers to a 'pre-human-

---

* It's a common misconception that they do this to catch fish. In fact, Eurasian beavers are herbivores, eating rhizomes of water plants and bark from tree branches, some of which they store underwater for winter – so they dam to extend their range, and their larder.

intervention' state. 'Most of the chalk streams would be re-braided;* they would have woodland growing through them. I'm told there's only one wet woodland in a chalk river left . . . surely those can be [restored] to stop floods when there's really high water. And, in low water, they're storing a lot of water within them.' Following flooding in Finchingfield, Essex, for example, a beaver reintroduction in the enclosed Spains Hall estate helped re-wet the woodland by slowing the flow of a brook. When the inevitable winter flooding returned in January 2021, drone shots showed much more flooding in the beaver-managed enclosure, and much less than usual downstream. Estate manager Archie Ruggles-Brise excitedly blogged: 'The beaver ponds are very full today! However, the nature of the beavers is that having seen this flood come through they will now build the dams higher and wider, increasing the volume after each successive flood event. A proportion of the water you can see remains for long periods, right through to the driest months of the year . . . [helping to] keep the river flowing later in the year when everything dries up.' This was evidence, he said, with just one pair of beavers, of 'a floodplain coming back into action as it should!'[4]

The UK government launched a public consultation in 2021–2, asking for views on changing the Wildlife and Countryside Act 1981 to move Eurasian beavers from 'Animals no longer normally present' to 'Native animals'. There was a lot riding on the outcome, Henry tells me. Due to their very recent Prodigal Son return, beavers didn't have native wildlife status. As soon as they are formally re-recognized as Brits, they'll be able to wander freely once again – then we'll really find out what these furry 'wetland engineers' are capable of, and will potentially see many of England's water

---

*   Another half-remembered geography lesson: a braided river is made up of lots of river channels flowing over a wide riverbed, creating many small, often temporary, islands in between, as opposed to a meandering river, which is one snaking water channel.

problems – specifically water storage, water quality and flood management – begin to be resolved, for free, by nature.

Farmers, however, aren't so keen. In response to the consultation, the National Farmers Union (NFU) said, 'beaver reintroductions can have negative impacts – potentially undermining riverbanks, damaging trees, impeding farmland drainage and causing low-lying fields to flood.' They do, however, proffer an easy solution: 'Where there is a financial impact on a farm business, adequate compensation must be made.'[5] Defra's environmental land-management services (Elms) payments to farmers are specifically designed to 'recognize the value of existing natural assets . . . protecting and enhancing these assets to achieve good environmental and climate outcomes' – it would be very easy to add 'land lost to beaver flooding' to the list. Water companies can privately supplement Elms payments too. Anglian Water and Severn Trent's 'Get River Positive' plan includes a pledge to reintroduce beavers, alongside a deal to incentivize regenerative farming practices and provide access to green financing[6] – again, this could easily include compensation for small losses to beaver-dammed flood defence and groundwater recharge.

This isn't a UK-centric lesson. While the diminutive North American beaver remains common, the continent can still relearn the importance of a wider-spread population. Prior to the arrival of European colonists, beavers inhabited creeks and wetlands from the Arctic tundra down to the deserts of northern Mexico. Alice Outwater's 1996 book *Water: A Natural History* notes that an estimated 'two hundred million beavers once lived in the continental United States, their dams making meadows out of forests, their wetlands slowly capturing silt.'[7] Today, just 10 per cent of that number are left.

The chair of UN Water, Gilbert F. Houngbo, wrote in the UN World Water Development Report 2018 that, 'For too long, the world has turned first to human-built, or "grey", infrastructure to improve water management.' The world's water economy must instead adopt a new approach, he said, of 'working with nature, rather than against it.'[8] Even in the lowland countries of Europe, literally built by hydrologically taming nature, this lesson holds true. In Belgium, says Marijke Huysmans, meandering rivers have previously been straightened and canalized to encourage drainage. 'But, in the current situation, this is a fairly bad idea. So now we actually need to do the reverse.'

As well as campaigning for abstraction reduction, Charles Rangeley-Wilson spends a lot of time on river restoration. He says a lot can be done with relatively little money. 'As an example, I've just moved the River Nar over 2 km, from a high-perched impounded and dredged channel, which was crap ecologically, back into the bottom of the floodplain, filled it with bits of timber . . . [and] re-wilded the river, essentially.' Water companies will rightly question the sense of returning flow to a chalk stream if it's 'basically a suspended canal'. Rewilding and restoring rivers and floodplains must, therefore, go hand in hand with reducing abstraction, to reconnect the rivers to the water tables and aquifers.

The opposite approach – dredging, weirs, canalization – still has its supporters. But Charles believes that mindset is on its way out. During catastrophic flood events, though, it can still rear its ugly head, as it did at the time of the flooding of the Somerset Levels (so called because it's a floodplain), in 2015. On his blog, Charles called for 'new thinking and new policy based on flood *management*, more than *defence*.' Floods are natural and cyclical events, he argued, and the ways in which we manage and engineer the landscape leave it susceptible to harm. 'Our two-tier, engineering-based approach of dredging and defence is mutually contradictory and has disastrous consequences . . . dredging makes flooding worse.' Dredging a river makes it deeper, allowing it to carry more water, adding more fuel to flood with. A natural river includes its own in-built flood defence and water storage: a floodplain. Building on floodplains and dredging rivers means that former town planners 'have inadvertently built natural time-bombs and then asked us to live beside them.'[9]

Floodplains were central to pre-industrial society. The meadows that grew on floodplains were known to be the most diverse, with wild flowers used for medicine and the most prized pasture for common grazing. Stewart Clarke at the National Trust writes: 'We tend to think of floodplains as land that gets flooded when things go wrong but we need to revise our collective understanding because, by doing so, we will manage them very differently.' Most river scientists would, he says, recognize the description of the river channel as the dry-weather channel and the floodplain as the wet-weather channel. If the floodplain is the space for water during wetter times of year, it makes no sense to use them for anything that might be vulnerable to flooding. Instead, 'Imagine a broad network of green corridors snaking through our countryside, linking existing natural habitats and reaching into our cities and towns. This is what alternative use of floodplains could deliver.' The proposed Nature Recovery Network for England plans to do just that and, writes Clarke, 'could also help reduce flood risk, store and lock up carbon and create space for people to enjoy nature.'[10]

In England, 95 per cent of all semi-natural grassland was lost in the twentieth century to industrial agriculture.[11] Now, just 4 square miles (10.4 km^2) of pristine, never-been-ploughed floodplain meadow is believed to remain. To my astonishment, I find some of it in my own river catchment, alongside the River Thames, near Oxford, less than 2 km away from Farmoor Reservoir. It surprised its owners when they first moved there too. In 2000, Oxford academics Catriona Bass and Kevan Martin bought a ruined house on an island in the Thames; the patch of land it came with was a secondary thought, but it turned out to be one the country's most ancient flood meadows, Long Mead, listed in the Domesday Book of 1086. Now, conserving it, preserving it and extending its boundaries is an obsession for Bass and Martin, and they have since set up several environmental initiatives, including the Thames Valley Wildflower Meadow Restoration Project, and have joined the Nature Recovery Network. A steady stream of scientists make a pilgrimage to their meadow. When I visit, in late May 2022, the meadow is just weeks away from its prime, Bass tells me. But it's already an extraordinary sight. Let an ordinary patch of field go fallow in this region and nature will quickly take over with tall grasses, nettles, dock, cow parsley and thistles. But there are none of those here. The surface is a dense tapestry of low-lying wild flowers with delightful names such as bird's-foot trefoil, ribwort plantain and meadow vetchling. Bass reaches down to one called meadowsweet, topped with a spray of delicate white flowers. It contains the ingredient used in aspirin, she says, and crushes a small stem between her fingers for me to smell. It smells exactly like the fizz from soluble aspirin.

According to the research Bass and Martin have done, it takes at least 150 years for a meadow to grow as densely as this. The thick, low-lying carpet is too close-knit for typical field weeds such as nettles to penetrate. They believe it's the most botanically diverse habitat left in Britain. I meet one pilgrim scientist out on the meadow, Mike Wilson of the National Museum of Wales, Cardiff, cataloguing the insect life by sweeping an adapted leaf blower in suction mode, to take back to the lab. Some of the insects are so specialist, he tells me, that they only live on certain rare plants. An Open University study (not yet published) looking at carbon sequestration here suggests that these meadows' carbon capture may be equivalent to that of peat bogs, known to exceed that of forests. The root system of this unassumingly pretty pasture goes down over two metres, where the carbon is stored in the soil below. Back in the barn, Martin shows me some soil samples they've taken, comparing Long Mead to a conventional wheat field and a recently restored meadow. The samples immediately take me back to the underpants test on James Alexander's farm, but the difference here is even more stark. The soil from the conventional field looks like a solid lump of clay – you could turn a pot out of it; the chance of water penetrating it seems slim. So light and soft is the Mead soil, meanwhile, and so dark the organic matter, that the samples look like they come from different countries, not neighbouring fields. It strikes me that I'm looking at the ultimate 'no-till' field. James had done well to keep the plough off his field for twenty-five years; but the Long Mead is over 1,000 years till-free and counting. And, importantly, it floods here. Regularly. A couple of winters ago, it was under water for almost three months. But this habitat is uniquely adapted to it.

When the water recedes, the plants and insects bounce back like nothing happened (quite how the insects survive is what Mike's Cardiff study hopes to find out), and the water goes overnight, 'Like a bathtub when you take the plug out', says Bass. The infiltration rates here are second to none. What a flood meadow is, then – and what the soil samples clearly show – is a giant sponge. The water table is always high beneath it, as it holds water and slowly releases it during the dry times. It is a habitat uniquely designed by nature to alleviate drought and absorb floods – two properties we now desperately need more than ever.

The Nature Recovery Network hopes to revive these meadows and create unbroken chains of them throughout the land. Bass and Martin are, with their own time and money, slowly persuading nearby farmers and buying up expired leaseholds, but it's slow, thankless work, constantly overshadowed by bigger, headline-catching environmental projects, such as nearby solar farms. I ask what Thames Water – the private company who will directly benefit from these ecosystem services, with Farmoor Reservoir just downstream – are doing to help. They look at me blankly: 'Nothing.' In fact, worse than nothing. One of the Thames Valley Wildflower Meadows – and one of the 10.4 km² left in the whole of England – Osney Meadow, in nearby Botley, is about to be bulldozed. And jointly holding the steering wheel will be Thames Water and the Environment Agency. They are building the Oxford Flood Allevi-ation Scheme – and, most gallingly of all, doing it in the name of 'nature-based solutions'. The Environment Agency's own website says that it 'will create a new stream with wetland wildlife corridor' that will 'bring additional environmental improvements to the area, including creating new wetland which will link up existing wildlife sites'. A lovely watercolour illustration shows the new stream along-side 'grazing cattle' and 'species-rich floodplain meadow'.[12] Nowhere does it say that much of a pristine habitat, undisturbed for 1,000 years, will be destroyed in the process.

*

The world has long acted under the misassumption, says the Stockholm International Water Institute's Torgny Holmgren, 'that grey infrastructure, purpose-built by humans, is superior to what nature itself can bring us in the form of mangroves, marshes and meadows.' Whereas it is, in fact, the grey infrastructure that has proven 'inflexible [and] less suitable in changing environments and increasingly uncertain times.'[13] Nature's ecosystems, he says, can offer softer, more malleable solutions, making water systems more efficient. SIWI is the unofficial UN of water – this is now the mainstream view, not one of fringe environmentalists. The World Water Council also argues that 'Healthy ecosystems, sufficient water and biodiversity play a critical role as infrastructure', and that 'the maintenance or restoration of ecosystems should be considered a priority for both public and private investments.'[14] The IPCC Mitigation Report (2022) finds that 'Landscape restoration is nearly always a [climate] mitigation action . . . Restoration of ecosystems is associated with improved water filtration, ground water recharge and flood control.'[15]

Even in the context of Pakistan – the country with the most megadams per head of population – the vice chancellor of Pir Mehr Ali Shah Arid Agriculture University, Rawalpindi, has said, 'Nature-based solutions have the potential to solve many of our water challenges. We need to do so much more with "green" infrastructure and harmonize it with "grey" infrastructure wherever possible by planting new forests, reconnecting rivers to floodplains, restoring wetlands for balancing [the] water cycle.' If those words are relevant for Pakistan, then they hold true for every country in the world.[16]

According to the NGO American Rivers, who campaign for dam removal, 'Removing a dam is the fastest, most effective way of restoring a river . . . few things have such a fundamental impact on a river as a dam.'[17] The days of the American West relying on large dams must come to an end for the simple reason that they cut off the connection to the water cycle. Helen Dahlke at UC Davis tells me that, before California had reservoirs, 'you had crazy flooding in the spring from snowmelt; the entire valley was basically a big wetland.

You flooded, you recharged [the groundwater], and then rivers could sustain from that water for a long, long time.' Even if California removed its dams, it would 'take decades, of course, to reach that state again.' And flooding isn't a vote winner, I suggest. Helen agrees: 'It's still the costliest disaster we have. It's more costly than droughts. But, in Germany, for example, we do controlled flooding. When we have big flood events, there are agreements with farmers to open the levees at certain locations to flood the fields behind them, because they know the river needs room.' Whereas, in California, this would require huge political effort, and likely much litigation.

However, even Republicans are beginning to advocate for dam removal. Idaho senator Mike Simpson wants to remove four dams on the Snake River. Initially sceptical, after three years in consultation with regional stakeholders, and with the Idaho salmon population down to a critical 4 per cent of its historic numbers, he concluded that restoring the river's natural flow was the best option: 'In the end, we realized there is no viable path that can allow us to keep the dams in place,' he explained.[18] His plan drew support from the Spokane and Shoshone-Bannock tribes and the Confederated Tribes of the Colville Reservation, Yakama Nation and Umatilla Reservation. Shannon Wheeler, chairman of the Nez Perce Tribe, told the local paper, 'The impacts of the dams as a whole have affected our people economically, culturally, spiritually and physically.'[19] Again, Senator Simpson is no looney outlier. The Iron Gate Dam, one of four on California's Klamath River, will be removed in 2023. Frankie Myers, vice chair of the Yurok Tribe, told Denver7 News, 'The lower four dams on the Klamath River – for Yurok people – are a monument to colonialism.' It will be the largest dam removal in US history.[20] Montana's senator Jon Tester is also pushing to place seventeen rivers under the Federal Wild & Scenic Rivers Act, and in the process block and remove several mines and dams – the sort of things that senators used to do anything to attract. By 2020, there were 90,000 dams in the US, 85 per cent of which were beyond their fifty-year intended lifespan.[21]

The Vjosa River in Albania is the last truly wild river in Europe, with no dams or weirs on its entire 270-km course. With the threat of its first dams planned, campaigners, supported by outdoor clothing brand Patagonia and Leonardo DiCaprio, led a #VjosaNationalParkNow campaign in 2021 to give the river protected National Park status. In January 2022, the Albanian government granted 'natural park' status – short of the full protection campaigners were hoping for, but a step in the right direction. Two earmarked hydropower plants had their permits revoked.[22] In March 2023, in a historic ceremony on the banks of the river, Albania's prime minister Edi Rama declared the Vjosa Europe's first wild river national park. It could yet become the talisman of the free-flowing movement that is sweeping the continent. The European Biodiversity Strategy aims to restore 25,000 km of free-flowing rivers in the EU by 2030. This can only be achieved by removing thousands of dams and weirs, most of which no longer fulfil their original purpose or are completely abandoned. The EU-funded AMBER project found 1.2 million barriers blocking Europe's rivers, of which almost 200,000 (around 17 per cent) were considered obsolete.*

Sandra Postel at the Global Water Policy Project has said that building a more secure and resilient water future means 'working with nature as more of a partner' – less command and control and more natural services and ecosystem services – 'looking at nature and working with nature as much as we can.' One clear example, she says, is our approach to flood control. By moving away from grey infrastructure and returning to nature-based solutions, 'we get these multiple co-benefits where we're controlling and mitigating floods, we're storing more carbon in the soil, we're recharging groundwater, we're holding and purifying water before it heads down stream.'[23] This doesn't require ripping up towns and starting again. But it does require ripping out some weirs and dams, and restoring rural and semi-urban floodplains and wetlands.

It is not, then, just about restoring fish-migration routes – as

*   For a comprehensive map of Europe's river barriers, including the UK, see www.amber.international/european-barrier-atlas

important as that is. From a purely water-infrastructure perspective, restored rivers, floodplains and replenished aquifers do a better job of supplying us with water than dams do.

The Great Valley Grasslands Floodplain Restoration Project in California, for example, is removing levees on the San Joaquin River to allow it to spread out during floods. Reconnecting the river to 48.5 ha of floodplain will help maintain both surface water and groundwater. The Central Valley Flood Protection Plan focuses on floodplain restoration to 'reduce the consequences of inevitable floods'. The front page of the Plan's website (multibenefitproject.org) tackles the obvious pushback front and centre: 'Millions of Central Valley residents live and work in high-risk flood zones. Taxpayers could be liable for billions of dollars in damages if a major flood damages private property.' Rather than make that risk more likely, a nature-based 'multi-benefit' approach reduces that risk. So far, this has included the purchase of thousands of hectares across the valley for floodplain restoration. According to Julie Rentner, president of River Partners, a non-profit involved in the scheme, these 'reconnected floodplains are thriving as wildlife habitat. The whole complex has endured deep flooding many times in the last twenty years, and it's served as a shock absorber, taking pressure off floodwater further downstream.' Scientists at Lawrence Livermore National Laboratory have said that the rivers of San Joaquin Valley once provided twice as much groundwater as they do today. 'That's not surprising, because of the dams built in the last hundred years,' commented Rentner on the study. 'It suggests that rivers could at least double what they put into the ground today. Allowing more water through river corridors and letting it soak in, especially in areas with groundwater deficits, could put water back in the ground through floodplain recharge.'[24] River Partners believe their work conserved a billion gallons of water in 2021 alone – that's around 12,300 m^3 a day, similar to a modest-size desalination plant (the Mossel Bay plant in South Africa, for example, produces 15,000 m^3 a day[25]), but with none of the running costs. Functioning floodplains, as I saw in Oxford, also sequester carbon rather than emit it.

Nature is more climate-resilient than concrete. This has been known for some time. A multi-sector paper in 2009, with input from the US Forest Service, recommends riparian restoration as a key response to climate change, saying, 'Riparian ecosystems are naturally resilient, provide linear habitat connectivity, link aquatic and terrestrial ecosystems, and create [refuge] for wildlife: all characteristics that can contribute to ecological adaptation to climate change.' It adds that, 'in order to restore hydrologic connectivity, reduction of upstream groundwater pumping and surface water augmentation is also necessary.'[26] Few people were listening in 2009, but they sure are now. The *Los Angeles Times* had, according to Marc Reisner, declared of the city's very first dam in 1905, 'The cable that has held the San Fernando Valley vassal for ten centuries to the arid demon is about to be severed by the magic scimitar of modern engineering skill.'[27] In 2017, the same newspaper ran an editorial stating that the Central Valley Floodplain Restoration dam-busting plan, 'that restores ecosystem health along the San Joaquin River and replenishes the groundwater,' is commended to readers as 'little-discussed but crucially important.'[28]

Alien hunters

One of the most striking examples of habitat conservation leading to increased water in a watershed can be found on the mountain-sides near Cape Town, South Africa. What you plant, or allow to grow, in and around your watershed matters. As Mark Dent at the Alliance for Water Stewardship, Pietermaritzburg, says: 'In South Africa, if you take 100 per cent of rainfall as the primary resource coming into the system, 92 per cent of that gets absorbed into the soil matrix and gets consumed and pumped back into the atmosphere by nature's little green pumps. Of the 8 per cent that gets into rivers, fully half to 60 per cent of that is pumped back onto the land to irrigate crops. You're looking at 95 to 96 per cent of the whole deal, goes back up through evapotranspiration . . . So, clearly, a big leverage point is on the plant and agriculture side of things.'

Talking to me from her home office in Scarborough, on the outskirts of Cape Town, Louise Stafford, Director for the Nature
Conservancy, South Africa, attributes Day Zero, in part, to the pro-
liferation of invasive species. 'Most of the problem in the catchments
here in Cape Town is that invasive species and invasive trees were
removing water out of the catchment that could have been available
downstream in the dams,' she says. There had long been a call to
remove damaging invasive species in and around Cape Town, but,
until Day Zero, few people had made the direct link between them
and diminishing water levels – few, that is, except Stafford. 'We
looked at what information is available; we refined the information –
and then we used some hydrological models to show the impact on
the water-supply system.' A spatial analysis by Stafford and her team
revealed a widespread invasive-plant infestation across two thirds
(69 per cent) of the twenty-four water catchments in the Western
Cape, with 14,400 ha (10 per cent) of that area suffering severe water
depletion as a result. These thirsty 'alien' plants, including Australian
acacia, eucalyptus and European pine, have roots that extend deep
into the soil and use up to 20 per cent more water compared to the
region's native plants, such as fynbos. Stafford's analysis demon-
strated that the restoration of priority catchments by removing these
water-hungry invasives would generate annual water gains of 50B L
(50M m^3) within five years – one sixth of the city's water supply. The
annual gains would double to 100B L, covering a third of the city's
water needs, within thirty years, as greater numbers of invasive,
thirsty trees were removed. Amazingly, this one action could almost
guarantee that Day Zero would never return to Cape Town's shores
again. While more than R8 billion (£406 million) in public funding
was being considered for boosting water supply through deep aquifer drilling or desalination, Stafford and her team believed that they
could do the job for R372 million (£19 million) – 95 per cent less –
giving an effective raw-water cost of R0.8 (4p) per m^3, compared to
R10–15 (51–76p) for wastewater recycling or desal. Her published
study finds that, 'catchment restoration would supply water at two
thirds to less than one tenth the cost per cubic metre as the other

supply alternatives considered in Cape Town . . . with improved supply showing as soon as the first winter rains. Furthermore, catchment restoration produces water yield gains into perpetuity if areas cleared of invasive alien plants are maintained.'[29] It would also create around 200 permanent jobs for a tree-removal workforce.

To pay for the initial efforts, a Greater Cape Town Water Fund was set up (based on the international Water Fund model, only the second to be established in Africa), bringing together the private and public sector and local communities. Simply clearing the catchment of these invasive species and letting the native species return was equivalent to *two additional months of water* a year for the city, Stafford tells me. 'This is buying us a lot of time.' Follow-up studies showed a return on investment of the invasive-tree removal of 351 per cent, she says. Once the city authorities saw these figures, they bought into it and released a new water strategy. 'Whereas, in the past, invasive trees were mentioned, but it was never quantified, once we had the sums, the city realized that it is actually in their interest to bring that into the mix. So, they permitted the fund and approved funding for work on nature-based solutions outside the city boundaries. This will be the first city in South Africa that does that.' A precedent, she says, now set for others to follow.

To understand how removing plants can deliver such massive water benefits, you need to understand just how large an area this is. The sub-catchments of the Wemmershoek, Theewaterskloof and Berg River dams alone cover 54,300 ha (543 km²) – that's larger than Barbados. This supplies 73 per cent of the water for Cape Town. 'Each of the major dams are in the mountains,' explains Stafford. 'So, when it rains, the water comes down and goes into the dam. We have a very short rainy season, it's only around three months, and then you get nine dry months where the rivers keep flowing.' Usually, some rainfall enters the rivers and fast flows into the dams, but then the native mountain vegetation – fynbos – and the soil structure beneath it acts as a sponge (much like the Oxford flood meadow), slowly releasing water during the summer months. Alien plants have a competitive advantage over indigenous ones, having no natural enemies and being bigger and taller, with windblown seeds able to spread widely. They use up the water instead of it being released into side streams. An image of these catchments from the air 'looks like hairs on a dog's back,' explains Stafford, with pine trees instead of the native plants 'literally as far as the eye can see.'

To remove so many trees over such a vast area, much of it accessible only by helicopter, is a mammoth task. Initially, the national government provided some wildfire crews to begin the clearances, but there just weren't enough teams to do it. So, a recruitment and training programme was developed by Stafford. Now, candidates undergo ten days' specialist training in high-altitude clearing and wilderness skills before being flown into a camp where they'll live and work for weeks at a time. It sounds similar to working in off-shore oil and gas fields, but for the frontline of water conservation instead of fossil-fuel extraction. The inaccessibility of the area is such that the trees are either ringbarked and left to die, or felled and left in situ. In more accessible areas, says Stafford, farmers use the woodchips on crops during winter, to prevent the soil from drying out. Given the scale of the catchment, the work will be ongoing for many years; Stafford envisages the skills gained by workers being highly sought after by other regions dealing with invasive species.

'Because the water-supply system catchments are so big, we had to start with the quickest wins, and the highest return on investment,' Stafford explains. They identified seven priority areas, but this was still 'about 52 per cent of the entire system. That's where about 76 per cent of the current water losses were happening.' The work began in 2019, and when we first spoke, in February 2021, about 15,000 ha (27 per cent) of the area had already been cleared. Stafford sent me a web page that updates progress in granular, sub-catchment by sub-catchment detail.* I found myself checking it

---

* Perhaps second only to the Jal Jeevan Mission in terms of granularity and accessibility of data; at the time of writing, it could still be found at: https://public.tableau.com/app/profile/waterfunds/viz/GCTWFDSSv1/PublicDSS

every now and again over the following months, imagining Stafford's brave team dangling from abseil lines over rocky crevasses, stripping bark and pulling saplings. In October 2022, I checked to find that 25,670 ha had been cleared (44 per cent complete), with four of the eight priority catchments already over the halfway point.

Speaking to Louise Stafford made me realize how overlooked invasive species are in the water conversation. We often see trees on a hillside as simply 'nature', without questioning whether those trees are native or how they're affecting the local water table. 'Invasive species should be considered as one of the reasons for water losses internationally', argues Stafford. Other examples include African buffelgrass in Arizona, Texas and Mexico, where teams of volunteers with names like the State Park Buffel Slayers and the Sonoran Desert Weedwackers have been fighting back for years; cotton-wood trees in Nevada; eucalyptus in Colombia; Himalayan balsam in the UK; and also, arguably, managed Sitka spruce plantations in Europe, including Ireland and Belgium. Stafford underlines how our land-use practices impact on our water supply. This holds a very important message for tree-planting initiatives for climate mitigation too, such as the EU's strategy to plant 3 billion additional trees by 2030, or the Trillion Trees campaign of the WWF and Restore Our Planet. Such nature-based carbon capture is crucial – but only if it's 'the right trees, in the right places', warns Stafford. 'What's the use of transforming areas that are supposed to be grassland to plant trees?' she asks. 'You've got to work with nature.'

When Chennai's Day Zero followed just a year after Cape Town's, restoration and conservation became key to rebuilding its water system. When Chennai lost its wetlands, it lost its water resilience. The link is that linear. The Tamil Nadu capital used to be surrounded by 8,000 ha of wetland, but, after decades of development, found itself with just 10 per cent of that left. 'Wetlands, as we all know, provide surface water', says Alpana Jain, at the Nature Conservancy (TNC) in India. Yet the Pallikaranai Marshland is among the last remaining 800 ha of wetland. The decline began under British colonial rule, Jain tells me. 'Why did the wetlands get encroached in the first place?

Because the British made rules that all wetlands are wastelands. And that rule never got to changed till very recently, about ten years ago.' While the coffin was built by the British redcoats, the nails were driven in by one of modern India's success stories: the IT boom. 'The whole IT industry has been built over that wetland,' she says. 'When the 2015 floods happened, the basements and the ground floors of these IT companies flooded. So, that's when the realization struck in a big way that we have to protect our wetlands if we want to protect ourselves.' The drought of 2019 posed an even bigger problem. The wet season in Chennai is from around October until early January, but April to June is very dry; for water from the monsoon to last the duration, the wetlands need enough capacity to hold the water and recharge the groundwater.

The IT industry is now too established to move. But the surviving Pallikaranai Marshlands have now been protected by the Indian Forest Department and, with the help of Jain and TNC, the remaining wetland plus 200 lakes within the city are being restored. TNC have taken on one lake to showcase best practice for others to copy. Lake Sembakkam, surrounded on all sides by residential developments, had become putrid with untreated sewage and solid-waste dumping. It then feeds into an equally polluted 40-ha urban wetland. Jain and her team used nature-based solutions to turn it from an eyesore and disease risk into a site fit for fish, birds and recreation. 'To begin with, we looked at a 5-km radius catchment of the lake to understand the hydrology of the area,' she explains. 'After that, we created a restoration plan, which dealt with improving biodiversity habitats and achieving water-quality standards. We aimed for the Central Pollution Control Board's standard D, which is basically to bring the water to a level which is good for wildlife and fisheries; you can't drink it or swim in it, but it's good for groundwater recharge.' If you're wondering why not standards A, B or C, consider the starting point. 'The water quality was really, really bad,' explains Jain. 'We had something like 7M L per day of wastewater coming into the lake, a mix of sewage and grey water. There's no formal sewage-disposal system in the area, so everybody is basically dumping their sewage in the storm-water drains, which end up in the lake.' Unsurprisingly, tests found extreme pollution levels, not just in the lake, but also in the groundwater below.

The restoration plan, designed in partnership with IIT Madras University, restored reed beds and wetlands to meet the wastewater as it comes in, and installed mechanical pumps and aerators, much as you'd find at a formal wastewater-treatment plant – though Jain is keen to point out that it's the reed beds that are 'the primary treatment process'. In addition to reed beds, TNC is replanting the native plants that used to be integral to the Pallikaranai Marshland. This is not only to help plants and animals return, but to help people return too. Turning a sewage dump into a nature and recreation area will help people to take pride in and ownership of it, says Jain, which is essential to the plan working longer term. I speak to Jain before the reed beds have been fully replanted, but 'already we are at the stage, after the current monsoons, that the lake is full, and it looks beautiful,' she says. 'Birds, flamingos, pelicans are coming back. And people are so happy about it; you see people actually taking a walk . . . We've cleaned it up, removed the invasives. It's breathing once again.' Initial restoration has increased the water-storage capacity of the lake by over 100,000 m³. A three-year monitoring plan is in place, with learning guidelines constantly updated to help the city authorities work on the other lakes and wetlands. The central government's Jal Shakti ministry in Chennai 'have spoken with us about our work and are very interested in understanding our template,' says Jain.

If Chennai fully restores its 200 lakes and wetlands, it may be enough to ensure water resilience for the city. Although ever-expanding population growth and climate change keep making the task harder, TNC's work offers a blueprint to follow. Chennai is working on its twenty-year Master Plan, to begin in 2026, and Jain says she's in 'advanced conversations' to make nature-based solutions and natural water infrastructure central to it.

Peru's Natural Infrastructure for Water Security project is attempting to make nature-based solutions a national strategy too.

It's stated goal is to scale up the conservation, restoration and sustainable use of ecosystems and indigenous knowledge to reduce water risks. Between 2014 and 2020, finance for natural infrastructure for water security in Peru grew rapidly, increasing by thirteen times overall. With the main aim of improving water infiltration into groundwater, the most common interventions have included reforestation or afforestation, followed by infiltration ditches and indigenous water and soil conservation. The *qochas* have gone full circle, from being the central means of indigenous water harvesting, to being derided and falling into disuse, to now becoming part of national policy.[30]

Which brings *us* full circle, too. Should we just re-naturalize – to use Niels Hartog's word – the entire system? Rewild and re-meander all rivers, reconnect floodplains and let nature take over the managed aquifer recharge for us? Niels pushes back on that idea a bit: 'You know, in some ways, we are locked-in, in the sense that we're using all the land. We have a function for that land. And so, yes, lots more porous pavements . . . for sure, that's part of the solution. But, you know, we cannot return fully to nature. So, my perspective is: go as far as you can with utilizing what nature is providing, but it's silly to think that we live in harmony with what nature provides – because we're just too many, we have too many demands.' I want to rail against this. But he has a point. Cities – entirely man-made and an inevitable part of our future – are not going to be given over to wolves and beavers any time soon. 'Rainfall, precipitation, is going to get more extreme, more water in shorter periods of time,' says Niels, as did everyone else I talked to for this book. 'And our current infrastructure is not capable of dealing with it. So, *anything* you can do to dampen these peaks, with the surface features . . . SuDS, green roofing, parks – they all play a role.' And, yes, plenty of beavers and nature-based solutions too, further up catchment, even giving some farmland back for that purpose. 'We have to be smarter with water' urges Niels.

CONCLUSION

Must Try Harder

According to some, the world's water situation is now so dire that 'water wars are the new oil wars'. Whether it's VPs of the World Bank[1] or WEF blogs[2] or US Vice President Kamala Harris's speech marking her first hundred days in office, saying, 'For years and generations, wars have been fought over oil. In a short matter of time, they will be fought over water.'[3] It's the view of many Western politicians and commentators that the Syrian civil war was one such water war. President Barack Obama said in a 2015 speech that climate-change-induced drought had 'helped fuel the early unrest in Syria, which descended into civil war'[4]; his secretary of state John Kerry also argued that 'it's not a coincidence that immediately prior to the civil war in Syria, the country experienced its worst drought on record . . . intensifying the political unrest that was just beginning.'[5] Dr Peter Gleick, cofounder of the Pacific Institute, which hosts the Water Conflict database, writes: 'One recent example of a major conflict with complicated but direct connections to water is the unraveling of Syria and the escalation of massive civil war there.' [6]

The same could be said – but rightly isn't – about Ukraine. The North Crimean Canal was built by the Soviet Union between 1957 and 1975 to bring much-needed water from the Dnieper River to Crimea, where the summer is hot and dry, and the winter arid. The North Crimean Canal soon provided 85 per cent of the peninsula's water. In 2014, when the Russian military annexed Crimea and Russian officials took over the canal, Ukraine responded by damming it up; Crimea's forty-years-strong water source disappeared over-night. After seven years of dispute, in July 2021, Russia lodged an application with the European Court of Human Rights against Ukraine, with the canal's closure central to their complaint. Russia requested that the court 'order the Ukrainian authorities to sus-pend the blockade' of the canal's water. The court rejected the request. Ukraine instead reinforced the initial sandbagged dam with permanent concrete. Later that month, the *Financial Times* described 'Moscow's struggle to supply Crimea's 2.4m residents with fresh water' as 'an undeclared war'. The article added, 'Kyiv fears that Moscow is plotting a military incursion to secure water flows.'[7] The Russian-appointed Crimean governor Sergei Aksyonov responded explicitly: 'There won't be any "water war".' Just six months later, on 24 February 2022, Russia launched a full-scale invasion of Ukraine. On that very first day, Russian forces destroyed the dam, unblocking the North Crimean Canal, and the waters of the Dnieper River flowed into Crimea once again.

So, should we be attributing recent wars to water scarcity? An academic collaboration between UK, US and German universities into the causes of the Syrian war cautions against such analysis: until a few years ago, say the authors, the 2003–5 war in Darfur was widely identified as the 'first climate war', but such claims have since been discredited by, among other things, the finding that Darfur's war neither occurred during nor was directly preceded by drought. Syria, too, is not as straightforward as it seemed: 'Rather than . . . displaying a long-term drying trend, the observed rainfall [in Syria] is better characterized as highly variable with an anomalous decade from 1999 to 2008, culminating in a severe drought (centred on 2007/8) after which rainfall returned to its pattern of high inter-annual variability (wet, dry, wet). Our analysis of raw station data from north-east Syria confirms . . . there is no evidence of progres-sive multi-decadal drying . . . contrary to what . . . others claim or imply.' The study also found that 'large-scale migration from and within rural Syria . . . was occurring well before the 2006/7–2008/9 drought'. The Syrian government's 'liberalization' programme had already led to a sharp rise in rural-to-urban migration between 2000 and 2005, leading the authors to conclude that, although there were doubtless socioeconomic consequences to the drought, most of the 'water war' claims since are 'overstated and unreliable'.[8]

Few are more steeped in 'water wars' research than Mark Zeitoun, co-founder of the University of East Anglia's Water Security Research Centre. He tells me, bluntly, 'There have been no water wars. And, having observed these things for a long time, I don't think there will be one anytime soon.' Zeitoun's career has been dedicated to water conflict. Starting out as an on-the-ground, dirty-boots civil engineer for the likes of the Red Cross, USAID and Oxfam, in Lebanon, Iraq, Gaza and the Left Bank, Congo and Chad, he later switched to an academic career, working under Tony Allan – the King's College London water academic of 'virtual water' fame. When we speak in December 2020, he argues that sensationalist messaging about 'water wars' is a distraction from the daily conflicts and injustices that flare up over water. Should we care less about clashes that don't tip over into war? Or do we just wait, ghoulishly, for a real 'war' to begin before we take notice? 'There's a lot of conflict around water which falls short of a war, but that deserves our attention,' he tells me. 'If you could find a snappy way of saying that, I'd be grateful, because that's my message. No: water wars. Yes: water conflicts.'

On the Syrian war, Zeitoun is his most bullish. 'I totally, 100 per cent disagree with those who say that the war in Syria is even partly related to the drought,' he scorns. According to his account, there were very many reasons to oppose that government and other groups within the state. 'Water and drought might have exacerbated it. But the people who say it led to the war in Syria haven't done the serious work to put that alongside the existing tensions.' Those existing tensions included, he says, 'class divisions, religious divisions, outright torture and discrimination, a failed economy, neo-liberalization of the agricultural sector . . .' He could go on, though the point is made. Instead, he offers a counter-factual: 'Consider Sudanese researchers going to Belfast in the winter of

1985, seeing people huddled around peat fires, and writing that the reason for the Troubles is because it's so cold in Northern Ireland. They would be laughed at. But that's what we white guys are doing, and we get away with it.' Few international media outlets describe the 2022 Russian invasion of Ukraine as a water war, despite the North Crimean Canal arguably being a more direct trigger than drought in Syria. Majority-white populations are seemingly judged differently – that it's brown-skinned countries who fight over water and famine; white countries do not. The Russian invasion of Ukraine obviously had multiple, decades-long, complicated factors that led up to it, beyond the situation with the North Crimean Canal. It clearly shouldn't be simplified, patronisingly, as a 'water war'. We should, then, extend the same intellectual courtesy to Syria and elsewhere.

In fact, the US can look closer to home for climate conflict. In June 2021, fears of a confrontation between law enforcement and right-wing militia were reported at Klamath Falls, Oregon. Armed protesters affiliated with the right-wing anti-government People's Rights Network were threatening to open the headgates of a dam. *The Guardian* US reported, 'The area is a hotbed of militia and anti-government activity and also hit by the mega-drought that has struck the American west and caused turmoil in the agricultural community as conflicts over water become more intense.'[9] Meanwhile in Utah, the nation's fastest growing population despite having the second driest climate, officials want to build a 225-kilometer pipeline to tap water from Lake Powell, irrespective of how that would affect downstream supplies, and regardless of the fact that Lake Powell is running out of water anyway. 'The Utah State Legislature is still mired in an era of unreality,' Daniel McCool, a political scientist at the University of Utah, told *Science* magazine, calling the move 'a declaration of water war, frankly.'[10]

You increasingly don't need a war, however, to drive waves of refugees. According to the United Nations High Commissioner for Refugees (UNHCR), an annual average of 21.5 million people have been forcibly displaced by weather-related events – such as floods, storms, wildfires and extreme temperatures – since 2008.[11] By 2020, of the 40.5m forcibly displaced that year, more people were forced from their homes by climate disasters than armed conflict by a factor of three to one (30.7m new climate refugees compared to 9.8m war refugees).[12] Environmental reporter Abrahm Lustgarten has covered the movement of climate refugees for *The New York Times* and *ProPublica*. Initially he looked south, writing 'As their land fails them, hundreds of millions of people from Central America to Sudan to the Mekong Delta will be forced to choose between flight or death. The result will almost certainly be the greatest wave of global migration the world has seen.' But by September 2020, living in California, Lustgarten no longer needed to travel to find climate refugees. It was now happening in his own country, the United States. He reported that in California, '900 [recent] blazes incinerated six times as much land as all the state's 2019 wildfires combined, forcing 100,000 people from their homes. Three of the largest fires in history burned simultaneously in a ring around the San Francisco Bay Area . . . Across the United States, some 162 million people – nearly one in two – will most likely experience a decline in the quality of their environment, namely more heat and less water.'[13] He noted the incongruence of 'more than 1.5 million people [having] moved to the Phoenix metro area' despite its dependence on the already rapidly declining Lake Mead, and temperatures that regularly hit 115 degrees Fahrenheit (46C). That year, in 2020, the population of California shrank for the first time in years. This could have been dismissed as a pandemic-related blip, but then it happened again in 2021. The trend wasn't due to deaths or reduced immigration, it was Californians fleeing. About 280,000 more people left California for other states than moved there in 2021.[14]

Not only is loose talk of 'water wars' often looking in the wrong direction, it also lets governments off the hook for their mismanagement of water resources, insufficient infrastructure investment, failed diplomacy and internal corruption. As Mr McCarthy so memorably sang to me in Ghana, politicians can blame anything on climate change, while avoiding hard decisions. Iran's former deputy vice-president, Kaveh Madani, agrees, writing in criticism of his country's policies: 'Climatizing extremes can be a gift to most decision-makers who don't want to accept liability for their bad decisions. They will tell you that there is not much they can do . . . they will remind you that they are the victims of a global catastrophe'. Whereas instead, he says, 'alongside that global struggle, we must remember that local decision-makers are liable for avoidable failures of environmental management.'[15] This goes for all nations. As *The New York Times* puts it, 'as the scientific consensus around climate change and climate migration builds . . . [the] United States has done very little at all.'[16]

For all the talk of conflict, water can just as readily be a broker for peace. In November 2021, the grand dream of EcoPeace Middle East – the Jordanian, Palestinian and Israeli charity that kindly hosted me in the Jordan Valley – finally came to fruition: Gidon Bromberg's long-held (lobbied for, negotiated and authored) 'water–energy nexus'. That water-scarce Jordan could sell solar power to land-and-renewables-scarce Israel and Palestine, in return for desalinated water from Israel. 'The Minister of Energy in Israel [Dr Yuval Steinitz] actually gave us a letter in support of piloting the purchase of renewable energy from Jordan,' Bromberg excitedly told me, in August 2020. A year and a half later, in November 2021, it was agreed by Israel, Jordan and the UAE, which brokered the deal. The Emirati government-owned firm Masdar will construct a large solar facility in Jordan to begin producing emissions-free electricity for Israel by 2026. In exchange, Israel will send 200M m^3 of desalinated water to Jordan annually, alongside annual payments of $180 million to Masdar, which will split the proceeds with Jordan. Palestine was notably absent from the deal. But Bromberg, who had worked on this for so long, had made an impossible dream come true. And, as the Brookings Institute commented at the time of the announcement, 'Good news is a rare commodity in the Middle East. So is practical progress on climate change.'[17]

Paying for our mistakes

As Adrian Sym at the water standard AWS told me, 'It's not actually about water. It's about politics.' The water academics Biswas and Tortajada similarly argue that providing clean water to people 'is not rocket science. The policies to make this possible have been known' for decades, 'as has been the technology.' The missing ingredient, they say, has always been 'sustained political will' and 'good governance.'[18]

A few years before Australia's Murray–Darling Basin Plan tried to combine sustainable irrigation with environmental conservation, Spain had attempted something very similar. Environmental flows – water allocation for nature to replenish rivers and wetlands – were also purchased by the state following the Special Plan for the Upper Guadiana Basin 2008 (Plan Especial del Alto Guadiana, or PEAG), to save the collapsing wetlands of La Mancha (see page 94). At the time, the depletion of the aquifer and peatland fires had achieved national and international infamy. UNESCO threatened to downgrade La Mancha from biosphere-reserve status and gave the Spanish government until 2011 to regenerate the wetland. The PEAG aimed to achieve that by acquiring 130 GL of water rights between 2008 and 2015 via the purchase (or 'buy-back') of private groundwater rights through a 'public water bank'. The sweetener for farmers was that up to 30 per cent of these rights would be reassigned to farmers currently without legal water rights – in other words, bringing illegal wells into the legal system in order to meter them. This was particularly significant in the Western Mancha aquifer, where no new pumping concessions had formally been granted since overexploitation was first identified in 1987. Pumping had subsequently increased anyway, but illegally and unmonitored. Of the state-purchased water, a minimum target for environmental flows was set for 35 GL a year by 2027, with a dedicated budget of almost €3 billion (both EU and nationally funded). Within a month of buy-back, well owners were required to close and seal their wells. Having sold their water rights, farmers could continue to farm, but only under dryland (rain-fed) conditions. Users with remaining rights were also not allowed to expand their irrigated area.[19]

The PEAG's seemingly faultless design was best-practice, gold-standard water law that we can learn from. There's also a lesson in why it began to crumble. A review paper co-authored by Nuria Hernández-Mora finds that the River Basin Authority assigned insufficient human resources 'to monitor meters or to visit plots. There were instances of meter tampering or direct sabotage. In addition . . . 95 per cent of the rights procured were located outside the priority area designated for purchase.'[20] In other words, they were being ripped off by spurious claims. Ultimately, under pressure from farming lobbies and local government, the original sweetener became its core purpose: 81 per cent of the rights purchased by the PEAG were reallocated to formerly illegal users, 'thus becoming a de facto regularization of previously illegal groundwater abstraction.' This remains the PEAG's main legacy.* Just 9 per cent of the total volume originally allocated as environmental flows were delivered, and many of the wells that should have been sealed continued to be used. The review noted, dryly, 'it did not achieve the original environmental goals (aquifer recovery) that justified its operation and financial costs.' In July 2013, a liquidating director was appointed to close the PEAG down.

Speaking to me from Madrid in 2021, Nuria Hernández-Mora tells me the PEAG could and should have been an international success story: 'It really was a lost opportunity,' she says. An incoming conservative government was in no mood to purchase water for the environment or take any away from farmers. Nuria has since heard from an old contact that the River Basin Authority have now 'reverted to their old positions of twenty-five years ago, as though

---

* Though, Nuria tells me, the emergence of solar pumps is a threat to even this, as illegal 'off grid' pumping is harder to monitor.

nothing happened. The wetland is lost.' With Australia's similarly designed Murray-Darling Basin Plan up for renewal in 2025, it feels like the same precipice has been reached – a few tweaks from a government with a different agenda, and the whole pack of cards could come tumbling down.

The world is littered with well-meaning, well-written water policy that became politically crippled. The South African National Water Act (NWA) of 1998 was also seen as revolutionary at the time, hailed by the international community as one of the most progressive pieces of water legislation attempted in the world. The country was neatly divided into nineteen water-management catchment areas, some of which were further split into upper, middle and lower catchments. Each was to have its own Catchment Management Agency with legally defined roles. 'Essentially, our National Water Act is perfectly situated for water stewardship,' Mark Dent at the Alliance for Water Stewardship, South Africa, told me. 'But that was twenty-five years ago, and not a heck of a lot of progress has been made on it since.' The nineteen Catchment Management Agencies have since been reduced to nine. Of those nine, by 2015, only two were operational. In 2017, those two were consolidated into one by the Department for Water. A 2020 inquiry by the Water Research Commission referred to 'a history of failure', in particular with farms holding historic irrigation rights showing 'resistance to the re-allocation of water'.

Barbara Schreiner, former adviser to the Minister of Water Affairs during the water reforms from 1995 to 2007, has described it as the 'Volkswagen vs. Rolls Royce issue'.[21] The South African Water Act was hailed internationally as 'the Rolls Royce of water-management practice.' However, given the nascent point in the country's history, it might have been better, argues Schreiner, to have attempted a more basic, mass-rollout Volkswagen model.

No country or state has got this right, yet. California, the world's fifth-largest economy, hopes that its Sustainable Groundwater Management Act (SGMA), will bring some Rolls-Royce to the asphalt of the Golden State. However, SGMA doesn't make any changes to a landowner's water rights, nor does it state what 'sustainable' means, it only defines what isn't sustainable (via a list of six 'undesirable results') – instead, in the Land of the Free, it's up to the many groundwater sustainability agencies (GSAs) to come up with their own definitions, and 'work collaboratively with local agencies, landowners, municipalities, groundwater users, interest groups and other interested parties to ensure all interests are taken into account.'[22] Fudging this issue of water rights has left a ticking time bomb within the legislation. Of the first nineteen GSA plans submitted, lawsuits were already filed against five of them by summer 2021.[23] The law firm Brownstein Hyatt Farber Schreck issued a 'client alert' in February 2022, saying, 'It deserves restating plainly that SGMA does not impair water rights, and only a court can adjudicate water rights. [The California Department of Water Resources'] role is limited to review of the [groundwater sustainability plan's] technical components.'[24] SGMA attempted to stop the car (Rolls-Royce or no) from driving over a cliff, in the dark, without its headlights on; SGMA may have turned on the headlights, but the car is still driving off the cliff. If you look at today's levels of Lake Mead, at Jay Famiglietti's groundwater depletion graphs, or listen to the conversations of farmers like Bart and Dwayne, it's clear that there won't be much water left to manage by SGMA's end-goal of 2042. I can't help but imagine ageing Californian legislators sat on their verandas in 2042, patting each other on the back for achieving SGMA and the Drought Contingency Plan Authorization Act, over glasses of desalinated water and imported wine, to the backdrop of tumble weed and once farmed, now fallowed, desert. Both were progressive, hard-won, landmark legislation. But, as Jay would say, 'It's just math.'

In Alexandra Kleeman's novel *Something New Under the Sun*, published in 2021, she imagines a near-future Los Angeles circled by wildfires and protester roadblocks. Following a collapse of the central water infrastructure, a synthetic form of water called WAT-R comes in bottles, while tankers fill up domestic tanks with it for a hefty fee. The central character stays in a cheap hotel with no running water – guests are provided with a bottle of WAT-R to wash and brush their teeth with. (I don't know if Kleeman was familiar with Singapore's NEWater, but the capitalization seems a knowing nod.) Such a collapse of the Lake Mead system and over-abstraction of groundwater, to be replaced with expensive, recycled or desalinated water, seems a highly likely near future.

In late 2021, Metropolitan Water District, CAP, Arizona Department of Water Resources and Southern Nevada Water Authority (SNWA) joined forces – a veritable Marvel's Avengers of the West's water world – to begin the Regional Recycled Water Program, to purify treated wastewater to produce a 'new, drought-proof water supply' for Southern California.[25] 'This project could help the entire South-west,' said a Metropolitan press release, by 'adding new supplies, like recycled water.' The projected $3.4-billion project would produce up to 167,900 acre-feet (207M m³) a year. By comparison, the CAP canal delivered 1.5M af (1850M m³) a year to Arizona alone. We've drained what nature gave us, so now we'll have to create it ourselves at great cost.

The Lower Colorado River states will not run out of water. They are too big to fail. But they will run out of cheap water, with wide-reaching environmental-justice consequences. Further water banking and conservation measures will help ease that pain. But, ultimately, as Marco Velotta told me at Vegas City Hall, 'You can't conserve what you don't have.' The West never had the water that it divvied out in such precise legal terms in the first place. 'It was all based on a lie,' laughs Velotta, nervously. A reckoning was always going to come. Climate change has simply served it up earlier and hotter than anyone, or any state, was prepared for. In California, salmon are now being transported by truck to the ocean because they can no longer survive the low, warm waters of its rivers.[26]

Talk to anyone long enough about water in the American West, however, and they will bring up the next big megaproject messiah that will save them (and the salmon): a pipedream pipeline from the Mississippi River to the Colorado River. Such a 1,300-km system of

pipelines, canals and reservoirs would cost an estimated $23 billion*
and provide 1.2 km³ of water. Joanna Allhands for *Arizona Republic*
bemoans: 'Every time I write about water, I get a similar email from
different folks. It argues that if we can build pipelines to move oil,
we should be able to capture and pipe enough floodwater from the
Mississippi . . . to the Colorado.' Arizona state legislators have even
formally requested that Congress consider it. But, as Allhands continues, 'it makes us look wildly out of touch . . . we've studied this
before, multiple times. Each solution has been projected to cost
multiple billions of dollars. Most would not produce enough water
to fix our problems. And trust me, someone's going to fight several
hundred kilometres of pipe being laid across their land to make this
happen.'[27] It *isn't* going to happen. But it *is* entirely in keeping with
the grand tradition of the American West, where the greatest engineering projects of the twentieth century made deserts bloom and
metropolises rise improbably from the sand. It worked – for a few
decades, at least. So, the obvious go-to solution must be more
megaprojects. If you already have a 540-km concrete canal winding
its way through the desert, delivering water to your swimming pool
and flooding your backyard citrus orchard, so the mindset goes,
what's one more pipeline?

Sandra Postel said on the *What About Water?* podcast, 'I come
back to something Einstein said – we can't solve problems using the
same kind of thinking we used when we created the problems. So,
thinking we can engineer our way out by doing more and bigger of
the same kinds of things is *really* not going to work.' More likely the
Bureau of Reclamation, once a builder of such megaprojects,
becomes – by necessity and by a self-planted time bomb – a builder

---

* Likely a very conservative estimate, if you were to start land purchases
and settling the inevitable lawsuits. The UK's high-speed rail line, HS2,
for example, was initially budgeted at £32.7 billion in 2012; by the time
land purchasing was largely completed and construction underway in
2021, the updated estimate was already £107 billion, three times more,
and rising.

of desalination and wastewater-treatment plants. During my visit, President Biden's Infrastructure Bill was making its way through Congress. The Bureau of Reclamation had previously received just $20 million a year from Congress for desalination projects and $65 million for water recycling; the new Bill included $250 million for desalination over five years, and $1 billion for water recycling.[28] It is both a rational response to the water crisis, and a capitulation, a white flag waved: *We got it wrong, nature; sorry we sucked you dry.*

Singapore is much further down this road, almost totally reliant on large reverse osmosis (RO) plants for desalination and NEWater. 'It is very energy intensive,' Professor Hu Jiangyong at the Centre for Water Research, Singapore, warns me. 'RO produces water of very good quality – almost all the pollutants can be rejected – but the energy involvement is just too much. So, we will need to have some other nature-based solutions [alongside] to reduce the energy consumption. We are very mindful about how much energy is involved in water reclamation.' The water ministry, PUB, is expecting its water demand to double by 2060. A big part of their plan is simply more desalination and more NEWater plants. The Keppel Marina East Desalination Plant – Singapore's fourth – opened in 2020. A fifth plant opened in 2021. But, increasingly, there is a recognition that 'we need solutions that cost less energy', repeats Hu.

Meanwhile, in 2022, Arizona's Republican Governor Doug Ducey proposed to spend $1 billion over three years to help 'secure Arizona's water future for the next hundred years' by desalinating water from the Gulf of Mexico. 'Instead of just talking about desalination, the technology that made Israel the world's water superpower,' he said, 'how about we pave the way to make it actually happen?' Never mind that Arizona is thirteen times bigger than Israel and is entirely landlocked. The two proposed plants south of Puerto Peñasco would each produce 100,000 af (123M m^3) a year, and would cost between $3 billion and $4 billion to build, plus annual operating costs, creating a price of between $2,000 and $2,200 per acre foot.[29] The current delivery rate for CAP water is just $157 per acre foot.[30] That's a 1,301 per cent mark-up. And bear in mind that the CAP was widely derided as being too expensive at the time. This offers a taste of just how different – and costly – the post-Colorado River and groundwater-pumping days will be.

America, as dystopian literature like Kleeman's warns, will likely see a rapidly growing gap of inequality, between the haves and have-nots. We have gotten used to widening discrepancies in housing, wages, healthcare, crypto coin. But a lack of water leads to very real poverty, very quickly. Pools, fountains and farms filled with desal-inated and treated wastewater will run alongside communities – as I saw in Ghana – spending half their household budget to fill a jerrycan. That's the reality of expensive water. But, as Bronson Mack told me as we drove past a luxury villa on the outskirts of Las Vegas, 'Every American thinks that they're just one day away from being one of those people. We're all millionaires in waiting.' If we're not careful, access to water will simply, slowly, depressingly become another thing to aspire to.

Culture shocks

The term 'Day Zero' was coined to shock a city out of complacency. Even so, Cape Town ultimately avoided Day Zero because it rained. And, as Louise Stafford's invasive-tree-removal work has shown, restoring the natural water system is arguably even more crucial than changes to personal water-use. But the two things are con-nected. Whether I have a low-flow shower or not doesn't make a massive difference to my local river catchment; my drains lead back to the river anyway. But what it does – and what all water-efficiency measures do – is raise my awareness. 'We don't think about where water comes from,' agrees Stafford. 'And it's our responsibility to give that message so that people can have an appreciation of where water actually comes from.' Ultimately, she says, 'You can have the biggest dam with the prettiest pipelines and water-treatment works in the world, but, if you don't have the water that goes in the system, that infrastructure is just a monument. We need to think beyond the dams and beyond the pipes and look at where the water comes from and invest there in order to have the water; otherwise, it's like having a beautiful car without any fuel.'

Another way of looking at this is asking what a well-managed water system *should* look like. 'Think of it this way,' says Jay Famiglietti. 'If we were to redesign what a water system should look like, it would look totally different to how it looks today. It would involve a lot more natural storage; it would involve a balance of surface water and groundwater use. The design would be within a watershed, rather than some monstrous thing that spans many states or provinces. It would be more like what John Wesley Powel envisioned: just using what's available within a watershed.'* Jay warns against tech-fix solutions, and he's speaking as a man who worked on NASA missions. 'The problem with "solutions people" is that they just want to sell a solution. Like the people who talk about pulling water from the air, right? Like, if you could just condense water from air, then this would be problem solved? What they're talking about *may* provide enough water for a small community. That's about it. Same thing with desalination, right? The desalination people never talk about what would happen if we put forty desalination plants up and down the West Coast of the United States. What would it *actually do* if you threw that much brine back into the ocean?' Instead, says Jay, we must start thinking more holistically: 'We understand a lot more about the Earth as a system now . . . If we were to redesign it today, it would involve elements of both hard infrastructure and nature-based solutions. It's not an either/or.'

Jordan is as drought and climate-compromised as they come. And yet half of all its municipal water is lost to leaks in the network:

* After the 1869 Powell Geographic Expedition, John Wesley Powell drew up recommendations for the US Congress of how to create a sustainable water supply. He proposed state boundaries based on watersheds, suggested homesteading and land-based water rights were nonsensical, and that only a tiny proportion of the arid West could be sustainably irrigated. He was ignored. The full remarkable story is told in Marc Reisner's *Cadillac Desert* (London: Penguin Books, 1986).

and that's with a network only running once or twice a week. His Excellency Saad Saleh Abu Hammour described water losses of 50 per cent to me as a 'national shame': 'Fixing leaks is the main problem for Jordan in the coming few years. Whenever we pump 100M m³, we lose 50M m³; that's a very big amount of water.' Reducing that to 20 per cent, he said, would 'gain a source of water bigger than the Disi aquifer.' If it was down to him, he would start from scratch, too, and divide Amman into districts, putting each district up for competitive tender to international water companies to fix the leaks, and divide any water-saving profits half and half between contractor and government.* Why hasn't this happened? 'Because . . . This is the case in Jordan.'

Leaks – also known as 'non-revenue water' by accounting and policy types – are a problem the world over, although some countries fare better than others. Both Singapore and Israel are down to single-figure percentage losses, while 25 per cent is more typical for countries like the UK. In Delhi, says Kangkanika Neog, 'almost 40 per cent of the water is lost this way.' Water companies endlessly complain about how hard it is to fix ageing underground infrastructure,† but it can be done. Phnom Penh, the capital of Cambodia, used to suffer from nearly 83 per cent water losses, but reduced it to a world-leading 8 per cent thanks to constant monitoring of pressure levels via software and a call hotline – leaks are repaired within three hours, while invisible leaks are identified at night using stethoscopes.[31] The subsequent cost savings made water bills cheaper for everyone in the city and free for the poorest households. (So, the next time your water company complains about their legacy pipes, tell them, 'If they can do it in Phnom Penh, why can't

---

* Perhaps unsurprisingly, 'paid by performance' is the World Bank's preferred approach to fixing leaks too.
† An Annual Leakage Conference has been held in the UK every year since 2000, where the water sector gathers to console each other about how hard it is, and occasionally sets targets for 2050, when they'll all be long gone.

you?'). That said, in areas reliant on groundwater, 'non-revenue water' isn't entirely lost. In Mexico City, where almost half its water is 'lost' to leaks, the grateful aquifer below soaks it back in, arguably protecting the city from further ground-level subsidence. I also have some sympathy with Andrew Tucker's point that, whenever water companies speak about water-saving, the press and public angrily shout back, 'Fix your leaks!' It kills off debate to the detriment of all other issues and available water solutions, namely nature-based solutions, aquifer recharge, water storage, rainwater harvesting and sewage treatment. We need to widen our focus, not restrict it.*

When we restrict our focus, says Rick Hogeboom, there's a real danger our water prospects will go in the wrong direction. 'We have worked a lot with the investor community and financial institutions in the past few years,' Hogeboom tells me. 'You see them waking up more and more to water issues, but they think in terms of financial indicators, in terms of risks: "Does it pose a risk to us? If there's water-scarcity risk, should we move on somewhere else?" That's a different question to, "Do our activities contribute to an unsustainable situation? And do we need to take responsibility for that as a business?"' See the commercial mining of Arizona groundwater, for example, or pension funds buying up Australian nut orchards. 'If you live on the savings account, as long as you don't have to pay the interest, then it's fine, you are spending someone else's money. I think that's the phase we are in now.' In short, even if water scarcity is acknowledged and calculated, the management decision can still be to extract, destroy and move on. There will always be another country or river basin willing and happy to embrace 'foreign direct investment'. If so, warns Hogeboom, 'investment activity will consume and pollute water for decades to come . . . this has the potential to go really wrong.'

Our huge water footprint, reliant on importing embedded or virtual water, had its fig leaf removed by the COVID-19 pandemic.

---

* But, even so, fix your leaks!

'We have now seen some of the vulnerabilities of the globalized system,' says Rick. We don't grow where the water is most sustainable; we grow 'where the labour is cheapest, where you have the land available, where you have regulations that are favourable, where you have trade negotiations. Water is just not factored into that.' It should be. International water stewardship written into every purchase agreement and trade negotiation would be far more effective than any number of water labels or low-flush toilets.

Is there a concerted international effort, or even appetite, for this? The international equivalent of India's water-for-all Jal Jeevan Mission is the UN's Sustainable Development Goal 6: 'Ensure access to water and sanitation for all by 2030'. And by 'all', it really means all, worldwide. The seventeen Sustainable Development Goals (SDGs) were adopted by the UN in 2015 as a universal call to action to end poverty, protect the planet and ensure that all people 'enjoy peace and prosperity' by 2030.[32] If that all sounds a bit too *Star Trek* Federation, the type of words that diplomats like to spout over fine wine and lobster at Davos, then Maggie White of SIWI attempts to put me right. The SDGs, she says, 'are very useful in that they are driving all development-corporation set-ups, at the moment. A country cannot ask for any funding support from the World Bank, from the IMF, from the French Development Agency, for example, without referring to the SDGs and how this or that project will deliver on the targets of the SDGs . . . Countries who are within the UN system have agreed to these global objectives, and hence, all of their reporting, all of their funding requests or grants or loans are dependent on how they comply or fulfil those objectives.' She believes – as do many who work in international development – that SDG 6 is not only important, but is the key to unlock the other sixteen SDGs too. But the UN's progress report for SDG 6 doesn't look good. In fact, it finds 'alarmingly inadequate progress'. Global water stress has gotten worse since 2015.[33] In 2020, 129 countries and territories were not on track to meet their target. At current rates of progress, UNICEF and the WHO estimate that, by 2030, rather than access for 'all', 1.6 billion people will still be without safe drinking water.

Are world leaders at the annual UN Climate Conferences taking sufficient heed of this? Water isn't mentioned once in the Paris Agreement (2015) or the Draft Agreement of COP26 in Glasgow.[34] The director-general of UNESCO, Audrey Azoulay, has admitted that, 'the word "water" rarely appears in international climate agreements, even though it plays a key role in issues such as food security, energy production, economic development and poverty reduction.'[35] In short, 'there's a lot of talk' and not enough action, sums up Rick Hogeboom. 'I view it a little bit like climate change. I mean, we have known for over fifty years that carbon poses a threat to the climate, and it took us forty years or so to get from that insight to action. That's how I see the water journey. But we don't have another forty years. We will simply run out of water.'

Back in credit

Something strange happened to my online water bill. It's not something I often check. Thames Water's customer interaction could be described as 'soft touch' at best. They email me a water bill once a quarter – sometimes I look at it, sometimes I don't. It doesn't affect what I pay, because, despite having a meter, I just pay a flat fee – £47 every month – irrespective of how much water my family actually use. The theory is that the amount will be less in winter, more in summer, and averages out at around £47. When I registered for an online account, hoping to see some detailed breakdowns of water usage, all I found were my previous bills in PDF form. Not exactly engaging. In February 2021, about halfway through writing this book, my quarterly bill arrived in my inbox. I clicked to open and found a surprise: £95 in credit. I was clearly paying too much. So, I rang customer services.

'It looks like your water usage is really not that high, in all honesty,' confirms a friendly Welshman called Alan at the call centre. This is good news. Part of my wake-up call for writing this book was a water bill in 2019 showing that our household consumption was 160 L per person – above the already high UK average. And one of the people in our house was a very little person – our then one-year-old daughter. She wasn't taking any twenty-minute showers. But our usage now, confirms Alan, is a much more respectable 110 L each. And Silvia was now three, and did, admittedly, occasionally take twenty-minute showers. I ask Alan what our direct debit should be adjusted to. 'Let me just get my calculator out . . . and, bearing in mind usage does change, and it has been a *little bit* high in the past . . . You're now looking at £23.' I'm shocked. My water bill has been cut in half. We haven't installed a Flow Loop shower or air-flush toilets – we even had the paddling pool out over summer. I think back to the water audit, and that alarming finding that over half our water use came from the shower, with a carbon footprint higher than that of our car. We subsequently fitted an aerator to the showerhead that changed the flow from 12 L to 7 L per minute – almost half. And fixed the leaky loo. And replaced the broken-down dishwasher with a low water-use model. Small changes, really; nothing life-changing. But in doing so we'd halved our water bill and fallen well below the national water use average in the process.

Towards the end of my original call with Thames Water's Aussie water-efficiency guy, Andrew Tucker, I remember suggesting that sometimes things (like my water bill) have to get really bad before they get better. In cities like Cape Town and Chennai, it took a Day Zero, literally running out of water, before populations truly addressed their water scarcity. Maybe southern England will have to experience this too, before we change our ways? 'I've been through that, Tim, and I wouldn't wish it upon my worst enemy,' said Tucker. And he told me his Australian genesis story. He was working as an environmental adviser at one of Australia's 'biggest, ugliest mines – the biggest in the southern hemisphere,' producing lead, copper and zinc. The mining town had 30,000 people and a large reservoir, bigger than any that serve London and fed by a single river. 'But we hadn't had rain for, like, three years. The dams were down to 15 per cent capacity – we were sucking mud.' This was, then, another Day Zero. The response of both the townspeople and Tucker shows, in microcosm, what needs to happen to solve a water crisis. It also shows how far removed this is from many of our daily lives. 'We put in very aggressive customer engagement,' recounts Tucker. 'Every single day, advising the mayor of the city, huge leakage-response plans, education on the news, every channel: no showers, no watering lawns, no washing cars, water rationing, things you could only do on odd dates or even dates, on-the-spot fines, priority uses only.' On the mine site, where water use would be about the same as the city itself, they completed a full audit. 'Every single tap, every single hose, every single water outlet was mapped and measured. What can we reuse? What leakage can we stop? We trained up 2,000 people on the mine to become water-efficiency experts, telling them, "In two months' time, this place is going to shut down unless we stop using water." We reduced water consumption by about 50 per cent, while still maintaining production.' Eventually, Tucker says, they mapped out the hierarchy for what would need to be switched off first – city businesses, then onsite mine operations – to keep homes, hospitals and schools with water. 'We were ready to shut down everything.' His family back home still take a bucket into the shower to save water for reuse, he says. When you've lived it, everyone gets on board.

The citizens of Santiago, Chile, similarly never believed they would suffer from a water shortage. As recently as 2019, Dr Pablo García-Chevesich at the University of Chile explained that the common viewpoint in the city was 'there is no value for the resource . . . it's very common to see gardens with high water consumption . . . 30-minute showers, and even people sweeping leaves off the street with a hose.' Just three years later, severe water rationing was imposed – forget manicured lawns: grass is now described as a 'luxury' in Santiago.[36] No one believes they can run out of water, until they do.

Ultimately, our response to the water crisis must work in conjunction with our response to the climate crisis. Every 1 degree temperature rise – every 0.1 of a degree – influences the water cycle and makes the water-management challenge harder, potentially untenable. A 2020 paper from Germany, using 254 years of climate data, predicted that the likelihood of severe drought in Europe increases sevenfold in 2051–2100 with a high-emissions future; it reduces 'by almost half' with mid-level emissions, and increases are 'almost negligible' under the most optimistic, net-zero pathway.[37] That's an awful lot to play for. A water-company boss would sell their grandma for a sevenfold reduction in their future supply risk. And yet that's where we're at: reduce carbon emissions, reduce water crisis.

Jennifer Francis, at the Woodwell Climate Research Center, describes extreme weather events – which have tripled in number since 1980 – as 'symptoms of the underlying disease'. It's time, she says, for emergency-room triage: 'a rapid and drastic reduction in our emissions of heat-trapping gases, while also bracing for worsening events to come'. We can't avoid the near-term escalation in the water cycle. That's already underway. But we can control 'how much worse they become', she says.

The IPCC's 2018 report outlined how water management should be modified in the face of climate change. The 'adaptive measures' it recommends include 'rainwater harvesting, conservation tillage, maintaining vegetation cover . . . improved pasture management, water reuse, desalination, and more efficient soil and irrigation water management'. The obvious anomaly in that otherwise nature-based and regenerative-farming list is desalination. Yet desalination is here to stay and is only going to grow. For some desert-based regions, it *is* the only option on the table. We can't ignore desalination or wish it away; we should embrace it if (and only if) all other options have been exhausted. And only when the water produced is affordable to the end user or the subsidizing government. Similarly, the restoration and protection of natural floodplains is crucial, say the IPCC, but the benefits are felt only 'where the infrastructure to capture that resource' is already in place downstream[38] – we must push ahead with nature-based solutions and rewilding, but in combination with the requisite boreholes, reservoirs and water-treatment plants to recoup the benefits. We can't move away from grey engineering entirely – in some places, such as southern England, we need more of it. Grey engineering can run alongside the blue and green, but, in almost every country in the world, we need to tip the balance more in favour of the latter.

Back in Tucson, Arizona, Dr Jennifer McIntosh has removed the swimming pool from her backyard. She laughs, saying they're the only people crazy enough to do it. 'Then we put in fake grass. Which some days I regret, when it's 115°F [46°C] outside.' But she knows there's a reckoning coming. 'How ethical is it to live in a place like Tucson? People are moving to the south-west because it's inexpensive. But, in terms of water supply and climate-change impacts, is that the right decision? And so, do we flee? Or do we stay and try to help the problem? I've kind of rooted in to staying and trying to help . . . and answer the question, what does sustainability *mean* in the south-west?' And, if it means no swimming pools and artificial grass for all, will as many people still want to live there? In June 2022, the same harsh reality finally hit Californians too. 'Get ready for short showers and brown lawns,' announced the *LA Times*. 'More than 6 million Southern Californians will be placed under new drought rules.' A Metropolitan Water District manager told the paper, 'We need to prioritize between watering our lawns and having water for our children and our grandchildren.'[39] The direst of warnings during my visits in 2021 were already coming to pass.

Within the problem also lies the solution, to paraphrase Charles Rangeley-Wilson. Many people, businesses and governments are currently so wasteful with water, that even small tweaks and changes can make a big difference. There's plentiful low-hanging fruit for water efficiency – from needlessly flood-irrigated back gardens to criminally wasted rainwater – just hanging there, waiting to be plucked. We must all try harder and act with urgency. Because the water cycle has already changed. The global food crisis has already arrived. And our twentieth-century water infrastructure is no longer fit for purpose. As Jake Rigg at Affinity Water said to me: 'The problem with all of this is we're running out of time, right?'

At the end of my conversation with Chloe Whipple, whose one-woman show saw her live off 15 L of water a day, she says, 'It sounds like you've had a similar journey of knowledge to me, of being like, "Oh shit. What? Our water system isn't OK? How is this not at the forefront of all news and planning, like globally and nationally? Who is solving this? How are we letting this just slip under the radar?!"' Yep, I agree, that about sums it up. 'It's really bonkers!' she exclaims. It *is* bonkers. Chloe finished her theatre show with the audience closing their eyes, holding hands, and chanting, 'WE LOVE WATER!' We all do. It's a universal. And yet we treat water as if we're in an abusive relationship with it – and we are the abuser. 'It's interesting, when we start to look at sustainable solutions,' Chloe continues. 'Once you start harvesting rainwater or reusing grey water, you're like, "Oh, I've got loads of it." And then you realize the abundance around us. It's like, if we carry on like this, then, yeah – it's all gonna run out. But, actually, if we just make some tweaks, then we have plenty.'

The 'last drop' doesn't mean waiting for the water to run out. It means *valuing* every last drop as precious. If we do that – and we hold politicians accountable for that – then there may yet be enough on this blue planet to go around.

Acknowledgements

My unending thanks and love to my wife Dr Patricia Brekke and my family for supporting me – in every definition of that word – during this book journey. The difficulties in its creation will forever be tied to the complications of a pandemic, illness, home-schooling, travel bans, shared stress and anxiety. Amid all that, my thanks to the NHS for supporting my family and the entire country through that most difficult of times.

Big thanks to my publisher, Picador, and editor extraordinaire, George Morley, for believing in me and this book from the start – and for those pandemic Zoom pep talks when I needed it most. And to my agent Jenny Hewson at Lutyens & Rubinstein for making this book a reality.

Thanks to the Society of Authors' K Blundell Trust grant, which helped fund some of the crucial overseas travel, without which the research for this book would have been much diminished.

My thanks to Becky Trotman, senior media relations officer at Thames Water, who helped set up numerous interviews and site visits, often in very difficult, middle-of-a-pandemic circumstances. It must have felt like letting the fox into the chicken coup at times, yet there was never any insistence on my towing a company line.

My thanks to the team at the Chalk Aquifer Alliance, in particular Sharon Moat, Libby Ranzetta, Mark Wilkinson and John Pritchard, for hosting me at their respective chalk streams (and my friends Emma and Carl Roche for offering me free board in Cambridgeshire that night!). The CAA are (alongside many others I met along the way, including Ash Smith) a humbling reminder that no change happens without dedicated, knowledgeable amateurs putting their time into the often thankless task of environmental campaigning.

Thanks to Virtalent and virtual PA Gina Wallace for stepping in to help with travel arrangements for the book, negotiating Covid travel red lists and PCR requirements – I couldn't have gotten through that without you (and the Californian insider tips were much appreciated!).

I am indebted to several people for their help and hospitality during my travels. In Amman, Jordan, to Dave Simpson and Emma Cliffe for welcoming me to their home with only the briefest of introductions (my thanks there to Dave Norton!), showing me the city and advising me on everything from SIM cards to the best Dead Sea soap! To EcoPeace Middle East, and in particular its Jordan director Yana Abu Taleb, for so generously hosting me at their field sites and offices – you went above and beyond, and I hope my text does some justice to your work.

In Ghana, my fixer Komi 'Fred' Vedomey made everything happen – even when, on occasion, all hope seemed lost, he always pulled it out of the bag somehow. And in Arizona, I'll forever be grateful to Lynda Person and her husband John Peterson for being so generous with their time and Tesla, giving me a deep insight into Arizona's water woes (plus a Mexican restaurant or two).

In short, this book – and any non-fiction book worth its salt – couldn't happen without people giving up their time of their own volition, with very little expected in return. Whether it's opening up their house or farm or sewage works to me, the easier option will always be to say 'no'. This book, then, is dedicated to those who say 'yes' – it's you who get to tell and shape the story, and inform others of what's really happening. I literally couldn't do anything without you. So, if you are quoted in this book, in whatever form – thank you for saying 'yes'. Your work will live on.

Notes

INTRODUCTION

1   'Karameh Dam, Jordan', Environmental Justice Atlas, 2 September 2018, https://ejatlas.org/conflict/karama-dam-jordan

2   'Parliamentary justice: the dream of a minister who cost the state treasury 80 million dinars', ammonnews.net, 12 January 2021, https://www.ammonnews.net/article/587592 [in Arabic]

3   OMICS International, 'Saad Saleh Abu Hammour, Arab Potash Company, Jordan', https://biography.omicsonline.org/jordan/arab-potash-company/saad-saleh-abu-hammour-224608 [accessed 21 February 2023]

4   Elias Mechael Salameh, 'The Tragedy of the Karama Dam Project/Jordan', *Acta Hydrochimica et Hydrobiologica* 32(3), September 2004, pp. 249–58, https://www.researchgate.net/publication/230328743_The_Tragedy_of_the_Karama_Dam_ProjectJordan

5   Laith Al-Junaidi, 'Jordan to buy 50 million m3 of water from Israel', Anadolu Agency, 12 October 2021, https://www.aa.com.tr/en/middle-east/jordan-to-buy-50-million-m3-of-water-from-israel/2389837

6   Rachel Ramirez, Pedram Javaheri and Drew Kann, 'The shocking numbers behind the Lake Mead drought crisis', *The Mercury News*, 17 June 2021, https://www.mercurynews.com/2021/06/17/the-shocking-numbers-behind-the-lake-mead-drought-crisis/

7   GWCL Company Profile (website), https://www.gwcl.com.gh/company-profile/

8   'Average precipitation in depth (mm per year)', World Bank (website), https://data.worldbank.org/indicator/AG.LND.PRCP.MM

9   Ibid.

10 The United Nations World Water Development Report 2022.

11 European Court of Auditors, Special Report no. 33, 'Combating
    desertification in the EU: a growing threat in need of more action',
    2018, https://op.europa.eu/webpub/eca/special-reports/
    desertification-33-2018/en/#chapter0

12 UN Water, 'Summary Progress Update 2021: SDG 6 – water and
    sanitation for all', 24 February 2021, https://www.unwater.org/
    publications/summary-progress-update-2021-sdg-6-water-and-
    sanitation-all

13 Emma Howard Boyd, UK Environment Agency, 'Water Security in a
    Changing Environment', 30 June 2021, https://www.gov.uk/govern-
    ment/speeches/water-security-in-a-changing-environment

14 Helen Davidson, 'China drought causes Yangtze to dry up, sparking
    shortage of hydropower', *The Guardian*, 22 August 2022, https://www.
    theguardian.com/world/2022/aug/22/china-drought-causes-yangtze-river-
    to-dry-up-sparking-shortage-of-hydropower

CHAPTER 1

1 'Disi water project comes on stream', Economist Intelligence, 29 July
   2013, http://country.eiu.com/article.aspx?articleid=1590783543#:~:text
   =with%20the%20disi%20water%20estimated%20to%20last%20for%20
   just%2050%20years%3B%20it%20was%20originally%20regarded%20
   as%20a%20holding%20measure%20until%20major%20desalination%20
   projects%20could%20be%20completed.

2 Johanna Montanari, 'Water woes pushing Jordan close to "Day Zero"
   scenario – experts', *Jordan Times*, 8 August 2019, https://jordantimes.
   com/news/local/water-woes-pushing-jordan-close-%E2%80%98day-
   zero%E2%80%99-scenario-%E2%80%94-experts

3 Petru Saal, 'This is how Cape Town's water collection points will work',
   Times Live, 28 January 2018, https://www.timeslive.co.za/news/
   south-africa/2018-01-28-this-is-how-cape-towns-water-collection-
   points-will-work/

4 Louise Stafford, 'Cape Town Faces "Day Zero"', Nature Conservancy
   (website), https://www.nature.org/en-us/about-us/where-we-work/
   africa/stories-in-africa/cape-town-faces--day-zero-/

5 The United Nations World Water Development Report 2020, 'Water
   and Climate Change'.

6 Asit K. Biswas and Cecilia Tortajada, Policy Forum, 6 July 2018.

7 Sujith Sourab Guntoju, Mohammad Faiz Alam and Alok Sikka,
   'Chennai water crisis: A wake-up call for Indian cities', *Down To Earth*,
   5 August 2019, https://www.downtoearth.org.in/blog/water/chennai-
   water-crisis-a-wake-up-call-for-indian-cities-66024

8   Rachel Salvidge, 'Countdown to Day Zero: what will happen when London runs out of water?', *ENDS Report*, 2020, https://www.endsreport.com/article/1690368/countdown-day-zero-will-happen-when-london-runs-water?s=09

9   NITI Aayog, 'Composite Water Management Index: A Tool for Water Management', June 2018, https://www.niti.gov.in/writereaddata/files/document_publication/2018-05-18-Water-index-Report_vS6B.pdf

10 Jaz da King of da South, Instagram, 3 March 2021, https://www.instagram.com/p/CL-XDwWADOM/

11 Will Stribling, 'Thousands of Jackson residents enter second week without running water', Mississippi Today, 1 March 2021, https://mississippitoday.org/2021/03/01/thousands-of-jackson-residents-enter-third-week-without-running-water/

12 Safia Samee Ali, 'Jackson, Mississippi, water crisis brings to light long-standing problems in city', NBC News, 3 March 2021, https://www.nbcnews.com/news/us-news/jackson-mississippi-water-crisis-brings-light-long-standing-problems-city-n1259376

13 Molly Schwartz, 'The Water Crisis in Jackson, Mississippi, Is a Dire Warning Sign', *Mother Jones*, 24 March 2021, https://www.motherjones.com/politics/2021/03/water-crisis-in-jackson-mississippi-is-a-dire-warning-sign/

14 Henry Fountain, 'Severe Drought, Worsened by Climate Change, Ravages the American West', *New York Times*, 19 May 2021, https://www.nytimes.com/2021/05/19/climate/drought.html

15 L. O'Hanlon, 'The Looming Crisis of Sinking Ground in Mexico City', *Eos*, 22 April 2021, https://eos.org/research-spotlights/the-looming-crisis-of-sinking-ground-in-mexico-city

16 Joshua Greene, 'Bottled water in Mexico: The rise of a new access to water paradigm', *WIREs Water* 5(4), July/August 2018, https://wires.onlinelibrary.wiley.com/doi/10.1002/wat2.1286

17 '50 years later, donkeys continue delivering water in Mexico City villages', Mexico News Daily, 2 April 2021, https://mexiconewsdaily.com/news/50-years-later-donkeys-continue-delivering-water-in-mexico-city-villages/

18 MTST – Movimento dos Trabalhadores Sem-Teto, Facebook post, 25 September 2014, https://www.facebook.com/mtstbrasil

19 Carlos Meneses, 'Impact of severe Brazil drought seen in low reservoirs, soaring power bills', *La Prensa Latina*, 1 October 2021, https://www.laprensalatina.com/impact-of-severe-brazil-drought-seen-in-low-reservoirs-soaring-power-bills/

20 'Brazil hydro reservoirs at highest level since 2016, allaying rationing fears', Reuters, 7 January 2022, https://www.reuters.com/markets/

commodities/brazil-hydro-reservoirs-highest-level-since-2016-
allaying-rationing-fears-2022-01-07/

21 Ana Ferraz, 'Reservoirs in six Brazilian states are below 2021 levels',
    The Brazilian Report, 19 Jan 2022, 'https://brazilian.report/liveblog/
    2022/01/19/reservoirs-below-2021-levels/

22 Julieta Pelcastre, 'NASA, ESA Sign Climate Change Agreement',
    Diálogo, 12 August 2021, https://dialogo-americas.com/articles/
    nasa-esa-sign-climate-change-agreement/

23 John Vidal, 'Cities in peril as Andean glaciers melt', *The Guardian,* 29
    August 2006, https://www.theguardian.com/environment/2006/
    aug/29/glaciers.climatechange

24 Lorena Guzmán, 'Chile's mega-drought rolls on', *Diálogo Chino,* 11
    October 2019, https://dialogochino.net/en/climate-energy/30820-
    chiles-mega-drought-rolls-on/

25 Andrea Becerra et al, 'A New Course: Managing Drought and
    Downpours in the Santiago Metropolitan Region', NRDC, July 2019,
    https://www.nrdc.org/sites/default/files/new-course-managing-
    drought-downpours-santiago-report.pdf

26 'Chile announces unprecedented water rationing plan: Here's how it
    will work', CNBC TV18, 12 April 2022, https://www.cnbctv18.com/
    environment/chile-announces-unprecedented-water-rationing-plan-
    heres-how-it-will-work-13131372.htm

27 Mindy Wright, 'Most Water-Stressed Countries In The World For
    2019', *CEO World Magazine,* 8 August 2019, https://ceoworld.
    biz/2019/08/08/most-water-stressed-countries-in-the-world-for-2019/

28 Marijke Huysmans, 'Minder oppompen of meer infiltratie? Hoe
    verhoog je de grondwaterstand het meest?', Mythes over grondwater
    (blog), 2 June 2020, https://grondwatermythes.blogspot.com/2020/05/
    meer-of-minder-grondwater-oppompen-meer.html

29 Zuhal Demir, 'Grondwaterstanden bij start zomervakantie 2021 hoger
    dan vorig jaar', Zuhal Demir (website), https://www.zuhaldemir.be/
    nieuws/grondwaterstanden-bij-start-zomervakantie-2021-hoger-dan-
    vorig-jaar?pq=nieuws&page=8#views-row-4

30 Zhang et al, 'Water Scarcity and Sustainability in an Emerging
    Economy: A Management Perspective for Future'. *Sustainability,* 2021,
    13, 144. https://dx.doi.org/ 10.3390/su13010144

31 'Water shortage of 31m acre feet expected by 2025', *Dawn,* 29 March
    2018, https://www.dawn.com/news/1398099

32 Alan Buis and Janet Wilson, 'Study: Third of Big Groundwater Basins
    in Distress', NASA, 16 June 2015, https://www.nasa.gov/jpl/grace/
    study-third-of-big-groundwater-basins-in-distress

33 Raghu Murtugudde, 'Pakistan world's top groundwater exporter, India
    third', Third Pole (website), 3 May 2017, https://www.thethirdpole.

net/2017/05/03/pakistan-worlds-top-groundwater-exporter-india-third/

34 Jacob Kurtzer, 'Pakistan's Deadly Floods Pose Urgent Questions on Preparedness and Response', Center for Strategic & International Studies, 14 September 2022, https://www.csis.org/analysis/pakistans-deadly-floods-pose-urgent-questions-preparedness-and-response

35 'Appeal: Devastating floods in Pakistan', Unicef, https://www.unicef.org/emergencies/devastating-floods-pakistan-2022

36 Mohsin Hafeez, 'How Pakistan can prevent another flood disaster', Al Jazeera (website), 11 October 2022, https://www.aljazeera.com/opinions/2022/10/11/how-pakistan-can-prevent-another-flood-disaster

37 Brian Eyler, *Last Days of the Mighty Mekong* (London: Zed Books, 2019).

38 'The Mekong Delta, the drama of all deltas', We Are Water Foundation (website), 3 May 2021, https://www.wearewater.org/en/the-mekong-delta-the-drama-of-all-deltas_339961

39 Ibid.

40 Brian Eyler et al, 'How China Turned Off the Tap on the Mekong River', Stimson Center (website), 13 April 2020, https://www.stimson.org/2020/new-evidence-how-china-turned-off-the-mekong-tap/

41 Vaibhav Chaturvedi et al, 'Reallocating Water for India's Growth', CEEW Report, December 2020, https://www.ceew.in/sites/default/files/CEEW-Reallocating-Water-for-Indias-Growth-20Nov20_0.pdf

42 Echo Xie, 'China on higher alert as floods hit – and climate change promises more extreme weather', *South China Morning Post*, 4 July 2020, https://www.scmp.com/news/china/society/article/3091831/china-higher-alert-floods-hit-and-climate-change-promises-more

43 'Sahel: A food crisis fueled by conflict is set to worsen during lean period', ReliefWeb (website), 12 May 2022, https://reliefweb.int/report/burkina-faso/sahel-food-crisis-fueled-conflict-set-worsen-during-lean-period

44 'Drought forces Mozambique capital to ration water', CGTN Africa (website), 15 February 2018, https://africa.cgtn.com/2018/02/15/drought-forces-mozambique-capital-to-ration-water/

CHAPTER 2

1 Jay Famiglietti biography, Steven Barclay Agency (website), https://www.barclayagency.com/speakers/jay-famiglietti

2 'The Water Institute 2016 RBC Distinguished Lecture: Jay Famiglietti',
   Water Institute (video), https://www.youtube.com/watch?v=
   8AXJMKdinBs

3 'Water Resources Management', World Bank (website), https://www.
   worldbank.org/en/topic/waterresourcesmanagement

4 'Replenishing a Broken Water Cycle, with Sandra Postel' (episode),
   What About Water? (podcast, season 3), 24 November 2021, https://
   www.whataboutwater.org/s03e04/

5 Nadine Schmidt, Schams Elwazer, Barbara Wojazer and Sharon
   Braithwaite, 'Germany mounts huge rescue effort after floods leave
   dozens dead and many more missing', CNN (website), 16 July 2021,
   https://edition.cnn.com/2021/07/15/europe/germany-deaths-severe-
   flooding-intl/index.html

6 Luisa Neubauer, Twitter post, 16 July 2021, https://twitter.com/
   Luisamneubauer/status/1415982920951554049

7 Minghao Zhou, Twitter post, 21 July 2021, https://twitter.com/
   Minghao_Zhou/status/1417719446941192193

8 Peter Gleick, Twitter post, 22 July, 2021, https://twitter.com/
   PeterGleick/status/1418338770542227457

9 'Catastrophic floods caused by extremely heavy rains claim lives of at
   least 63 people in Henan, China', Watchers News (website), 21 July
   2021, https://watchers.news/2021/07/21/henan-zhengzhou-gongyi-
   china-flood-july-2021-casualties/

10 Hayley Smith and Lila Seidman, ' "Storm of the season" dumps
   record-breaking rainfall on SoCal and snow in the mountains', *Los
   Angeles Times*, 14 December 2021, https://www.latimes.com/
   california/story/2021-12-14/storm-of-the-season-delivers-rain-snow-
   and-wind-to-socal

11 Jennifer Francis, 'Another week, another weather tantrum: We are
   changing the atmosphere and global weather patterns', *The Hill*, 21 July
   2021, https://thehill.com/opinion/energy-environment/564077-
   another-week-another-weather-tantrum-we-are-changing-the/

12 Andrea Becerra et al, 'A New Course: Managing Drought and
   Downpours in the Santiago Metropolitan Region', NRDC, July 2019,
   https://www.nrdc.org/sites/default/files/new-course-managing-
   drought-downpours-santiago-report.pdf

13 'Human Activities Are Drying Out the Amazon', NASA Earth
   Observatory (website), 2019, https://earthobservatory.nasa.gov/
   images/145834/human-activities-are-drying-out-the-amazon

14 Ibid.

15 'Climate Change 2022: Mitigation of Climate Change', Working Group
   III contribution to IPCC's Sixth Assessment Report, https://report.
   ipcc.ch/ar6wg3/pdf/IPCC_AR6_WGIII_FinalDraft_FullReport.pdf

16 Y. Kamae et al, 'Atmospheric Rivers Bring More Frequent and Intense Extreme Rainfall Events Over East Asia Under Global Warming', *Geophysical Research Letters* 48(24), 28 December 2021, https://agupubs.onlinelibrary.wiley.com/doi/10.1029/2021GL096030

17 Stuart Layt, ' "River in the sky" left south-east Queensland drenched, experts say', *Brisbane Times*, 28 February 2022, https://www.brisbanetimes.com.au/national/queensland/river-in-the-sky-left-south-east-queensland-drenched-experts-say-20220228-p5a0am.html

18 United Nations World Water Development Report 2020: Water and Climate Change, Paris, UNESCO.

19 Meiping Sun, Shiyin Liu, Xiaojun Yao et al, 'Glacier changes in the Qilian Mountains in the past half century: Based on the revised First and Second Chinese Glacier Inventory[J]', *Acta Geographica Sinica* 70(9), 2015, pp. 1402–14.

20 'Uganda: Floods and Landslides – Sep 2021', ReliefWeb (website), https://reliefweb.int/disaster/fl-2021-000153-uga

21 Aditya Tarar, 'Fire raining from the sky in Pakistan! Mercury crosses 51 degrees, Jacobabad becomes the hottest area in the world', Hindustan News Hub (website), 15 May 2022, https://hindustannewshub.com/world-news/fire-raining-from-the-sky-in-pakistan-mercury-crosses-51-degrees-jacobabad-becomes-the-hottest-area-in-the-world/

22 Al Shaw, Abrahm Lustgarten and Jeremy W. Goldsmith, 'New Climate Maps Show a Transformed United States', ProPublica (website), 15 September 2020, https://projects.propublica.org/climate-migration/

CHAPTER 3

1 'Municipal water consumption', Britannica (website), https://www.britannica.com/technology/water-supply-system/Municipal-water-consumption

2 David Owen, *Where the Water Goes: Life and Death Along the Colorado River* (New York: Riverhead Books, 2017).

3 A. Park Williams, Benjamin I. Cook and Jason E. Smerdon, 'Rapid intensification of the emerging southwestern North American megadrought in 2020–2021', *Nature Climate Change* 12, 14 February 2022, pp. 232–4, https://www.nature.com/articles/s41558-022-01290-z.epdf

4 Steven R. Fassnacht, 'Snow can disappear straight into the atmosphere in hot, dry weather', The Conversation (website), 27 July 2021, https://theconversation.com/snow-can-disappear-straight-into-the-atmosphere-in-hot-dry-weather-162910

5   Andrew Schwartz, 'I'm a scientist in California. Here's what worries me most about drought', *New York Times*, 4 April 2022, https://www. nytimes.com/2022/04/04/opinion/environment/california-drought-wildfires.html

6   'Fact Sheet: Drought Contingency Plan, Arizona Implementation', Central Arizona Project, February 2022, https://library.cap-az.com/ documents/departments/planning/colorado-river-programs/CAP-FactSheet-DCP.pdf

7   Bureau of Reclamation, 'Inflow for Lake Powell 1964 to 2009', River Simulator (website), http://www.riversimulator.org/Resources/Graphs/ InflowLakePowell1964to2009.pdf

8   'Glen Canyon Dam', Bureau of Reclamation (website), 15 November 2022, https://www.usbr.gov/uc/water/crsp/cs/gcd.html

9   Stephanie Elam, 'Lake Mead plummets to unprecedented low, exposing original 1971 water intake valve', CNN (website), 29 April 2022, https://edition.cnn.com/2022/04/27/us/water-intake-exposed-lake-mead-drought-climate/index.html#

10 Lynda Person, 'Can agriculture use less water?' *Arizona Capitol Times*, 8 October 2021, https://azcapitoltimes.com/news/2021/10/08/ can-agriculture-use-less-water/

11 'Town Lake Water Quality', City of Tempe, AZ, https://www.tempe. gov/government/community-services/tempe-town-lake/how-town-lake-works/town-lake-water-quality

12 David Owen, *Where the Water Goes* (New York: Riverhead Books, 2017), p. 142.

13 'Central Arizona Project: Highlighting key features from the CAP system, an engineering marvel', https://storymaps.arcgis.com/stories/ b7b28dd4c36a413e8d533ba540f998cb

14 David Owen, *Where the Water Goes* (New York: Riverhead Books, 2017), p. 142.

15 Mark Silverstein, 'Las Vegas, we have a problem', *Nevada Independent*, 19 October 2021, https://thenevadaindependent.com/article/ las-vegas-we-have-a-problem

16 W. Michael Hanemann, 'Working Paper No. 937: The Central Arizona Project', California Agricultural Experiment Station, Giannini Foundation of Agricultural Economics, October 2002, https://gspp. berkeley.edu/assets/uploads/research/pdf/cap.pdf

17 'Irrigation Non-Expansion Areas' (info document), Arizona Department of Water Resources, 8 June 2017, https://infoshare. azwater.gov/docushare/dsweb/Get/Document-10190/Irrigation% 20Non-Expansion%20Areas%20(INAs).pdf

18 Keith Schneider, 'Unsafe Yield', Circle of Blue (website), https://www. circleofblue.org/2022/world/unsafe-yield/

19 'Ground-Water Depletion Across the Nation', US Geological Survey, November 2003, https://pubs.usgs.gov/fs/fs-103-03/#pdf

20 'Human Effects on the Hydrologic System of the Verde Valley, Central Arizona, 1910–2005 and 2005–2110, Using a Regional Groundwater Flow Model', Scientific Investigations Report 2013–5029, USGS, 2013, https://pubs.usgs.gov/sir/2013/5029/sir2013-5029.pdf

21 'GWSI Well Information', Arizona Department of Water Resources, https://gisweb.azwater.gov/gwsi/Detail.aspx? SiteID=332308111345001

22 'GWSI Well Information', Arizona Department of Water Resources, https://gisweb.azwater.gov/gwsi/Detail.aspx?SiteID=331324111370001

23 Keith Schneider, 'Unsafe Yield', Circle of Blue (website), https://www.circleofblue.org/2022/world/unsafe-yield/

CHAPTER 4

1 'Iran: Deadly Response to Water Protests', Human Rights Watch (website), 22 July 2021, https://www.hrw.org/news/2021/07/22/iran-deadly-response-water-protests#

2 Richard Spencer, 'Iran fast running out of water, says exiled expert after deadly protests', *The Times*, 24 July 2021, https://www.thetimes.co.uk/article/iran-fast-running-out-of-water-says-exiled-expert-after-deadly-protests-mt3gdndc3

3 World Commission on Dams, *Dams and Development: A New Framework for Decision-Making* (London: Earthscan Publications, 2000), p. xxviii, https://archive.internationalrivers.org/sites/default/files/attached-files/world_commission_on_dams_final_report.pdf

4 Ibid, pp. xxx–xxxi.

5 Ibid, p. 16.

6 Ana Swanson, 'How China used more cement in 3 years than the U.S. did in the entire 20th Century', *Washington Post*, 24 March 2015, https://www.washingtonpost.com/news/wonk/wp/2015/03/24/how-china-used-more-cement-in-3-years-than-the-u-s-did-in-the-entire-20th-century/

7 Echo Xie, 'Explainer: Why are the floods so severe in China this year?', *South China Morning Post*, 19 July 2020, https://www.scmp.com/news/china/society/article/3093713/global-warming-and-illegal-land-reclamation-add-severe-floods

8 BBC News, 'Ethiopia's Abiy Ahmed: The Nobel Prize winner who went to war', 11 October 2021, https://www.bbc.co.uk/news/world-africa-43567007

9   By *Egypt Today* staff, 'GERD: A year since talks collapsed', *Egypt Today*, 04 Apr 2022, https://www.egypttoday.com/Article/1/114516/GERD-A-year-since-talks-collapsed

10   Charlotte Bruneau and Ahmed Rasheed, 'As its rivers shrink, Iraq thirsts for regional cooperation' Reuters, 8 September 2021, https://www.reuters.com/business/environment/its-rivers-shrink-iraq-thirsts-regional-cooperation-2021-09-06/

11   UN Environment Programme, 'Options for decoupling economic growth from water use and water pollution', Report of the International Resource Panel Working Group on Sustainable Water Management, 2015.

12   'Lake Powell Reservoir: A Failed Solution', Glen Canyon Institute, https://www.glencanyon.org/lake-powell-reservoir-a-failed-solution

13   The Sonoma Resource Conservation District and the Resource Conservation District of Santa Cruz County, 'Slow it. Spread it. Sink it. Store it!' LandSmart, June 2015, https://dailyacts.org/wp-content/uploads/2015/12/Slow-it-Spread-it-Sink-it-Store-it.pdf

14   J. S. Famiglietti, Min-Hui Lo, James Bethune et al, 'Satellites Measure Recent Rates of Groundwater Depletion in California's Central Valley', *Geophysical Research Letters* 38(3), February 2011, https://www.researchgate.net/publication/237998500_Satellites_Measure_Recent_Rates_of_Groundwater_Depletion_in_California's_Central_Valley

15   Leonard F. Konikow, 'Long-Term Groundwater Depletion in the United States', *Groundwater* 53(1), January–February 2015, pp. 2–9, https://www.documentcloud.org/documents/1674356-konikow-2015-groundwater.html

16   'Groundwater: "Go Deep or Go Dry" is Unsustainable', *What About Water?* (podcast), season 2, episode 5, https://www.whataboutwater.org/s02e05/

17   Debra Perrone and Scott Jasechko, 'Deeper well drilling an unsustainable stopgap to groundwater depletion', *Nature Sustainability* 2, 22 July 2019, pp. 773–82, https://www.nature.com/articles/s41893-019-0325-z

18   Scott Jasechko and Debra Perrone, 'Global groundwater wells at risk of running dry', *Science* 372(6540), 23 April 2021, pp. 418–21, https://www.science.org/doi/abs/10.1126/science.abc2755

19   Elad Levintal, Maribeth L. Kniffin, Yonatan Ganot et al, 'Agricultural managed aquifer recharge (Ag-MAR) – a method for sustainable groundwater management: A review', *Critical Reviews in Environmental Science and Technology*, 28 March 2022, https://www.tandfonline.com/doi/full/10.1080/10643389.2022.2050160

20   GuruJal Information Borchure: An integrated water management initiative, 2020. https://gurujal.org/reports/

21 Alvar Closas, François Moll and Nuria Hernández-Mora, 'Sticks and carrots to manage groundwater over-abstraction in La Mancha, Spain', *Agricultural Water Management*, Volume 194, December 2017, https://doi.org/10.1016/j.agwat.2017.08.024

22 https://www.internationalrivers.org/wp-content/uploads/sites/86/2021/10/UN-CBD-COP15-Freshwater-Expert-Letter-Signatories-Oct-2021.pdf

23 'Dead Sea sinkhole expansion recorded by scientists' camera', *Jerusalem Post*, 1 February 2021, https://www.jpost.com/israel-news/dead-sea-sinkhole-expansion-recorded-by-scientists-camera-watch-657437

24 Figures from interview with HE Abu Hammour.

CHAPTER 5

1 S. Ayoob and A. K. Gupta, 'Fluoride in Drinking Water: A Review on the Status and Stress Effects', *Critical Reviews in Environmental Science and Technology* 36(6), November 2006, pp. 433–87, https://www.researchgate.net/publication/233357494_Fluoride_in_Drinking_Water_A_Review_on_the_Status_and_Stress_Effects

2 Richard Damania, Sébastien Desbureaux et al, *Quality Unknown: The Invisible Water Crisis* (Washington, DC: World Bank Group, 2019), https://openknowledge.worldbank.org/bitstream/handle/10986/32245/9781464814594.pdf

3 Tushaar Shah, Chittaranjan Ray and Uma Lele, 'How to clean up the Ganges?', *Science* 362(6414), 2 November 2018, p. 503, https://www.science.org/doi/10.1126/science.aav8261

4 Richard Damania, Sébastien Desbureaux et al, *Quality Unknown: The Invisible Water Crisis*, https://openknowledge.worldbank.org/bitstream/handle/10986/32245/9781464814594.pdf

5 Stephanie L. Wear, Vicenç Acuña et al, 'Sewage pollution, declining ecosystem health, and cross-sector collaboration', *Biological Conservation* 255, March 2021, https://www.sciencedirect.com/science/article/pii/S0006320721000628

6 J. Mateo-Sagasta et al, *More people, more food, worse water? A global review of water pollution from agriculture* (Rome: Food and Agriculture Organization of the United Nations, 2018), https://www.fao.org/3/CA0146EN/ca0146en.pdf

7 *Pharmaceuticals in Drinking-Water* (Geneva: World Health Organization, 2012), https://apps.who.int/iris/bitstream/handle/10665/44630/9789241502085_eng.pdf

8 Karel Allegaert, Mariska Y. Peeters, Bjorn Beleyn et al, 'Paracetamol pharmacokinetics and metabolism in young women', *BMC*

Anesthesiology 15, 13 November 2015, https://bmcanesthesiol. biomedcentral.com/articles/10.1186/s12871-015-0144-3

9 José Angosto et al, 'Removal of Diclofenac in Wastewater Using Biosorption and Advanced Oxidation Techniques: Comparative Results', *Water* 12(12), 2020, p. 3567, https://www.mdpi.com/ 2073-4441/12/12/3567

10 James Meador, Andrew Yeh, Graham Young et al, 'Contaminants of emerging concern in a large temperate estuary', *Environmental Pollution* 213, June 2016, pp. 254–67, https://sci-hub.3800808.com/ 10.1016/j.envpol.2016.01.088

11 Ibid.

12 Reiner Hengstmann, *Scoping Study on the European Union Standards in Textile and Leather Sectors of Pakistan*, WWF-Pakistan, 2020, https://wwfasia.awsassets.panda.org/downloads/final_eu_scoping_ study_1.pdf

13 'Fresh Water', Forest & Bird (website), https://www.forestandbird.org. nz/what-we-do/fresh-water

14 Janet McConnaughey, 'Gulf of Mexico's "dead zone" much smaller this year', Associated Press (website), 4 August 2020, https://apnews. com/article/gulf-of-mexico-mississippi-river-weather- louisiana-9d1bcc60a66b54a96976ca17cf9ce3b3

15 Jennifer Collins, 'Ocean "dead zones" cover an area larger than the UK', Deutsche Welle (website), 8 March 2017, https://www.dw.com/ en/ocean-dead-zones-cover-an-area-larger-than-the-united-kingdom/ a-39941558

16 Richard Damania, Sébastien Desbureaux et al, *Quality Unknown: The Invisible Water Crisis*, https://openknowledge. worldbank.org/bitstream/handle/10986/32245/ 9781464814594.pdf

17 City of Toledo, Facebook post, 2 August 2014, https://www.facebook. com/cityoftoledo/posts/urgent-notice-to-residents-of-toledo- lucas-county-who-receive-water-from-the-cit/738905586173078/

18 'Toxic Algae in Lake Erie', NASA Earth Observatory (website), 2013, https://earthobservatory.nasa.gov/images/82165/toxic-algae- in-lake-erie

19 Michigan Environmental Council, 'Another lackluster plan won't fix Lake Erie', 17 December 2021, https://www.environmentalcouncil.org/ another_lackluster_plan_won_t_fix_lake_erie

20 Fremont News-Messenger, USA TODAY Network, '2021 Lake Erie algal bloom more severe than scientists predicted', 3 November 2021, https://eu.thenews-messenger.com/story/news/local/2021/11/03/ lake-erie-algal-bloom-more-worse-year-than-scientists-predicted/ 6232058001/

21 J. Mateo-Sagasta et al, *More people, more food, worse water? A global review of water pollution from agriculture*, https://www.fao.org/3/CA0146EN/ca0146en.pdf

22 Richard Damania, Sébastien Desbureaux et al, *Quality Unknown: The Invisible Water Crisis*, https://openknowledge.worldbank.org/bitstream/handle/10986/32245/9781464814594.pdf

23 'The Harrowing Trail of Toxic Nutrients in Farm Country Water', Circle of Blue (website), http://www.circleofblue.org/nitrate/

24 Richard Damania, Sébastien Desbureaux et al, *Quality Unknown: The Invisible Water Crisis*, https://openknowledge.worldbank.org/bitstream/handle/10986/32245/9781464814594.pdf

25 J. Mateo-Sagasta et al, *More people, more food, worse water? A global review of water pollution from agriculture*, https://www.fao.org/3/CA0146EN/ca0146en.pdf

26 Brian Bienkowski, 'Peak Pig: Read our full series on the fight for the soul of rural America', *Environmental Health News*, 13 November 2017, https://www.ehn.org/peak-pig-the-fight-for-the-soul-of-rural-america-2504375655/day-3

27 Natalia Milojevic and Agnieszka Cydzik-Kwiatkowska, 'Agricultural Use of Sewage Sludge as a Threat of Microplastic (MP) Spread in the Environment and the Role of Governance', *Energies* 14(19), 2021, https://www.mdpi.com/1996-1073/14/19/6293/htm

28 Damian Carrington, 'Nanoplastic pollution found at both of Earth's poles for first time', *The Guardian*, 21 January 2022, https://www.theguardian.com/environment/2022/jan/21/nanoplastic-pollution-found-at-both-of-earths-poles-for-first-time

29 Graham Readfearn, 'WHO launches health review after microplastics found in 90% of bottled water', *The Guardian*, 15 March 2018, https://www.theguardian.com/environment/2018/mar/15/microplastics-found-in-more-than-90-of-bottled-water-study-says

30 Damian Carrington, 'Microplastics may be linked to inflammatory bowel disease, study finds', *The Guardian*, 22 December 2021, https://www.theguardian.com/society/2021/dec/22/microplastics-may-be-linked-to-inflammatory-bowel-disease-study-finds

31 Mark Olalde, 'The Haunting Legacy of South Africa's Gold Mines', *Yale Environment 360*, 12 November 2015, https://e360.yale.edu/features/the_haunting_legacy_of_south_africas_gold_mines

32 Sara E. Pratt, 'All the glitters . . . Acid mine drainage: The toxic legacy of gold mining in South Africa', *Earth Magazine*, 5 January 2012, https://www.earthmagazine.org/article/all-glitters-acid-mine-drainage-toxic-legacy-gold-mining-south-africa/

33 S. Luzipo (chairperson), 'Derelict and Ownerless Mine interventions'
    (meeting), South Africa Parliamentary Monitoring Group (website),
    20 November 2019, https://pmg.org.za/committee-meeting/29381/

34 Stephen Tuffnell, 'Acid drainage: the global environmental crisis
    you've never heard of', The Conversation (website), 5 September 2017,
    https://theconversation.com/acid-drainage-the-global-environmental-
    crisis-youve-never-heard-of-83515

35 Sharon Lerner, '3M knew about the dangers of PFOA and PFOS
    decades ago, internal documents show', The Intercept (website), 31 July
    2018, https://theintercept.com/2018/07/31/3m-pfas-minnesota-
    pfoa-pfos/

36 'Information on PFAS | Historical Use', DuPont (website), https://
    www.dupont.com/pfas/historical-use.html [accessed April 2022]

37 The Threat of PFAS: The Forever Chemicals, Canadian Environmental
    Law Association, https://cela.ca/wp-content/uploads/2019/10/
    PFAS-Fact-Sheet.pdf

38 Richard Damania, Sébastien Desbureaux et al, Quality Unknown: The
    Invisible Water Crisis, https://openknowledge.worldbank.org/
    bitstream/handle/10986/32245/9781464814594.pdf

CHAPTER 6

1 'England faces "serious risk of running out of water within 20 years"',
   UK parliament (website), 10 July 2020, https://committees.parliament.
   uk/work/317/water-supply-and-demand-management/news/115817/
   england-faces-serious-risk-of-running-out-of-water-within-20-years/

2 Emma Howard Boyd / Environment Agency, 'Environment Agency
   Chair Emma Howard Boyd speaks about water scarcity at the WWT
   event "Water Security in a Changing Environment"', GOV.UK
   (website), 30 June 2021, https://www.gov.uk/government/speeches/
   water-security-in-a-changing-environment

3 Gareth Davies , Report by the Comptroller and Auditor General,
   National Audit Office, Water supply and demand management, 18
   March 2020

4 The Development of the Water Industry in England and Wales, Ofwat /
   Defra, 2006, https://www.ofwat.gov.uk/wp-content/uploads/2015/11/
   rpt_com_devwatindust270106.pdf

5 'Nine water bosses pocket £70 million as bills spiral', GMB Union
   (website), 10 June 2019, https://www.gmb.org.uk/news/nine-
   water-bosses-pocket-70-million-bills-spiral

6 Joe Watts, 'Michael Gove launches searing attack on water company
   bosses over tax avoidance and executive pay', Independent, 1 March

2018, https://www.independent.co.uk/news/uk/politics/michael-gove-water-company-bosses-speech-attack-audience-tax-executive-pay-a8235531.html

7 Editorial, 'Sewage spills highlight decades of under-investment at England's water companies', *Financial Times*, 28 December 2021, https://www.ft.com/content/b2314ae0-9e17-425d-8e3f-066270388331

8 Michael Robinson, 'How Macquarie bank left Thames Water with extra £2bn debt', BBC News, 5 September 2017, https://www.bbc.co.uk/news/business-41152516

9 Gareth Davies, *Water supply and demand management*, National Audit Office, 11 June 2020, https://www.nao.org.uk/wp-content/uploads/2020/03/Water-supply-and-demand-management.pdf

10 GARD, 'News Release: Thames Water reveals 10 year Abingdon Reservoir construction blight', 12 April 2010, https://www.abingdonreservoir.org.uk/downloads/news-release-04-10.pdf

11 RAPID, *Strategic regional water resource solutions guidance for gate two* (Birmingham: Ofwat, April 2022), https://www.ofwat.gov.uk/publication/strategic-regional-water-resource-solutions-guidance-for-gate-two/

12 Anglian Water and Affinity Water, *Strategic Solution Gate One Submission: Preliminary Feasibility Assessment – South Lincolnshire Reservoir*, 5 July 2021, https://www.anglianwater.co.uk/siteassets/household/about-us/strategic-solution-gate-one-submission-preliminary-feasibility-assessment-south-lincolnshire.pdf

13 Anglian Water and Cambridge Water, *Strategic Solution Gate One Submission: Preliminary Feasibility Assessment – Fens Reservoir*, 5 July 2021, https://www.anglianwater.co.uk/siteassets/household/about-us/strategic-solution-gate-one-submission-preliminary-feasibility-fens-reservoir.pdf

14 RAPID, *Strategic regional water resource solutions: Standard gate one final decision for Grand Union Canal Transfer* (Birmingham: Ofwat, December 2021), https://www.ofwat.gov.uk/wp-content/uploads/2021/12/Standard-gate-one-final-decision-for-Grand-Union-Canal-Transfer.pdf

15 RAPID, *Standard gate one final decision for South East Strategic Reservoir Option* (Birmingham: Ofwat, 2022), https://www.ofwat.gov.uk/wp-content/uploads/2022/01/Standard-gate-one-final-decision-for-South-East-Strategic-Reservoir-Option_Final.pdf

16 Chris Yandell, 'Southern Water scraps plan for new plant near Ashlett Creek', *Southern Daily Echo*, 30 September 2021, https://www.dailyecho.co.uk/news/19613608.southern-water-scraps-plan-new-plant-near-ashlett-creek/

17 Charles Rangeley-Wilson, *Chalk Stream Restoration Strategy 2021*, CaBA Chalk Stream Restoration Group.

18 Agreement under Section 20 of the Water Resources Act 1991 Between Southern Water Services Limited and the Environment Agency, 2018, https://www.southernwater.co.uk/media/2665/sw-drought-permit-appendices-1-17-0719.pdf

19 Sandra Laville, 'England's privatised water firms paid £57bn in dividends since 1991', *The Guardian*, 1 July 2020, https://www.theguardian.com/environment/2020/jul/01/england-privatised-water-firms-dividends-shareholders

20 *Hertfordshire's State of Nature* (St Albans: Herts & Middlesex Wildlife Trust, 2020), https://www.hertswildlifetrust.org.uk/stateofnature

21 Affinity Water, *Designing Our Future Together: Our Strategic Direction Statement 2025–2050*, December 2021, https://www.affinitywater.co.uk/docs/corporate/plans/strategic/AW0031_Strategic-direction-statement.pdf

22 'Affinity Water launches Chalk Stream restoration in London and the Eastern Region', Affinity Water (website), 28 September 2020, https://www.affinitywater.co.uk/news/restoring-chalk-streams

23 Sandra Laville and Niamh McIntyre, 'Exclusive: water firms discharged raw sewage into England's rivers 200,000 times in 2019', *The Guardian*, 1 July 2020, https://www.theguardian.com/environment/2020/jul/01/water-firms-raw-sewage-england-rivers

24 Sandra Laville, 'Water firms discharged raw sewage into English waters 400,000 times last year', *The Guardian*, 31 March 2021, https://www.theguardian.com/environment/2021/mar/31/water-firms-discharged-raw-sewage-into-english-waters-400000-times-last-year

25 Rachel Salvidge, 'Water firm faces sentencing for 51 sewage pollution offences', *ENDS Report*, 6 July 2021, https://www.endsreport.com/article/1721425/water-firm-faces-sentencing-51-sewage-pollution-offences

26 Southern Water Annual Report and Financial Statements for the year ended 31 March 2021, https://southernwater.annualreport2021.com/media/unml5lxq/30055_southern-water-ar2021_full.pdf

27 Rachel Salvidge, 'Ignore reports of low-impact pollution events, Environment Agency tells staff', *The Guardian*, 10 January 2022, https://www.theguardian.com/environment/2022/jan/10/ignore-low-impact-pollution-events-environment-agency-staff

28 James Bevan / Environment Agency, 'Water is the most important thing there is', GOV.UK (website), https://www.gov.uk/government/speeches/water-is-the-most-important-thing-there-is

29 House of Commons Environmental Audit Committee, *Water Quality in Rivers: Fourth Report of Session 2021–22*, 5 January 2022, https://

committees.parliament.uk/publications/8460/documents/88412/
default/

30 Rebecca Duncan, 'New EA water quality statistics show failure at a
national scale', Rivers Trust (website), 18 September 2020, https://
theriverstrust.org/about-us/news/
new-ea-water-quality-statistics-show-failure-at-a-national-scale

31 Defra, EA and Rebecca Pow MP, 'Taskforce sets goal to end pollution
from storm overflows', GOV.UK (website), https://www.gov.uk/
government/news/taskforce-sets-goal-to-end-pollution-from-
storm-overflows

32 House of Commons Environmental Audit Committee, *Water Quality
in Rivers: Fourth Report of Session 2021–22*, 5 January 2022, https://
committees.parliament.uk/publications/8460/documents/88412/
default/

33 *Rivercide*, Spanner Films, 2021, available to view on YouTube: https://
www.youtube.com/watch?v=kSPtVkJ_Uxs

34 Charles Rangeley-Wilson, *Chalk Stream Restoration Strategy 2021*,
CaBA Chalk Stream Restoration Group.

35 Charles Rangeley-Wilson, 'To remove P or not to remove P, that is the
question', ChalkStreams.org (website), 22 January 2021, https://
chalkstreams.org/2021/01/22/to-remove-p-or-not-to-remove-p-that-
is-the-question/

CHAPTER 7

1 'Managing Alfalfa Under Pivot Irrigation', *American Dairymen*, 1 April
2014, https://www.americandairymen.com/articles/
managing-alfalfa-under-pivot-irrigation

2 'Alfalfa Irrigation', Valley irrigation equipment (website), https://www.
valleyirrigation.com/science-engineering/alfalfa

3 Brian Leib and Tim Grant, 'Understanding Center Pivot Application
Rate', University of Tennessee Institute of Agriculture, https://
extension.tennessee.edu/publications/Documents/W809-F.pdf

4 *NASA Earth Observatory maps by Lauren Dauphin, using data from
Deines, Jillian, et al. (2017). Story by Adam Voiland*, https://
earthobservatory.nasa.gov/images/92387/satellites-investigate-
irrigation-in-a-stressed-aquifer

5 Jillian Deines, Meagan Schipanski, Bill Golden et al, 'Transitions from
irrigated to dryland agriculture in the Ogallala Aquifer: Land use
suitability and regional economic impacts', *Agricultural Water
Management* 233, 30 April 2020, https://www.sciencedirect.com/
science/article/abs/pii/S0378377419318062#

6   E. Elhadj, 'Camels Don't Fly, Deserts Don't Bloom: an Assessment of SaudiArabia's Experiment in Desert Agriculture', Occasional Paper Paper 48, Water Issues Study Group, SOAS/KCL, University of London, 2004

7   'Saudi Arabia ends domestic wheat production program', World-Grain.com (website), 18 March 2016, https://www.world-grain.com/articles/6275-saudi-arabia-ends-domestic-wheat-production-program

8   'Almarai acquires huge farmland in Arizona', *Arab News*, 11 March 2014, https://www.arabnews.com/news/537336

9   The Metropolitan Water District of Southern California, Board of Directors Water Planning and Stewardship Committee board meeting information, 14 September 2021, https://mwdh2o.legistar.com/View.ashx?M=F&ID=9801165&GUID=B624E1D8-7809-4D23-8A35-0706EB31C0D2

10 John Bishop, 'Rushall Organic Farm: Education in the Countryside', Rushall Farm (website), http://www.rushallfarm.org.uk/blog/february-2020/

11 Guy Singh-Watson, 'Guy's news: Praying for thunder', *Wicked Leeks*, 28 November 2018, https://wickedleeks.riverford.co.uk/opinion/farming/guys-news-praying-thunder

12 Gian Singh, 'Punjab Assembly Elections 2022: Ignoring the groundwater depletion problem', *Down to Earth*, 27 January 2022, https://www.downtoearth.org.in/blog/water/punjab-assembly-elections-2022-ignoring-the-groundwater-depletion-problem-81286

13 Vaibhav Chaturvedi, Kangkanika Neog, Sujata Basu et al, *Reallocating Water for India's Growth: Sectoral Withdrawals, Water-efficient Agriculture, and Institutional Mechanisms* (New Delhi: Council on Energy, Environment and Water, 2020), https://www.ceew.in/publications/reallocating-water-to-grow-india

14 Abrahm Lustgarten, 'Use It or Lose It Laws Worsen Western U.S. Water Woes', *Scientific American*, 9 June 2015, https://www.scientificamerican.com/article/use-it-or-lose-it-laws-worsen-western-u-s-water-woes/

15 Arizona State Senate, Fifty-Fifth Legislature, First Regular Session, 'Fact Sheet for H.B. 2056/S.B. 1368: water conservation notice; no forfeiture', February 2021, https://www.azleg.gov/legtext/55leg/1R/summary/S.2056-1368NREW_ASENACTED.pdf

16 Timothy Egan, 'War on Peruvian Drugs Takes a Victim: U.S. Asparagus', *New York Times*, 25 April 2004, https://www.nytimes.com/2004/04/25/us/war-on-peruvian-drugs-takes-a-victim-us-asparagus.html

17   CRS Report for Congress, US–Peru Economic Relations and the
      US–Peru Trade Promotion Agreement, November 2007, https://www.
      files.ethz.ch/isn/118410/2007-11-06_US-Peru_Trade.pdf
18   Rachel Dring, 'Asparagus – draining Peru dry', Sustainable Food Trust,
      April 21, 2014. [Original URL extant but available via: https://www.
      resilience.org/stories/2014-04-21/asparagus-draining-peru-dry/]
19   N. D. Hepworth, J. C. Postigo, B. Güemes Delgado and P. Kjell, *Drop
      by Drop: Understanding the impacts of the UK's water footprint
      through a case study of Peruvian asparagus* (London: Progressio,
      CEPES and Water Witness International, 2010), https://www.
      progressio.org.uk/sites/default/files/Drop-by-drop_Progressio_Sept-
      2010.pdf
20   Ian James, 'The costs of Peru's farming boom', *Palm Springs Desert
      Sun*, 10 December, 2015, https://eu.desertsun.com/story/news/
      environment/2015/12/10/costs-perus-farming-boom/
      76605530/
21   Fabiola Torres and Cristian Díaz, 'The company that absorbed the
      home of the Lima bean', *Ojo Público*, 21 May 2018, https://ojo-publico.
      com/especiales/acuatenientes/the-company-that-absorbed-the-home-
      of-the-lima-bean.html
22   Sasha Chavkin, 'News drones reveal big companies are draining local
      water supplies in Peru, Colombia', International Consortium of
      Investigative Journalists (website), 4 June 2018, https://www.icij.org/
      inside-icij/2018/06/news-drones-reveal-big-companies-draining-
      local-water-supplies-peru-colombia/
23   Gaspar Nolte and Dwight Wilder, 'Voluntary Report: Asparagus
      Exports Down for the First Time', United States Department of
      Agriculture: Foreign Agricultural Service, 16 February 2021, https://
      apps.fas.usda.gov/newgainapi/api/Report/DownloadReportByFileNam
      e?fileName=Asparagus%20Exports%20Down%20for%20the%20First%
      20Time_Lima_Peru_02-16-2021
24   Andrea E. Becerra et al, 'A New Course: Managing Drought and
      Downpours in the Santiago Metropolitan Region', NRDC, July 2019,
      https://www.nrdc.org/sites/default/files/new-course-managing-
      drought-downpours-santiago-report.pdf
25   World Avocado Organization, brand manual, 2020, https://
      avocadofruitoflife.com/wp-content/uploads/2020/06/wao-bm.pdf
26   Asit K. Biswas and Cecilia Tortajada, 'The slow drip of water policy',
      Policy Forum, 6 July 2018, https://www.policyforum.net/the-slow-
      drip-of-poor-water-policy/
27   'Professor Tony Allan discusses water and food security in the Middle
      East and North Africa', YouTube: https://www.youtube.com/
      watch?v=9S5sWUEGviA [accessed 08 Mach 2021]

28 Hepworth et al, 'Drop by drop: Understanding the impacts of the UK's water footprint through a case study of Peruvian asparagus', Progressio in association with Centro Peruano de Estudios Sociales and Water Witness International, September 2010, https://www.progressio.org.uk/sites/default/files/Drop-by-drop_Progressio_Sept-2010.pdf

29 Rick Hogeboom, 'Your water footprint revealed', TED Talk, TEDxTwenteU, May 2018, https://www.ted.com/talks/rick_hogeboom_your_water_footprint_revealed

CHAPTER 8

1 Steve Smith, 'Farmers in Cornwall are being encouraged to soil their underpants', Plymouth Live (website), 31 January 2018, https://www.plymouthherald.co.uk/news/plymouth-news/farmers-cornwall-being-encouraged-soil-1145173

2 Natural Resources Conservation Service, 'Soil your undies challenge' (website), https://www.nrcs.usda.gov/conservation-basics/conservation-by-state/oregon/soil-your-undies-challenge [accessed April 2022]

3 Merlin Sheldrake, *Entangled Life: How Fungi Make Our Worlds, Change Our Minds and Shape Our Futures* (London: Bodley Head, 2020)

4 Brian Bienkowski, 'Can Farming Practices in Oklahoma Solve Climate Change?', *Scientific American*, 15 October 2015, https://www.scientificamerican.com/article/can-farming-practices-in-oklahoma-solve-climate-change/

5 Severn Trent Plc, Sustainability Report 2021, https://www.severntrent.com/content/dam/stw-plc/interactive-sustainability report/publication/contents/templates/Severn_Trent_plc_Sustainability_Report_2021.pdf

6 Ibid.

7 'Replenishing a Broken Water Cycle, with Sandra Postel' (episode), *What About Water?* (podcast, season 3), 24 November 2021, https://www.whataboutwater.org/s03e04/

8 Rachel Becker, 'California enacted a groundwater law 7 years ago. But wells are still drying up – and the threat is spreading', CalMatters (website), 18 August 2021, https://calmatters.org/environment/2021/08/california-groundwater-dry/

9 Stephanie Osler Hastings, 'A Test for California's Groundwater Regulations in the Megadrought', *Bloomberg Law*, 7 September 2021, https://news.bloomberglaw.com/us-law-week/a-test-for-californias-groundwater-regulations-in-the-megadrought-17

10 'Gov. Brown appoints UCI's Jay Famiglietti to state water board', UCI
    News (website), 9 January 2014, https://news.uci.edu/2014/01/09/
    governor-brown-appoints-ucis-jay-famiglietti-to-state-water-board/
11 Elad Levintal, Maribeth L. Kniffin, Yonatan Ganot et al, 'Agricultural
    managed aquifer recharge (Ag-MAR) – a method for sustainable
    groundwater management: A review', *Critical Reviews in
    Environmental Science and Technology*, 28 March 2022, https://www.
    tandfonline.com/doi/full/10.1080/10643389.2022.2050160
12 Augustine A. Ayantunde, Mawa Karambiri et al, *Multiple Uses of
    Small Reservoirs in Crop-livestock Agro-ecosystems of the Volta River
    Basin with an Emphasis on Livestock Management* (Colombo, Sri
    Lanka: International Water Management Institute, 2016), http://www.
    iwmi.cgiar.org/Publications/Working_Papers/working/wor171.pdf
13 'Study: Colorado Set To Lose Half Of Its Snow In 60 Years', CBS
    Colorado, 8 June 2022, https://www.cbsnews.com/colorado/news/
    colorado-snowpack-study-runoff/
14 Nisha Marwaha, George Kourakos et al, 'Identifying Agricultural
    Managed Aquifer Recharge Locations to Benefit Drinking Water
    Supply in Rural Communities', *Water Resources Research* 57(3), March
    2021, https://agupubs.onlinelibrary.wiley.com/doi/abs/10.1029/
    2020WR028811
15 Christoph Sprenger, Niels Hartog et al, 'Inventory of managed aquifer
    recharge sites in Europe: historical development, current situation and
    perspectives', *Hydrology Journal* 25(6), March 2017, https://www.
    researchgate.net/publication/315110753_Inventory_of_managed_
    aquifer_recharge_sites_in_Europe_historical_development_current_
    situation_and_perspectives
16 Seth Owusu, Olufunke O. Cofie, Paa Kofi Osei-Owusu et al, *Adapting
    Aquifer Storage and Recovery Technology to the Flood-prone Areas of
    Northern Ghana for Dry-season Irrigation* (Colombo, Sri Lanka:
    International Water Management Institute, 2017), http://www.iwmi.
    cgiar.org/Publications/Working_Papers/working/wor176.pdf
17 Christoph Sprenger, Niels Hartog et al, 'Inventory of managed aquifer
    recharge sites in Europe: historical development, current situation and
    perspectives', *Hydrology Journal* 25(6), March 2017.
18 CAWCD Board of Directors, 2022 Strategic Plan, Central Arizona
    Project, https://library.cap-az.com/documents/board/2022-Strategic-
    Plan.pdf
19 Keith Schneider, 'Unsafe Yield', Circle of Blue (website), https://www.
    circleofblue.org/2022/world/unsafe-yield/
20 'Alma Road Rain Gardens, London – case study', Susdrain, https://
    www.susdrain.org/case-studies/case_studies/alma_road_rain_
    gardens_london.html

21 ' "Sponge city": China tries new tack after centuries of trying to curb
    floods', *Japan Times*, 14 August 2020, https://www.japantimes.co.jp/
    news/2020/08/14/asia-pacific/sponge-city-china-floods/#.Xzvj6ehKjcs
22 L. E. Oates, L. Dai, A. Sudmant and A. Gouldson, *Building Climate
    Resilience and Water Security in Cities: Lessons from the sponge city of
    Wuhan, China* (London and Washington, DC: Coalition for Urban
    Transitions, 2020), https://www.academia.edu/43992773/BUILDING_
    CLIMATE_RESILIENCE_AND_WATER_SECURITY_IN_CITIES_
    LESSONS_FROM_THE_SPONGE_CITY_OF_WUHAN_CHINA?
    email_work_card=view-paper
23 Andrea Becerra et al, 'A New Course: Managing Drought and
    Downpours in the Santiago Metropolitan Region', NRDC, July 2019,
    https://www.nrdc.org/sites/default/files/new-course-managing-
    drought-downpours-santiago-report.pdf
24 Oates et al, 'Building Climate Resilience and Water Security in Cities:
    Lessons from the sponge city of Wuhan, China', *Coalition for Urban
    Transitions*, 2020, https://urbantransitions.global/publications
25 Shagun Garg, Mahdi Motagh, J. Indu and Vamshi Karanam, 'Tracking
    hidden crisis in India's capital from space: implications of
    unsustainable groundwater use', *Scientific Reports* 12, 2022, https://
    www.nature.com/articles/s41598-021-04193-9
26 'City of 1,000 Tanks, Chennai: Holistic urban strategy to combat
    floods, droughts and pollution through blue-green strategies', Ooze
    (website), http://www.ooze.eu.com/en/urban_strategy/
    city_of_1000_tanks_chennai/
27 'Chennai Metropolitan Area: Towards a regional urban landscape
    structure', City of 1000 Tanks (website), https://www.cityof1000tanks.
    org/projectproposals#/chennai-metropolitan-area-towards-a-
    regional-urban-landscape-structure/
28 Seattle Office of Sustainability & Environment, 'Green Stormwater
    Infrastructure in Seattle, 2015-2020 Implementation Strategy', http://
    www.seattle.gov/documents/departments/ose/gsi_strategy_nov_2015.pdf
29 'The Road to 700 Million Gallons: A Natural Approach to Stormwater
    Management', Green Stormwater Infrastructure, 2020 Progress
    Report, Seattle Public Utilities and King County, https://www.seattle.
    gov/documents/Departments/SPU/Documents/GSI-
    ProgressReport2020.pdf

CHAPTER 9

1 Cecilia Tortajada, Yugal Joshi and Asit K. Biswas, *The Singapore Water
   Story* (Abingdon: Routledge, 2013).

2 'Water Agreements', Ministry of Foreign Affairs, Singapore (Website), https://www.mfa.gov.sg/SINGAPORES-FOREIGN-POLICY/Key-Issues/Water-Agreements

3 'NEWater', *Blueprint For Survival* (documentary), series 1, episode 2, CNA, 17 January 2018, https://www.channelnewsasia.com/watch/blueprint-survival/newater-1560961

4 Cecilia Tortajada, Yugal Joshi and Asit K. Biswas, *The Singapore Water Story* (Abingdon: Routledge, 2013).

5 'NEWater', *Blueprint For Survival* (documentary), series 1, episode 2, CNA, 17 January 2018.

6 'Sunfruits: building water stewardship solutions in the Ica Valley', Peru. 1. Project / Program name: Rehabilitation & Operation of the Community of San Jose de Los Molinos.

7 Mayor Eric Garcetti, 'L.A.'s Green New Deal: Sustainable City pLAn 2019', https://plan.lamayor.org/sites/default/files/pLAn_2019_final.pdf

8 David Garrick, 'San Diego launching Pure Water, largest infrastructure project in city's history', *San Diego Union-Tribune*, 20 August 2021, https://www.sandiegouniontribune.com/news/politics/story/2021-08-20/san-diego-officially-launching-pure-water-largest-infrastructure-project-in-city-history

9 Royal Haskoning DHV / European Investment Bank, *Report: Red Sea Dead Sea Water Conveyance Study*, 6 December 2017, https://www.eib.org/attachments/registers/82425040.pdf

10 'Jordan allocated $1.8bln for water project', ZAWYA (website), 1 April 2022, https://www.zawya.com/en/projects/utilities/jordan-allocates-18bln-for-water-project-w7ekn1f7

11 Andrew Roscoe, 'Developers weigh up Jordan's Aqaba–Amman water project', Energy & Utilities (website), 1 April 2022, https://energy-utilities.com/developers-weigh-up-jordan-s-aqabaamman-water-news116979.html

12 Accountant General, 'Background – Seawater Desalination in Israel', Ministry of Finance / Israeli government website, https://www.gov.il/en/departments/general/project-water-desalination-background

13 Soteris A. Kalogirou, 'Seawater desalination using renewable energy sources', *Progress in Energy and Combustion Science* 31(3), 2005, pp. 242–81, https://www.sciencedirect.com/science/article/abs/pii/S0360128505000146

14 'Does size matter? Meet ten of the world's largest desalination plants', Aquatech (website), 19 April 2021, https://www.aquatechtrade.com/news/desalination/worlds-largest-desalination-plants/

15 Gitika Bhardwaj, interview with Loïc Fauchon, 'Exploring the Looming Water Crisis', Chatham House (website), 28 November 2019,

https://www.chathamhouse.org/2019/11/exploring-looming-
water-crisis

16 'Five things to know about desalination', UN Environment Programme
    (website), 11 January 2021, https://www.unep.org/news-and-stories/
    story/five-things-know-about-desalination

17 Angelakis et al, 'Desalination: From Ancient to Present and Future',
    Water 2021, https://doi.org/10.3390/w13162222

18 Michael Birnbaum, 'Desalination can make saltwater drinkable – but
    it won't solve the U.S. water crisis', *Washington Post*, 28 September
    2021, https://www.washingtonpost.com/climate-solutions/2021/09/28/
    desalination-saltwater-drought-water-crisis/

19 'Reverse Osmosis Desalination: Brine disposal', Lenntech (website),
    https://www.lenntech.com/processes/desalination/brine/general/
    brine-disposal.htm

20 'UN Warns of Rising Levels of Toxic Brine as Desalination Plants
    Meet Growing Water Needs', United Nations University (website), 14
    January 2014, https://unu.edu/media-relations/releases/un-warns-of-
    rising-levels-of-toxic-brine.html

21 'Seawater Desalination: Ocean Plan Requirements for Seawater
    Desalination Facilities', California Water Boards (website), https://
    www.waterboards.ca.gov/water_issues/programs/ocean/
    desalination/

22 Alon Tal, 'Addressing Desalination's Carbon Footprint: The Israeli
    Experience', *Water* 10(197), 12 February 2018, https://www.mdpi.com/
    2073-4441/10/2/197/pdf#:~:text=All%20told%2C%20Israel's%20
    water%20supply,the%20energy%20demand%20%5B22%5D

23 IPCC, *Special Report: Global Warming of 1.5°C*, Chapter 4:
    'Strengthening and implementing the global response', 2018, https://
    www.ipcc.ch/sr15/chapter/chapter-4/

24 Elena Pappas, 'Quenching the world's thirst with off-grid water
    desalination', *Horizon*, 30 September 2021, https://ec.europa.eu/
    research-and-innovation/en/horizon-magazine/quenching-
    worlds-thirst-grid-water-desalination

25 Renee Cho, 'A 1,000 Year Drought is Hitting the West. Could
    Desalination Be a Solution?', State of the Planet (news website of
    Columbia Climate School), 26 August 2021, https://news.climate.
    columbia.edu/2021/08/26/a-1000-year-drought-is-hitting-the-
    west-could-desalination-be-a-solution/

26 'Unique Floating Wind Turbine Set for Middle East Debut',
    Offshore Engineer, 17 February 2021, https://www.oedigital.com/
    news/485362-unique-floating-wind-turbine-set-for-middle-
    east-debut

27 'Ghana Water losing GHc6m monthly to Teshie desalination plant', Citi FM (website), 2 October 2017, https://citifmonline.com/2017/10/ghana-water-losing-ghc-6m-monthly-to-teshie-desalination-plant/

28 'GWCL shuts down Teshie Desalination plant again', GhanaWeb (website), 3 October 2019, https://www.ghanaweb.com/GhanaHomePage/NewsArchive/GWCL-shuts-down-Teshie-Desalination-plant-again-785807

29 Timothy Ngnenbe, 'GWCL pays GH¢1.4m monthly four months after desalination plant closes', Graphic Online, 18 April 2018, https://www.graphic.com.gh/news/general-news/gwcl-pays-gh-1-4m-monthly-four-months-after-desalination-plant-closes.html

30 Vincent E. A. Post, Jacobus Groen, Henk Kooi et al, 'Offshore fresh groundwater reserves as a global phenomenon', Nature 504, 5 December 2013, https://sci-hub.yncjkj.com/10.1038/nature12858

31 John Dennehy, 'Rain enhancement pioneers win UAE's $5 million award at Abu Dhabi Sustainability Week', The National, 17 January 2018, https://www.thenationalnews.com/uae/environment/rain-enhancement-pioneers-win-uae-s-5-million-award-at-abu-dhabi-sustainability-week-1.696102

32 John Dennehy, 'Seeding flights in UAE "boosting rainfall from clouds by a third"', The National, 11 January 2018, https://www.thenationalnews.com/uae/environment/seeding-flights-in-uae-boosting-rainfall-from-clouds-by-a-third-1.694589

33 Marianne Bray, 'Beijing to shoot down Olympic rain', CNN (website), 9 June 2006, https://edition.cnn.com/2006/WORLD/asiapcf/06/05/china.rain/index.html

34 Arwa Mahdawi, 'China is scaling up its weather modification programme – here's why we should be worried', The Guardian, 15 December 2020, https://www.theguardian.com/commentisfree/2020/dec/15/china-scaling-up-weather-modification-programme-we-should-be-worried

35 'China to forge ahead with weather modification service', State Council, People's Republic of China (website), 2 December 2020, http://english.www.gov.cn/policies/latestreleases/202012/02/content_WS5fc76218c6d0f7257694125e.html

36 Katja Friedrich, Kyoko Ikeda, Sarah A. Tessendorf et al, 'Quantifying snowfall from orographic cloud seeding', PNAS 117(10), 24 February 2020, https://www.pnas.org/doi/10.1073/pnas.1917204117

37 Roy M. Rasmussen, Sarah A. Tessendorf, Lulin Xue et al, 'Evaluation of the Wyoming Weather Modification Pilot Project (WWMPP) Using Two Approaches: Traditional Statistics and Ensemble Modeling', Journal of Applied Meteorology and Climatology 57(11), 1

November 2018, https://journals.ametsoc.org/view/journals/apme/57/11/jamc-d-17-0335.1.xml

38   Aparna Roy, ' "Weather War": A latest addition to the Sino-India conundrum?', Observer Research Foundation (website), 22 August 2018, https://www.orfonline.org/expert-speak/43534-weather-war-a-latest-addition-to-the-sino-india-conundrum/

39   'Farmers blame cloud seeding program for drought', E&E News (website), 7 August 2017, https://subscriber.politicopro.com/article/eenews/1060058470?show_login=1&t=https%3A%2F%2Fwww.eenews.net%2Fgreenwire%2F2017%2F08%2F07%2Fstories%2F1060058470

40   Iain Woessner, 'Cloud seeding scrutinized at public meeting', Dickinson Press, 8 August 2017, https://www.thedickinsonpress.com/business/cloud-seeding-scrutinized-at-public-meeting

41   Oliver Milman, 'Make it rain: US states embrace "cloud seeding" to try to conquer drought', The Guardian, 23 March 2021, https://www.theguardian.com/environment/2021/mar/23/us-stated-cloud-seeding-weather-modification

42   Gabe Allen, 'Cloud Seeding in Colorado Could Make Waves in the West', Discover magazine (website), 27 October 2021, https://www.discovermagazine.com/technology/cloud-seeding-in-colorado-could-make-waves-in-the-west

CHAPTER 10

1   Richard Kingsford, 'The tragedy of the Murray–Darling river system is man-made', Sydney Morning Herald, 25 July 2017, https://www.smh.com.au/opinion/the-tragedy-of-the-murraydarling-river-system-is-manmade-20170725-gxi35b.html

2   'MDBA Chief Executive Phillip Glyde Cotton Collective Water Forum, Griffith' (speech transcript), Australian government Murray–Darling Basin Authority (website), 24 July 2019, https://www.mdba.gov.au/media/mr/mdba-chief-executive-phillip-glyde-cotton-collective-water-forum-griffith

3   Murray–Darling Basin water markets inquiry: Final Report (Canberra: Australian Competition and Consumer Commission, February 2021), https://www.accc.gov.au/system/files/Murray-Darling%20Basin%20water%20markets%20inquiry%20-%20final%20report%20-%20Overview.pdf

4   'Announcement: Retirement of MDBA Chief Executive Phillip Glyde', Australian government Murray–Darling Basin Authority (website), 3 December 2021, https://www.mdba.gov.au/news-media-events/

newsroom/media-centre/announcement-retirement-mdba-
chief-executive-phillip-glyde

5   'Retirement of MDBA Chief Executive Phillip Glyde', National
    Irrigators' Council (website), 8 December 2021, https://www.irrigators.
    org.au/retirement-of-mdba-chief-executive-phillip-glyde/

6   Mike Foley, 'Labor wades into Murray–Darling water wars to woo
    Adelaide voters', *Sydney Morning Herald*, 8 April 2022, https://www.
    smh.com.au/politics/federal/labor-wades-into-murray-darling-water-
    wars-to-woo-adelaide-voters-20220408-p5abx3.html

7   Kath Sullivan, 'Oil and gas lobby boss to head Murray–Darling Basin
    Authority, as former LNP politician set to chair river study', ABC
    News (website), https://www.abc.net.au/news/rural/2022-04-04/
    oil-and-gas-lobbyist-to-head-up-mdba/100964404

8   'Nationals put "upstream mates" in charge of River Murray', *Herald
    Sun*, https://www.heraldsun.com.au/news/south-australia/national-
    party-under-fire-for-putting-upstream-mates-in-charge-of-river-
    murray-review/news-story/528898863fb7152a644072d2f7595728

9   Parameswaran Iyer (ed.), *The Swachh Bharat Revolution:
    Four Pillars of India's Behavioural Transformation* (HarperCollins
    India, 2019).

10 *Rivercide*, Spanner Films, 2021, available to view on YouTube: https://
    www.youtube.com/watch?v=kSPtVkJ_Uxs

11 'TIME 100 Talks With Matt Damon And Gary White, Water.org and
    WaterEquity Co-founders', YouTube: https://www.youtube.com/
    watch?v=yUzqcLd-ErQ&ab_channel=TIME https://www.waterwise.
    org.uk/wp-content/uploads/2021/02/WESSG-A-Review-of-Water-
    Neutrality-in-the-UK-Feb-2021.pdf

12 Seth Siegel, *Let There Be Water: Israel's Solution for a Water-Starved
    World* (New York: St Martin's Press, 2015).

13 'Water Funds by Region', Nature Conservancy (website), https://
    waterfundstoolbox.org/regions

14 'EU slams plan to expand water use near Spain's Doñana park',
    Independent, 9 February 2022, https://www.independent.co.uk/news/
    world/europe/spain-barcelona-europe-andalusia-united-nations-
    b2011377.html

15 Mark Lowen, 'Spain's olive oil producers devastated by worst ever
    drought', BBC News (website), 29 August 2022, https://www.bbc.
    co.uk/news/world-europe-62707435

16 'Independent review of the costs and benefits of water labelling
    options in the UK', Energy Saving Trust, February 2019, https://www.
    waterwise.org.uk/wp-content/uploads/2019/02/Water-Labelling-
    Summary-Report-Final.pdf

17 'Use Cases: See how Hydraloop is used', Hydraloop website, https://
www.hydraloop.com/#:~:text=water%2520recycling%2520system-,Dav
id%2520Mertens%2520and%2520his,per%2520person%2520per%2520
day.,-I%2520work%2520for

18 'Power Generation: Gas Based Power Stations', NTPC (website),
https://www.ntpc.co.in/en/power-generation/gas-based-power-
stations

19 'Renewable future: Gujarat govt to set up 100 MW solar power
project atop Narmada canal', *Financial Express*, 22 February 2019,
https://www.financialexpress.com/india-news/renewable-
future-gujarat-govt-to-set-up-100-mw-solar-power-project-atop-
narmada-canal/1494866/; Kalpana Sunder, 'The "solar canals" making
smart use of India's space', Future Planet, BBC (website), 4 August
20220, https://www.bbc.com/future/article/20200803-the-solar-
canals-revolutionising-indias-renewable-energy

CHAPTER 11

1 Peter Melville-Shreeve, 'Exploring Smart Rainwater Management
Systems in a Small Island Economy', presented at the 6th Water
Efficiency Conference, University of the West of England, Bristol, 3
September 2020, https://www.youtube.com/watch?v=-P_RuRyXqyg

2 Robert Edwards, 'Fawley Desalination Plant: The New Forest
campaigners who have worked "tirelessly" opposing Southern Water
plans', Hampshire Live (website), 1 October 2021, https://www.
hampshirelive.news/news/hampshire-news/fawley-desalination-plant-
new-forest-5992271

3 Thorsten Schuetze, 'Rainwater harvesting and management – policy
and regulations in Germany', *Water Supply*, 2013, https://doi.org/
10.2166/ws.2013.035

4 'Sedema and Milpa Alta seek to create a network of rain harvesters',
Secretaría del Medio Ambiente, Gobierno de la Ciudad de México
(website), 20 January 2022, https://www.sedema.cdmx.gob.mx/
comunicacion/nota/buscan-sedema-y-milpa-alta-crear-red-de-
cosechadores-de-lluvia

5 Nuevas Esperanzas (website), https://nuevasesperanzas.org/projects

6 Shehfar, 'Beris to the rescue', *Down To Earth*, 31 January 2013, https://
www.downtoearth.org.in/coverage/beris-to-the-rescue-40065

7 'PM launches "Jal Shakti Abhiyan: Catch the Rain" campaign on
the occasion of World Water Day', Prime Minister's Office (website),
22 March 2021, https://pib.gov.in/PressReleasePage.aspx? PRID=
1706580

8 Sana Jamal, 'Lahore sets up first underground rainwater storage system', Gulf News (website), 20 July 2020, https://gulfnews.com/world/asia/pakistan/lahore-sets-up-first-underground-rainwater-storage-system-1.72655677

9 Afzal Talib, 'Rainwater harvesting project delayed', *Express Tribune*, 13 July 2021, https://tribune.com.pk/story/2310215/rainwater-harvesting-project-delayed

10 'Stormwater harvesting in Queen Victoria and Alexandra Gardens', Urban Water (website), https://urbanwater.melbourne.vic.gov.au/projects/water-capture-and-reuse/stormwater-harvesting-in-queen-victoria-and-alexandra-gardens/

11 'World-class Underground Discharge Channel', Trends in Japan (website), March 2013, https://web-japan.org/trends/11_tech-life/tec130312.html

12 Giovani De Feo, Sabino De Gisi, Carmela Malvano et al, 'The Greatest Water Reservoirs in the Ancient Roman World and the "Piscina Mirabilis" in Misenum', *Water Science & Technology: Water Supply* 10(3), January 2013, https://www.researchgate.net/figure/The-Piscina-Mirabilis-in-Miseno-in-the-bay-of-Naples-Southern-Italy_fig1_234025426

13 Charles R. Ortloff, 'Hydraulic Engineering at 100 BC–AD 300 Nabataean Petra (Jordan)', *Water* 12(12), 2020, https://www.mdpi.com/2073-4441/12/12/3498/htm

14 Kalpana Sunder, 'The remarkable floating gardens of Bangladesh', Future Planet, BBC (website), 11 September 2020, https://www.bbc.com/future/article/20200910-the-remarkable-floating-gardens-of-bangladesh

15 Archana Yadav, 'Resort to heritage', Down to Earth (website), 15 April 2016, https://www.downtoearth.org.in/news/agriculture/resort-to-heritage-53346

16 Keith Drew, 'Spain's ingenious water maze', BBC Travel (website), 21 February 2022, https://www.bbc.com/travel/article/20220220-valencias-la-huerta-spains-ingenious-water-maze?s=03

17 Vicuña Mackenna (1860), quoted by Carolina Domínguez, Maria Teresa Ore and Andres Verzijl, 'Pozas: A short story with some musings on human-groundwater relations along the coast of Peru', Transformations to Groundwater Sustainability (website), https://www.t2sgroundwater.org/narrative-reports-peru

18 Alberto Ñiquen Guerra, 'In the Peruvian Andes, residents sow water for the future', *Earth Journalism Network*, 13 December 2019, https://earthjournalism.net/stories/in-the-peruvian-andes-residents-sow-water-for-the-future

19 Amin Mahan, Reihaneh Khorramrouei, Ahmad Nasiri, 'Restoring the
    Qanats as a Traditional Water Transfer System: A Sustainable
    Approach', *International Journal of Architecture and Urban
    Development* 9(1), Winter 2019, https://ijaud.srbiau.ac.ir/article_14011_
    531192961bc388f1b3511a68b574cfeb.pdf
20 'Tradition valued', GIAHS, Food and Agriculture Organization of the
    United Nations (website), https://www.fao.org/giahs/news/
    archives/2014/05-traditionl-valued/en/
21 Gary Lewis, UN Resident Coordinator in the Islamic Republic of Iran,
    'The need to restore Iran's Qanats – the "Veins of the desert"' (speech
    given 25 June 2014)
22 'Ancient qanat in Tehran to be restored, revived', *Tehran Times*, 27
    January 2021, https://www.tehrantimes.com/news/457443/
    Ancient-qanat-in-Tehran-to-be-restored-revived
23 Approved Tariffs, Ghana Water Company Limited (website), https://
    www.gwcl.com.gh/approved-tariffs/
24 Bill Gates, 'Update: what ever happened to the machine that turns
    feces into water?', GatesNotes (blog), 11 August 2015, https://www.
    gatesnotes.com/Development/Omni-Processor-Update

CHAPTER 12

1 Martin Gaywood, Jeanette Elisabeth Hall, Andrew P. Stringer et al,
   Beavers in Scotland: A report to the Scottish Government (Inverness:
   Scottish Natural Heritage, June 2015), https://www.researchgate.net/
   publication/279751287_Beavers_in_Scotland_A_report_to_Scottish_
   Government
2 James Ashworth, 'Beavers reintroduced to London after 400 years',
   Natural History Museum (website), 17 March 2022, https://www.nhm.
   ac.uk/discover/news/2022/march/beavers-reintroduced-to-london-
   after-400-years.html#:~:text=Beavers%20have%20returned%20to%20
   London,council%20and%20Capel%20Manor%20College
3 Severn Trent Plc, Sustainability Report 2021, https://www.severntrent.
   com/content/dam/stw-plc/interactive-sustainability-report/index.
   html#page=1
4 Archie Ruggles-Brise, 'Flooding 14th Jan 2021', Spains Hall Estate
   (website), 17 January 2021, https://www.spainshallestate.co.uk/
   flooding-14th-jan-2021
5 'Beavers given legal protection without clear management plan',
   NFU (website), 22 July 2022, https://www.nfuonline.com/updates-
   and-information/beaver-reintroduction-consultation-have-
   your-say/

6   'Anglian Water joins forces with Severn Trent to revitalise the health of Britain's rivers', Anglian Water (website), 14 March 2022, https://www.anglianwater.co.uk/news/anglian-water-joins-forces-with-severn-trent-to-revitalise-the-health-of-britains-rivers/

7   Alice Outwater, *Water: A Natural History* (New York: Basic Books, 1996).

8   UN-Water, 'The United Nations World Water Development Report 2018: Nature-Based Solutions for Water', Paris, UNESCO, 2018.

9   Charles Rangeley-Wilson, 'Why Dredging Makes Flooding Worse', Charles Rangeley-Wilson (blog), 19 December 2015, https://charlesrangeleywilson.com/2015/12/19/why-dredging-makes-flooding-worse/

10   Stewart Clarke, 'Why we need to talk about floodplains', *The Green Alliance Blog*, 27 August 2021, https://greenallianceblog.org.uk/2021/08/27/why-we-need-to-talk-about-floodplains/

11   Clare Lawson, Emma Rothero, David Gowing et al, 'The Natural Capital of Floodplains: management, protection and restoration to deliver greater benefits', Valuing Nature: Natural Capital Synthesis Report VNP09, 2018, https://www.floodplainmeadows.org.uk/sites/www.floodplainmeadows.org.uk/files/VNP09-NatCapSynthesisReport-Floodplains-A4-16pp-144dpi.pdf

12   Environment Agency, Policy Paper: Oxford Flood Scheme, GOV.UK (website), https://www.gov.uk/government/publications/oxford-flood-scheme/oxford-flood-scheme

13   Torgny Holmgren, 'Balancing Green & Grey this World Water Day', Inter Press Service News Agency (website), March 2022, https://www.ipsnews.net/2018/03/balancing-green-grey-world-water-day/

14   The World Water Council, Water and Green Growth, 23 July 2012

15   Jim Skea, Priyadarshi R. Shukla, Andy Reisinger et al, 'Climate Change 2022: Mitigation of Climate Change – Working Group III Contribution to the IPCC Sixth Assessment Report', IPCC, 2022, https://report.ipcc.ch/ar6wg3/pdf/IPCC_AR6_WGIII_FinalDraft_FullReport.pdf

16   'Water shortage of 31m acre feet expected by 2025', Dawn (website), 29 March 2018, https://www.dawn.com/news/1398099

17   'Restoring Damaged Rivers', American Rivers (website), https://www.americanrivers.org/threats-solutions/restoring-damaged-rivers/, [accessed 11 February 2022]

18   Mike Simpson, 'What if? Simpson on Salmon Recovery' (promotional video), 7 February 2021, https://www.youtube.com/watch?v=z5_fr.7UGsw4&ab_channel=CongMikeSimpson

19   Orion Donovan-Smith, 'Fate of Republican Mike Simpson's plan to remove Snake River dams lies with Democrats and Biden infrastructure package', *Spokesman-Review*, 8 March 2021, https://

www.spokesman.com/stories/2021/mar/07/
fate-of-republican-mike-simpsons-plan-to-remove-sn/

20 Bo Evans, 'Largest dam removal in US history set to begin', ABC
Denver7 News (website), 24 February 2022, https://www.
thedenverchannel.com/news/national/largest-dam-removal-in-us-
history-set-to-begin#:~:text=HORNBROOK%2C%20Calif.,dam%20
removal%20in%20U.S.%20history.

21 Ibid.

22 Alice Taylor, 'Albanian government falls short of giving Vjosa River
maximum protection', Euractiv (website), 28 January 2022, https://
www.euractiv.com/section/politics/short_news/albanian-
government-falls-short-of-giving-vjosa-river-maximum-protection/

23 'Replenishing a Broken Water Cycle, with Sandra Postel' (episode),
What About Water? (podcast, season 3), 24 November 2021, https://
www.whataboutwater.org/s03e04/

24 Julie Rentner (interviewed by Sarah Bardeen), 'California's Rivers
Could Help Protect the State from Flood and Drought', Public Policy
Institute of California (website), 18 April 2022, https://www.ppic.org/
blog/californias-rivers-could-help-protect-the-state-from-flood-
and-drought/

25 'Mossel Bay desalination usage quandary for council', Water,
Desalination + Reuse (website), https://www.desalination.biz/
desalination/mossel-bay-desalination-usage-quandary-for-
council/

26 Nathaniel E. Seavy, Thomas Gardall, Gregory H. Golet et al, 'Why
Climate Change Makes Riparian Restoration More Important Than
Ever: Recommendations for Practice and Research', *Ecological
Restoration* 27(3), August 2009, https://www.researchgate.net/
publication/228665764_Why_Climate_Change_Makes_Riparian_
Restoration_More_Important_Than_Ever_Recommendations_for_
Practice_and_Research

27 Marc Reisner, *Cadillac Desert: The American West and its
Disappearing Water* (London: Penguin Books, 1986).

28 'Editorial: Houston's floods are a warning for California to shore up its
water systems. The good news: We're getting started', *Los Angeles
Times*, 2 September 2017, https://www.latimes.com/opinion/
editorials/la-ed-drought-flood-plans-20170902-story.html

29 Stafford et al, 'Greater Cape Town Water Fund: Business Case,
Assessing the return on investment for ecological infrastructure
restoration', The Nature Conservancy, April 2019, https://www.nature.
org/content/dam/tnc/nature/en/documents/GCTWF-Business-Case-
April-2019.pdf

30 'Project: Natural Infrastructure for Water Security', Forest Trends
    (website), https://forest-trends.org/infraestructura-natural-en-
    peru

CONCLUSION

1 Gideon Rachman, 'The threat of conflict over water is growing',
   Financial Times, 1 November 2021, https://www.ft.com/content/
   b29578f1-c05f-4374-bbb4-68485ef6dbf7
2 Paul Ratner, 'Where will the "water wars" of the future be fought?'
   World Economic Forum (website), 23 October 2018, https://www.
   weforum.org/agenda/2018/10/where-the-water-wars-of-the-future-
   will-be -fought
3 The Hill, Twitter post, 7 April 2021, https://twitter.com/thehill/
   status/1379603832888381443
4 Pete Kasperowicz, 'Obama: Climate change contributed to rise of
   Boko Haram, Syrian civil war', *Washington Examiner*,
   May 20, 2015
5 Tom S. Elliott, 'John Kerry: Global Warming Played a Role in Syrian
   Civil War', *National Review*, 17 October 2015
6 Peter H. Gleick, 'Water, Drought, Climate Change, and Conflict in
   Syria', *Weather, Climate, and Society* 6(3), 1 July 2014, https://journals.
   ametsoc.org/view/journals/wcas/6/3/wcas-d-13-00059_1.xml
7 Roman Olearchyk and Max Seddon, 'Crimea 'water war' opens new
   front in Russia-Ukraine conflict', *The Financial Times*, 29 July
   2021, https://www.ft.com/content/5eda71fc-d678-41cd-ac5a-
   d7f324e19441
8 Jan Selby, Omar S. Dahi, Christiane Fröhlich, Mike Hulme, 'Climate
   change and the Syrian civil war revisited', *Political Geography*,
   September 2017, https://doi.org/10.1016/j.polgeo.2017.05.007
9 Jason Wilson, 'Amid mega-drought, rightwing militia stokes water
   rebellion in US west', *The Guardian*, 8 Jun 2021, https://www.
   theguardian.com/us-news/2021/jun/08/klamath-falls-oregon-
   protests-ammon-bundy
10 Erik Stokstad, 'A voice for the river', *Science*, 1 July 2021, https://www.
    sciencemag.org/news/2021/07/colorado-river-shrinking-hard-choices-lie-
    ahead-scientist-warns
11 UNHCR, 'Displacement linked to climate change is not a future
    hypothetical – it's a current reality', 6 November 2016, https://www.
    unhcr.org/uk/news/latest/2016/11/581f52dc4/frequently-asked-
    questions-climate-change-disaster-displacement.html

12 Global Report on Internal Displacement 2021, IDMC, https://www.
    internal-displacement.org/sites/default/files/publications/documents/
    grid2021_idmc.pdf

13 Abrahm Lustgarten, 'How climate migration will reshape America',
    The New York Times, 15 September 2020, https://www.nytimes.com/
    interactive/2020/09/15/magazine/climate-crisis-migration-america.
    html?

14 'California's population shrinks for second year in a row', *The
    Guardian*, 2 May 2022, https://www.theguardian.com/us-news/2022/
    may/02/california-population-decline-trend-covid

15 Kaveh Madani, 'Iran's decision-makers must shoulder the blame for its
    water crisis', *The Guardian*, 5 August 2021, https://www.theguardian.
    com/commentisfree/2021/aug/05/iran-environmental-crisis-climate-
    change

16 Abrahm Lustgarten, 'The Great Climate Migration', *The New York
    Times* Magazine, 23 July 2020,https://www.nytimes.com/
    interactive/2020/07/23/magazine/climate-migration.html

17 Bruce Riedel and Natan Sachs, 'Order From Chaos: Israel, Jordan, and
    the UAE's energy deal is good news', Brookings (website), 23
    November 2021, https://www.brookings.edu/blog/order-from-chaos/
    2021/11/23/israel-jordan-and-the-uaes-energy-deal-is-good-news/

18 Asit K Biswas and Cecilia Tortajada, 'Untapped water', *The
    Kathmandu Post*, 14 March 2016

19 Alvar Closas, François Moll and Nuria Hernández-Mora, 'Sticks and
    carrots to manage groundwater over-abstraction in La Mancha, Spain',
    Agricultural Water Management, Volume 194, December 2017, https://
    doi.org/10.1016/j.agwat.2017.08.024

20 Ibid.

21 Barbara Schreiner, 'Viewpoint – Why has the South African National
    Water Act Been so Difficult to Implement?', *Water Alternatives*, 2013,
    https://www.researchgate.net/publication/285838406

22 Tara Moran and Dan Wendell, 'The Sustainable Groundwater
    Management Act of 2014: Challenges and Opportunities for
    Implementation', Water in the West, Stanford University, https://
    localwitw.stanford.edu/sites/default/files/WitW_SGMA_Report_
    08242015_0.pdf

23 Stephanie Osler Hastings, 'A Test for California's Groundwater
    Regulations in the Megadrought', Bloomberg Law (website), 7
    September 2021, https://news.bloomberglaw.com/environment-
    and-energy/a-test-for-californias-groundwater-regulations-in-the-
    megadrought-17

24 'Kicking Off 2022 with Significant Implications for Agriculture and
    SGMA: GSP Assessments, Submittals and Alternative Five-Year

Updates', Client Alert, Brownstein (website), 4 February 2022, https://www.bhfs.com/insights/alerts-articles/2022/kicking-off-2022-with-significant-implications-for-agriculture-and-sgma-gsp-assessments-submittals-and-alternative-five-year-updates

25 'Pure Water Southern California', Metropolitan Water District of Southern California (website), https://www.mwdh2o.com/planning-for-tomorrow/building-local-supplies/regional-recycled-water-program/

26 'CDFW Takes Proactive Measures to Increase Salmon Smolt Survival', California Department of Fish and Wildlife News (website), 28 April 2021, https://cdfgnews.wordpress.com/2021/04/28/cdfw-takes-proactive-measures-to-increase-salmon-smolt-survival/

27 Joanna Allhands, 'Arizona needs more water, but tapping rivers in other states is no way to do it', *Arizona Republic*, AZCentral (website), https://eu.azcentral.com/story/opinion/op-ed/joannaallhands/2021/05/26/mississippi-river-pipeline-wont-save-arizona-but-these-ideas-might/7436318002/

28 Coral Davenport and Christopher Flavelle, 'Infrastructure Bill makes first major U.S. investment in climate resilience', 6 November 2021, https://www.nytimes.com/2021/11/06/climate/infrastructure-bill-climate.html

29 Minute No. 323 Binational Desalination Work Group, 'Binational Study of Water Desalination Opportunities in the Sea of Cortez', April 2020, https://library.cap-az.com/documents/departments/planning/colorado-river-programs/Binational-Desal-Study-Executive-Summary.pdf

30 'Central Arizona Project: Final 2019–2024 Rate Schedule', 7 June 2018, https://library.cap-az.com/documents/departments/finance/Final-2019-2024-CAP-Water-Rate-Schedule-06-07-18.pdf

31 'Phnom Penh: Upgrading a Major Water Treatment Plant', Agence Française de Développememt (website), 20 March 2020, https://www.afd.fr/en/actualites/phnom-penh-station-des-eaux-chamcar-morn

32 'What are the Sustainable Development Goals?' United Nations Development Programme (website), https://www.undp.org/sustainable-development-goals

33 'Goal 6: Ensure availability and sustainable management of water and sanitation for all', United Nations Department of Economic and Social Affairs (website), https://sdgs.un.org/goals/goal6

34 Draft negotiating texts, UN Climate Change (website), https://unfccc.int/documents?f%5B0%5D=document_type%3A4168&search2=&search3=&page=%2C%2C0

35 'Foreword', *The United Nations World Water Development Report 2020: Water and Climate Change* (Paris: United Nations Educational,

Scientific and Cultural Organization, 2020), https://unesdoc.unesco.
org/ark:/48223/pf0000372985.locale=en

36   'In Chile's capital city, drought is making grass 'unacceptable' for
     many', NBC News (website), 26 April 2022, https://www.nbcnews.
     com/news/latino/chiles-capital-city-drought-making-grass-
     unacceptable-many-rcna26098

37   Vittal Hari, Oldrich Rakovec, Yannis Markonis et al, 'Increased future
     occurrences of the exceptional 2018–2019 Central European drought
     under global warming', *Scientific Reports* 10, 6 August 2020, https://
     www.nature.com/articles/s41598-020-68872-9#Sec2

38   Blanca E. Jiménez Cisneros, Taikan Oki, Nigel W. Arnell et al,
     'Freshwater Resources', *Climate Change 2014: Impacts, Adaptation,
     and Vulnerability* (Cambridge: Cambridge University Press, 2014),
     https://www.ipcc.ch/site/assets/uploads/2018/02/WGIIAR5-Chap3_
     FINAL.pdf

39   Hayley Smith, 'Unprecedented water restrictions hit Southern
     California today: What they mean to you', *Los Angeles Times*, 1 June
     2022, https://www.latimes.com/california/story/2022-06-01/southern-
     california-new-drought-rules-june-2022

Index

.